NOW IS THE TIME TO COLLECT

NOW IS THE TIME TO COLLECT

Daniel Giraud Elliot,
Carl Akeley, *and the*
Field Museum African
Expedition *of* 1896

PAUL D. BRINKMAN

THE UNIVERSITY OF ALABAMA PRESS | TUSCALOOSA

The University of Alabama Press
Tuscaloosa, Alabama 35487-0380
uapress.ua.edu

Typeface: Plantin

Cover image: Watercolor painting by Peggy Macnamara
Cover design: Lori Lynch

Published with a grant from the Field Museum of Natural History

Cataloging-in-Publication data is available from the Library of Congress.
ISBN: 978-0-8173-2199-4 (cloth)
ISBN: 978-0-8173-6148-8 (paper)
E-ISBN: 978-0-8173-9517-9

For Bill Stanley

Now is the time to collect. A little later it will cost a great deal more, and the collector will get a great deal less.

—*William T. Hornaday, 1891*

CONTENTS

PREFACE

This project started with a tantalizing email from a friend and colleague, Christine D. G. Weis, then head of the Marie Louise Rosenthal Library at Chicago's Field Museum. Christine informed me, in October 2014, that a London bookseller was peddling a previously unknown manuscript journal allegedly kept by zoologist Daniel Giraud Elliot on the museum's first zoological expedition to Africa, in 1896. The library already owned an assortment of letters, photographs, notes, and other archival materials that afforded a sketchy history of this formative expedition, but it lacked a day-to-day record. The asking price for the journal was hefty, but Christine was interested in acquiring it for the library. She wanted advice. Could I recognize Elliot's handwriting if she sent me a sample? What was my opinion of the potential research value of the journal? She shared a low-resolution scan of page 210 from the journal—dated July 18, 1896—and I was immediately and irrevocably hooked.

The library bought the journal with the generous aid of the Louann Hurter Van Zelst Purchase Fund. I saw it for the first time the following month. It was beautiful. It had 316 quarto-sized pages—including two pages numbered 290—with faded, difficult-to-decipher handwriting on its rectos. It was filled with a constellation of ink blots, smudges, inserts, and strike-throughs. The journal—now housed in a black, purpose-made Solander box—was still bound in its original cloth-backed, marbled limp boards, and was in very good condition. As I was then the world's only authority on Elliot's handwriting, Christine invited me to work on an annotated transcription of the journal. That idea then morphed into a more ambitious project to write a narrative history of the expedition, and the rest, as they say, is history. I cannot thank Christine enough for sharing this opportunity with me and for helping see it through to completion, despite some enormous challenges, both personal and professional.

Naturally, the lion's share of the research for this project was done at the Field Museum, utilizing the enormous body of historical resources housed there. Everyone in the library was uniformly helpful and accommodating of my work. However, I must single out Armand Esai, for facilitating access to a wealth of supplementary archival records at the museum; Nina Cummings and Rebecca Wilke, for handling my historical photograph requests; Gretchen Rings, for providing access to materials in the Mary W. Runnells Rare Book Room and for making arrangements for a generous publication subsidy from the Field Museum; and Angelo Giannoni, for laboriously scanning the journal. Angelo's fingerprints are literally all over this project. Museum zoologists were likewise helpful. Julian Kerbis Peterhans shared an important set of taxidermist Carl Akeley's African field notes with me. Adam Ferguson provided information about mammal specimens in the zoology collection. Finally, the late Bill Stanley was a strong supporter of this project from its inception.

My friend Sarah O'Brien hosted me on a dozen or more pizza-fueled visits to Chicago while I was doing this research. I'm grateful for her unflagging hospitality.

I'd also like to thank the many curators, collection managers, librarians, and archivists at several other libraries and museums where additional research was done. These include the American Museum of Natural History, the Milwaukee Public Museum, the Milwaukee County Historical Society, the Museum of Comparative Zoology, the Natural History Museum in London, the Chicago Academy of Sciences / Peggy Notebaert Nature Museum, the University of Florida, the Oak Park Public Library, and the Smithsonian Institution Archives. I'd like to thank Pamela Henson for helping me obtain a Smithsonian Institution Short-Term Visiting Fellowship, which made research in Washington, DC, more affordable.

Friends and colleagues at the North Carolina Museum of Natural Sciences supported this project in myriad ways. Librarian Janet Edgerton filled numerous interlibrary loan requests. Roy Campbell helped me pick and choose among the best historical images from the expedition. At various stages of this project, I had the help of two hardworking volunteer transcribers, Herman Perkins and Saul Schiffman. Together, they did the bulk of the initial transcription work on Elliot's journal. The former also transcribed a large collection of Elliot's correspondence. At the same time, I also had the benefit of another volunteer, Richard A. Webb, who did indispensable genealogical research for me.

Mark Barrow, Matthew James, Mark Alvey, Matthew Laubacher, Daniel Lewis, Gary Rosenberg, Mary Anne Andrei, and David Sepkoski were very generous in sharing their own research on American zoology, extinction, Carl Akeley, and other relevant topics with me. This project also benefited from a helpful dialog with Megan Raby.

Bits and pieces of this research have appeared elsewhere. Longer and more detailed versions of chapters 1 and 2 appeared in the *Journal of the*

History of Collections—published by Oxford University Press—and *Colligo*, respectively, and are used with permission. Meanwhile, sections of this book dealing with director Frederick J. V. Skiff's difficult relationships with the Field Museum's scientific staff have also been the subject of earlier works in *Archives of Natural History* and the *Museum History Journal*. I thank the editors and publishers of those journals for their support and the opportunity to expand this story here.

I'm grateful to a long list of people who read all or parts of the book-length manuscript and who made many helpful suggestions. This includes Sally Gregory Kohlstedt, Rich Bellon, Tara Greaver, Andrew Leman, Sarah Holshouser, Steve Emrick, Liz O'Brien, Sam Alberti, Abby Collier, Will Kimler, members of Eremotherium Club, and several anonymous reviewers.

I'd like to thank everyone at the University of Alabama Press, but especially my editor, Claire Lewis Evans, who shepherded my book through the publication process.

Finally, I thank my entire family, but especially my mother, who encouraged my interest in the history of science even when a career seemed impossibly remote, and Sarah Holshouser, who generously shared my time with Elliot and Akeley for the better part of six years.

NOTE ON THE FIELD MUSEUM

Chicago's Field Museum houses more than thirty million natural science and anthropology objects of all shapes and sizes: sixteen million insects, two million pressed plants, the type specimen of the enigmatic *Tullimonstrum gregarium*, a nineteenth-century Maori meeting house. Staggering in size, this collection aspires to be global and comprehensive in scope. Since the museum's founding in 1893, the museum's curators and scientists have used these objects to investigate, describe, and document the world's biological and cultural diversity.

Most of these objects were collected on expeditions by museum personnel. For nineteenth- and early twentieth-century museum zoologists, for example, expeditions usually involved hunting and trapping animals—by the hundreds and by the thousands—in their wild habitats, then bringing the specimens back to the museum for study and long-term preservation. Expeditions ran at great cost to the museum, and sometimes at the peril of the collector. In 1909, a Field Museum anthropologist, William Jones, was murdered on an expedition to the Philippines. Mammalogist William T. Stanley died of a heart attack while collecting mammals on an expedition to Ethiopia in 2015.[1]

The work of museum zoologists is vitally important for understanding the taxonomy, ecology, and conservation biology of Earth's animal life. The specimens they collect and care for in perpetuity at the museum serve as a permanent record of biodiversity. This is an especially important mission today, when natural habitats are shrinking around the globe, and so many of their denizens face the imminent threat of extinction.[2]

Some zoology specimens—especially the rarest, largest, and showiest animals like rhinos, elephants, and giraffes, the charismatic megafauna, as they're sometimes called—are used in exhibits to educate museum visitors. In the nineteenth and early twentieth centuries (before the proliferation of motion-picture photography, television, and the internet), object-based museum exhibits were considered the best and most effective means to learn about the zoology of impossibly remote places. Many educators and science communicators still prefer an object-based approach for teaching natural science.

Take the Field Museum's display of the Somali wild ass, *Equus asinus somalicus*, for example. In a single, glass-enclosed case in the museum's Carl E. Akeley Memorial Hall, four specimens pose in naturalistic, eerily lifelike positions. The perfectly coiffed hair of their manes, the bulging veins of their muzzles, and the vigilant stare of the tallest male specimen give this group of meticulously taxidermied specimens the appearance of suspended animation. Even a cursory examination of this display teaches the visitor something of the animals' anatomy, morphology, behavior, and natural habitat. Many of the museum's zoology exhibits convey the animals' conservation status, as well. The Somali wild ass is now listed as critically endangered. Current wild populations of this animal are estimated at a paltry twenty-three to two hundred mature individuals.[3]

Authoritative knowledge about diversity and abundance is arguably the single most important contribution of natural science museums to creating and maintaining a sustainable future. After all, how can anyone know what plants and animals to conserve without knowing what one has in the first place? Yet knowledge alone doesn't necessarily lead to action. And merely expressing concern about diminishing species shouldn't be confused with an act of conservation. Recognizing this, the Field Museum began a bold experiment in applied conservation biology in 1994, when it launched its Office of Environmental and Conservation Programs (ECP). The ECP experiment yielded actionable intelligence on biodiversity that has happily resulted in the conservation of vast tracts of species-rich land in Peru, Ecuador, Cuba, and elsewhere. Thanks to the success of this pilot program, environmental conservation is now a central pillar of the museum's scientific mission.[4]

The Field Museum still makes and maintains zoological (and other) collections. Indeed, responsible collecting remains the most reliable way to discover and to document biodiversity. Yet the ethos of collecting has changed significantly over the last century. Where animals were once hunted and trapped prolifically by museum scientists, they're now taken in more limited numbers. The so-called charismatic megafauna are seldom, if ever, hunted by museum scientists. Animals like these are now acquired from roadkill and through natural deaths at zoos or wildlife refuges. Where specimens were once taken freely from the colonial hinterland to be housed in Western museums, they are often now left in their country of origin or shared equitably among institutions.

This book examines in detail the activities and attitudes of a pair of Field Museum zoologists of the late nineteenth century, focusing primarily on their high-profile collecting expedition to British Somaliland in 1896. The hunter-naturalists who manned this expedition collected animal specimens with abandon. What they did in Africa, and why, is the story told here.

NOTE ON SOURCES

Elliot's journal and—to a lesser extent—his correspondence and other unpublished records, supplemented by a small body of Akeley's published writing, account for the majority of sources consulted in writing this book. For that reason, the point of view of this history is predominantly Elliot's. No firsthand sources for the African participants on the expedition could be located. Elliot's names and spellings for the people, tribes, and sites of Somaliland used in the journal have been retained in this history. Indeed, Elliot's spelling was so idiosyncratic that some of the places he named cannot now be located on a modern map of Somalia. Likewise, the scientific names Elliot used and published after the expedition have also been retained here, although some have by now been relegated to synonyms, or are no longer in use.

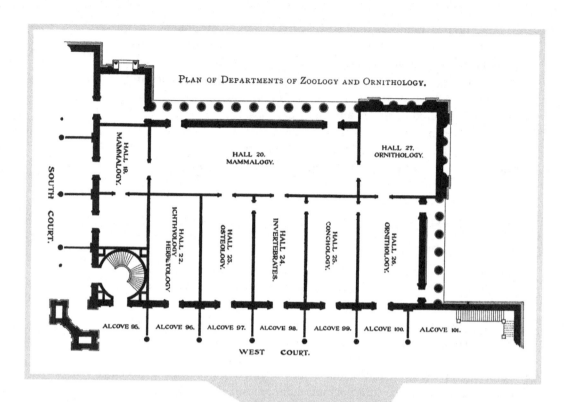

PLAN OF DEPARTMENTS OF ZOOLOGY AND ORNITHOLOGY.

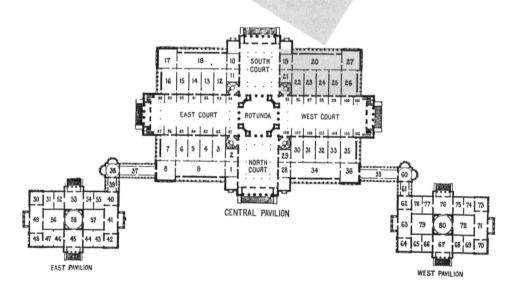

An 1896 floor plan of the Field Columbian Museum with a detail (*above*) showing the Departments of Zoology and Ornithology. Figure compiled by Haley McCay from originals in the *Guide to the Field Columbian Museum*.

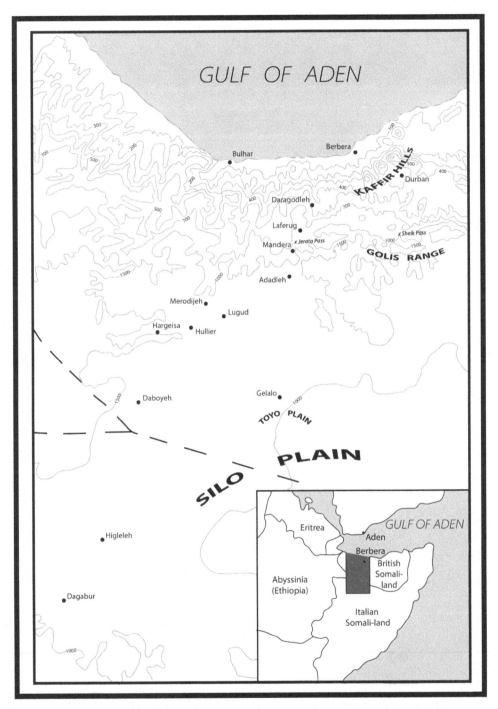

A map of Somaliland, showing the approximate route of travel of Daniel Giraud Elliot's 1896 expedition. The spelling of place names is based on Elliot's journal and letters. Many of the place names he used can no longer be found on modern maps of Somalia.

NOW IS THE TIME TO COLLECT

INTRODUCTION

S ickly and exhausted from a pernicious bout of fever, Daniel Giraud Elliot reclined in his tattered camp chair in the midmorning shade of a solitary tree. His khaki-colored safari suit was damp with sweat. A pith helmet straddled his lap, while the dome of his bald head glistened. His full, white beard and walrus mustache, often meticulously groomed, were a rat's nest of dirty tangles. Curator of zoology at Chicago's Field Columbian Museum, Elliot was leading a collecting expedition to the Horn of Africa in the spring and summer of 1896. His aim was the continent's living fauna, especially its showiest mammals. Having roughly equal cause for satisfaction and concern on the expedition, thus far, the bone-weary curator shut his eyes tight and tried to doze. He'd endured a late night plagued with the pain of fever and the anxieties of leading a safari, and all he wanted now was peace and quiet.

Elliot and his party were encamped on the dry and featureless Toyo Plain in the remote interior of what was then known as British Somaliland. The bulk of his caravan had been left behind at a previous campsite. Lack of water and fresh grass in this parched part of the country made it impractical to travel with all the expedition's animals, men, and material. A mere skeleton crew of essential safari personnel, including headman, guards, gun bearers, porters, and cooks, was with Elliot in camp. While the curator rested, his men attended to myriad duties in their usual fashion, singing and talking noisily among themselves.

One of Elliot's assistants, Edward Dodson, a young English taxidermist and explorer, had left camp on horseback at three o'clock that morning on a critical errand. Elliot had sent him to investigate a report that some of the expedition's specimens, stored in brine-filled barrels buried at the other campsite, had been swept away in a disastrous flash flood. The curator was deeply troubled by the probable loss of these valuable and hard-won specimens. Most—if not all—were not yet represented in the museum's collections. Moreover, many of the species were rare and getting rarer; some, Elliot feared, faced imminent extinction. To lose these priceless specimens was a calamity. "The news has disheartened us very much," he lamented in his journal.[1]

A promising young American naturalist, Carl Akeley, was chief taxidermist at the Field Columbian Museum and Elliot's right-hand man on the expedition. Though he lacked significant field experience, he'd proved himself to be a maniacal worker and a master of his craft—he was a virtuoso with dead animal skins. Earlier that morning, using Elliot's own Lee-Metford rifle, the talented taxidermist had shot a magnificent bull hartebeest, one of the largest and most distinctive antelopes of Africa. The curator was delighted, describing his companion's conquest as "the finest specimen we have yet obtained, very dark in color, & marked with very black patches on shoulders & hind quarters, & a fine knotted pair of horns."[2] Akeley, stripped to the waist, his hands and forearms caked in gore, was laboring to preserve the skin and skeleton of his hartebeest in the stingy shade of Elliot's tree.

While the men were thus occupied, a mounted herald suddenly rode into camp and announced that Sultan Nur, ruler of the Habr Yunis (a powerful Somali tribe), along with his royal family and an escort of warriors, was coming to grant Elliot an unexpected audience. The headman for the expedition, Dualla Idris, an old acquaintance of Sultan Nur, translated for the English-speaking zoologist. Accustomed to visits with local sovereigns, Elliot expressed his satisfaction at the news and sank back into his camp chair to await the sultan's arrival. Sixty-one years old and an experienced campaigner, Elliot had a calmness and solemnity that masked the tension that spread through camp as word of the approaching visitors passed from man to man in anxious whispers.

Soon, a body of some two or three dozen well-armed horsemen was seen riding over the shimmering plain "in a stately manner, the heads of their spears & gaily decked bridles & saddles flashing in the sun." The herald raced back to the main body to announce that Elliot was waiting. Immediately, "the pace of their horses was quickened, and the whole band dashed up at full speed & halted in front of my chair," Elliot described in a letter to his museum superior, director Frederick J. V. Skiff. "They were a fine body of men," he wrote, "warriors every one, & I envied them the splendid ponies they were riding." Their leader, Sultan Nur, "a very celebrated man, & a tried warrior of a hundred battles, was in the centre facing me, a very tall spare man who sat his horse as erect as was the spear in his hand."[3]

No one from the sultan's party made a motion to dismount. Elliot, too, sat motionless, his own men gathered timidly behind his chair. Then one of the escorts began to chant, "expressing the King's delight at my coming to his country, & that as soon as he had learned where I was, he had hastened to see me, & that the Europeans & his people must be firm friends

& then they would be invincible & fear no one, & so on for quite a long while, the whole sung in a low rhythmical chant." Elliot listened politely, all the while scanning the crowd of warriors, noting their weapons and taking their measure. "During the delivery of this effusion," he continued, "each man sat his horse like a statue, never moving a muscle & I had a fine chance to run my eyes over them." Each man carried two spears—one for throwing and another for stabbing—a short double-edged sword, and a small, round shield made of rhino hide and hung from the left arm. Their bridles were decorated with brightly colored tassels and shiny metal. "Altogether they made a brave show. Evidently a tough crowd to fight."[4]

When the song ended, the sultan dismounted with a flourish. Elliot rose to greet him and shake his hand, expressing his great pleasure in being able to welcome such a distinguished guest to his camp. The tension evaporated. The men began to relax as they realized at last that the sultan's intentions were truly friendly. The curator sent someone to fetch his only high-back chair and invited his guest to sit in it. He then ordered coffee to be served.

Sultan Nur made a magnificent first impression. He was a man approaching sixty, Elliot estimated. He was slight in frame and more than six feet tall. He had a "very pleasant expression" and was "dignified and gentle" in his bearing. He was remarkably composed and serene in Elliot's

FIGURE I.1.

Sultan Nur (*seated, with spear*) visiting with Elliot (*seated, with pith helmet*) in camp. The other seated man is probably Dualla, who served as interpreter. Photograph courtesy of the Field Museum, CSZ6081.

company. "Indeed," Elliot wrote, "I have seen many a so-called swell among a select circle in a drawing room show to much less advantage & credit to himself, than did this ignorant savage. . . . He never lost his self-possession or dignity & behaved like a thoroughbred gentleman."[5]

Once Elliot and his guest had taken their seats, an elaborate equestrian exhibition—called a *dibaltig*—began. One or two of the sultan's horsemen would shoot out suddenly from the circle and ride his horse through "all kinds of evolutions returning at full speed, drawing his horse on his haunches in front of my chair crying 'Mot,' or Welcome." This continued until each horseman had taken at least one turn. A cousin of the sultan then commenced another chant, and the entire body divided into two parties and staged a mock battle. Elliot was impressed by the spirited acting, with "many near spear thrusts, that might easily have required the surgeons care" had they struck home. Elliot regretted that most of his own caravan and the greater part of his armed force had been left behind at the previous campsite, otherwise he would have been pleased to offer a counterdemonstration with a rifle-firing drill.[6]

The reception ceremony over, Elliot and his visitor began to talk through a translator. Elliot soon suspected that the main purpose of the sultan's visit was to receive medical assistance. "The poor man was evidently ill," Elliot explained, "& he said he heard I was not well, which I suppose he thought would create a bond of sympathy between us." The sultan then cataloged his injuries at length, which Elliot recounted in his letter to Director Skiff: "Not long ago there had been a great battle [and] Sultan Nur was in the thick of it." He'd suffered some horrific wounds, one near his right ankle, "which must have been a daisy." The wound had been fired with a red-hot spear, and the sultan's leg looked like "a gigantic pepper box," with gruesome holes where the iron had seared his flesh. "He was otherwise in a bad way in sundry parts of his body & would like me to do something for him." Elliot advised his guest that "he was not as young as he once was, & he could not go about getting hard knocks & recovering from them as he once could." The sultan laughed and said, "quite true."[7]

Elliot examined the sultan's leg closely. The wound had healed, and there was nothing to be done short of surgery to repair the damaged bone. But this was an option of last resort, which Elliot intended to exercise in the field only to save a man's life. Meanwhile, the sultan was rife with other symptoms, which Elliot diagnosed as rheumatism and dyspepsia. For these he prescribed some medicines that would alleviate but probably not cure the man's chronic suffering. According to Elliot, the sultan was grateful to receive what he called "true medicine that he hoped would benefit him."[8] Apparently, Sultan Nur had some knowledge of and even a little faith in Western science and medicine.

The sultan remained a long time talking with Elliot. Finally, with the hot afternoon advancing, they rose and shook hands, the sultan announcing that he and his party would take a rest in the shade of a nearby clump of trees. Elliot intended to make a gift of some trifles he had around camp, but his headman advised him not to, explaining that the sultan would likely return or send one of his sons for gifts at a later date. Elliot therefore explained that he'd left many things at his previous camp, but that he would send for cloth. He was only sorry, he explained, that he couldn't present his visitor with any gifts in person. The interview over, Elliot retired to the shaded sanctuary of his tent and left the royal guests in the care of his headman. The sultan, with his escort and some of the expedition's hired men, went to sleep under the trees, and the camp "relapsed into a peaceful quiet that was most refreshing."[9]

Elliot later characterized the visit as "quite a ceremony . . . which was interesting to see." A curious man by nature, Elliot made note of the sultan's regal comportment and the warrior qualities of his escort. Habitual acquisitiveness drew his attention particularly to their stuff—dress, weapons, and rhino-hide shields. Sultan Nur, meanwhile, took the opportunity to scrutinize the American naturalists and their strange practices. He'd almost certainly seen big-game hunters in his country before, but probably none with the careful, methodical—even obsessive—work habits of Akeley, who'd been skinning his hartebeest when the royal party arrived, and who resumed his work as soon as it was polite to do so. The sultan watched the industrious taxidermist work and seemed "much interested."

In a postscript in his letter to Skiff, Elliot related that he explained to the sultan very matter-of-factly that the animal skin Akeley was so painstakingly treating "would be set up again just as in life" at a "great Institution in Chicago one of the great cities in the World."[10]

According to Elliot, Sultan Nur was "greatly surprised" to learn of the expedition's purpose. Yet there's no indication that the curator reflected on how strange this purpose must've sounded in translation. Rather, he took it for granted that the sultan immediately grasped the meaning of what he'd been told. "Hope he took it all in," Elliot wrote. "Verily the Field Columbian Museum will be known in the midst of darkest Africa." The sultan, Elliot concluded, had been "quite satisfied" with his visit.[11] But this is a difficult statement to credit. Elliot's guest departed with a bellyful of coffee, a modicum of medicine, and a vague promise of cloth. If his real purpose was to reconnoiter the American hunters, then he must've left with more questions than answers. The curator had told Sultan Nur something about *what* the expedition was doing in his country, but he hadn't bothered to explain *why*.

SALVAGE ZOOLOGY

In the nineteenth century, Western naturalists anticipated the extinction of large game animals in Africa and elsewhere as a consequence of European colonial expansion, and as human activities—especially land-intensive practices like farming, ranching, and logging—encroached more and more on wild animal habitats. Many naturalists viewed extinction as the inevitable, if lamentable, byproduct of humanity's steady advance. The permanent demise of some species was a pity, but the loss was a small price to pay to maintain the pace of progress.[12]

Museum zoologists of this era assumed the role of salvaging the remnants of these threatened species—skins, skeletons, eggs, nests, and whole animals—while they could still be acquired and preserving them as museum specimens for all time. The rationale behind *salvage zoology* was clear: certain animals, including large mammals, flightless birds, and island species, were doomed to extinction by the unrelenting spread of Western civilization. Zoologists, therefore, were obligated to harvest these specimens by any means necessary—including aggressive hunting and trapping—and to keep them forever in museum collections. "To museums must be largely assigned the work of conserving the remains of such forms ere they are absolutely lost," wrote American Association of Museums president Oliver C. Farrington. Museums would have to act expeditiously, for the pace of extinction was quickening and the window of opportunity for collecting specimens wouldn't stay open for long. "The time is near at hand," Elliot warned, "when, in certain lines of Zoology, especially in the large mammals of the world, it will be forever impossible to procure specimens.

They are certain, most of them, to become as extinct as the Mastodon or Dodo are to-day."[13]

Somewhat perversely, the scarcity of threatened animals tended to make their specimens more valuable as museum commodities. This was a common point of view among nineteenth-century museum zoologists. For example, to determine the "true market value" of various animal species, argued zoologist Gambier Bolton in 1899, "one must consider . . . whether they have already or are about to become extinct." What this meant, according to Bolton, was that some rare species were "worth their weight in gold."[14]

If this approach seems antithetical to the conservation ethos that's become de rigueur in museums of the twenty-first century, it is. Conservation is so central to the mission of natural science museums in the present day that it's easy to assume that museum zoologists of the nineteenth century collected specimens in order—somehow—to save animals from extinction. On the contrary, salvage zoologists raced to collect specimens before extinction put them permanently out of reach. Some prominent museum zoologists advocated passionately for conservation at the same time, but never at the expense of scientific collecting. The practice of salvage—to set out to shoot the last living example of a species to put it in a museum—is almost unthinkable today, yet it makes more sense when one realizes that naturalists of Elliot's era were motivated more by the potential loss of scientific information than they were by the loss of species. In 1887, when defending the salvage collecting of northern elephant seals, Smithsonian curator Frederic A. Lucas wrote, "No one deplores the destruction of animals more than does the present writer, and yet he deems the slaughter . . . not only justifiable but commendable."[15] Museum zoologists must have their specimens, no matter what the cost.

Elliot recognized the exotic fauna of Africa as an opportunity to make a conspicuous contribution to salvage zoology. The declining numbers of many of Africa's mammals were already widely acknowledged by hunters and travelers in the 1890s, yet few museums had a representative sample of these large and charismatic animals in their collections. Chicago's museum, for example, was starting almost from scratch. A new institution (founded in 1893 and opened in 1894), the Field Columbian Museum held only a smattering of African animals. Elliot sought to plug this gap. Arriving in Africa, he and his hunters proceeded to collect the local animals in prolific numbers, skin and skeletonize them in the field, pickle them in brine, and then ship them to Chicago. Duplicates would augment the museum's growing comparative zoology research collection, while excess specimens would be used in exchange with rival museums to acquire other rare animals. Meanwhile, the showiest specimens would be mounted in a series of elaborate, lifelike group displays at the museum. Akeley,

Elliot's superlatively talented taxidermist, would spearhead that initiative. It was an audacious undertaking—no American museum had ever sent a collecting expedition to Africa. Nor had many museums adopted such an ambitious and expensive strategy of zoological exhibition.[16]

There was a great deal of similarity between the practice of salvage zoology and big-game sport hunting. Both activities, for example, required traditionally masculine virtues like courage, endurance, strength, and steadfastness. Both involved the commodification of animals. Yet, Elliot professed to dislike hunting for mere sport, often taking pains to distinguish his work as a scientist from sport hunting. Akeley, too, had issues with certain kinds of sport hunters. "There are sportsmen and so-called sportsmen of all shades and degrees," he wrote in 1920. There was the hunter who was keen to get a trophy head of every species of game animal, men like Akeley's famous friend Theodore Roosevelt. "No one can take exception to him while there is plenty of game left," Akeley wrote forgivingly. "On the other hand, there is the man who hunts for record heads and with him I have little patience." Another class of so-called sportsmen was made up of hunters whose ambition was to kill to their limit. "Having paid for a license which allows them to kill a given number of animals of each species, they are never content until they have killed the full number." Another hunter that Akeley disapproved of was the man who went to Africa "determined to kill every available species within three months." Such a hunter had no "appreciation of Africa's charm" or any "real interest in its animals." The true sportsman, Akeley wrote prophetically, wants "to remain [in Africa] as long as possible."[17]

Like Elliot, Akeley was convinced of the inevitability of extinction. "The game must eventually disappear as the country is settled," he wrote, "and with it will be wiped out the charm of Africa." He blamed the spread of Western society: "The dark chapters of African history are only now being written by the inroads of civilization."[18]

THE OBSOLESCENCE OF NATURE

The present book is a narrative history of the Field Columbian Museum's 1896 expedition to Africa, which serves as a case study in salvage; it is not a book about salvage zoology per se. It draws liberally from a recently rediscovered field journal kept by Elliot, which provides—for the first time—a daily record of the expedition's activities. Its microhistorical narrative, reconstructed from the journal and other contemporary manuscript sources, paints a detailed picture of the practice of late nineteenth-century museum zoology. It describes and explains the operation of a zoological expedition. How, for example, did zoologists choose what to hunt? How did they find and acquire elusive game animals? How did they process and preserve specimens in the field? How did they solve their thorniest supply problems, including water, food, salt, ammunition, and specimen barrels? How did they adapt to the trying environmental conditions, including heat and sun, dust storms, drought, flood, and disease? How did they negotiate the cultural challenges of working with—and depending on—African auxiliaries with different languages, values, and customs? These and other operational issues are the subject of this book. It is precisely in the quotidian details that the salvage zoology agenda of the expedition shines through: in the planning and the elaborately and repeatedly argued justification of expenses; in the management of expectations against same; in the relentless pursuit of certain rare game; in the emphasis on exhibit-quality (and high-quantity) specimens to populate the museum's habitat groups; and in the running commentary on salvage coupled with the conspicuous lack of concern for wildlife conservation in Elliot's journal and correspondence.

This book also constitutes an early institutional history of zoology at the Field Columbian Museum. From its founding in 1893, its patrons had ambitions to build a world-class museum with zoology as an important component of its purview. The late nineteenth century, however, was a time of grave concern for the long-term viability of certain animal species. To build a world-class zoology collection under these circumstances required a salvage zoology approach. Elliot knew this. He worked constantly to get his museum superiors to understand the urgency of his salvage mission. Patrons, however, had to balance Elliot's agenda against the needs of the whole institution. Zoology sometimes had to take a back seat to other, more pressing museum priorities. Of paramount importance to patrons, for example, was the long-term financial health of the museum, and the maintenance of its temporary home.

Finally, this book tells the professional stories of the two American naturalists on the expedition. Elliot's twelve-year tenure at the Field Columbian Museum overlapped neatly with the heyday of salvage zoology in the 1890s, and this book will show that he was an unapologetic proponent of this practice. Elliot's sidekick, Akeley, meanwhile, was destined to change the public face of museums with the exquisitely lifelike habitat groups he built after the expedition's return. A committed hunter-naturalist early in his career, he would make a dramatic turn toward wildlife conservation later in his life. The story told here of their career arcs in Chicago, and especially their day-to-day experiences on the African expedition of 1896, provides insight into some of the forgotten motives and methods of museum zoologists at the turn of the twentieth century.

Elliot's purpose in Africa may seem anathema to the reader with a modern environmentalist sensibility. Yet, in amassing a large collection of animal specimens, Elliot was anticipating the obsolescence of nature. His long-term ambition was to make the Field Columbian Museum into "the Mecca for all naturalists." He wanted his institution to become the place where future naturalists would be required to make a pilgrimage to see the artifacts of nature after large game animals had been exterminated in the wild. He intended to make an auspicious start with the charismatic mammals and birds of Africa. "When the African collection is on exhibition [naturalists] certainly will have to come," he gloated, "for they will see nothing like it elsewhere."[19] It's important to note that Elliot's expedition didn't drive any African species to extinction. Indeed, scientific collecting cannot be shown conclusively to have caused the demise of any species.[20] Yet, from Elliot's point of view, extinction was a necessary—even *desirable*—adjunct to his ambitious museum plans.

The point here is subtle but important: Elliot and Akeley set out to salvage the remnants of Africa's wildlife, not to save it. Wildlife conservation wasn't their agenda. Through aggressive, systematic hunting and trapping, Elliot and Akeley stocked their museum with rare specimens and built elaborate dioramas that—more than a century later—still feature proudly and conspicuously at the Field Museum. In fact, ten of Akeley's original habitat groups, as well as one new, state-of-the-art diorama, all made from Somali specimens collected in 1896, are still on exhibit in Chicago. Interestingly, the cost of developing the new diorama—constructed in 2016—was met through a wildly successful crowdsourcing campaign. This suggests that dioramas are still relevant, still popular, still inspiring museum visitors.[21]

But what are they inspiring? Dioramas and habitat groups are complex objects, part science, part art. They're different things to different people: scientific specimens; museum icons; representations of species or environments; a snapshot of exotic nature; souvenirs of remembrance; even overt symbols of racism, sexism, and imperialism. The stakes can be high when representing animals in museums, Nigel Rothfels has argued. Curators and taxidermists who control that representation, and the uses they make of animal displays, are profoundly important. Yet the intended messages don't always square with the ones carried away by museum visitors.[22]

More often than not, dioramas are used by museum zoologists in the present day to tell a conservation story. Yet, the conservation messages one now reads on museum labels were often appended to dioramas later in the twentieth century, once visitors began asking awkward questions about when and how museums acquired their specimens. Because dioramas are among the most spectacular of museum exhibits, it's important for museum constituents, including patrons, curators, visitors, students and teachers, zoologists, and conservationists, to understand why they were built in the first place. Many of America's earliest dioramas were made to culminate the practice of salvage, to replace natural habitats that were being chopped down, dug up, and plowed under by the march of progress.

1

ONE OR TWO GOOD MEN

Chicago's Field Museum was a legacy of the World's Columbian Exposition of 1893. President Benjamin Harrison declared that the exposition had "surpassed all its predecessors in size, in splendor, and in greatness." The fact that it was held in Chicago—a brash boomtown of stockyards and smokestacks—made the Beaux-Arts spectacle of the fair even more astonishing. Easterners, especially those from older American cities that'd wrangled unsuccessfully for the honor of hosting the fair, had feared Chicago would put on a crude cattle show that would embarrass the nation. Instead, it was a monumental success. The leaders of Chicago's business community, many of whom were instrumental in planning and financing the fair, were exceedingly proud of their achievement. They regarded the fair as their personal triumph over East Coast elitism. A *Chicago Tribune* editorial trumpeted this sentiment, proclaiming: "It is a World's Fair as far as . . . exhibits are concerned. It is a Chicago Fair [in] energy, public spirit, enterprise, courage, and determination. . . . Chicago deserves the credit." *Shepp's World's Fair Photographed*, a popular souvenir booklet, went so far as to attribute divine grace to the builders of the fair. "Those who wrought these miracles," it proclaimed, "must have been very near to God."[1]

The exposition was intended to last only six short months, but proud Chicagoans, loathe to see it draw to a close, sought to perpetuate their achievement. To this end, in September 1893, a group of local businessmen conspired to establish a museum as a permanent memorial of the fair. Like the fair, the new museum was intended to be the finest of its kind: immense, spectacular, and all encompassing. Art, archaeology, science, industry, and history were all aspects of the museum's broad purview. Not only was the scope of the new museum determined by the fair, but its first collections and displays came directly from world's fair exhibitors, whether by purchase, donation, or, in some cases, salvage. With a liberal acquisitions policy, a generous budget, and a lot of hard work and hustle, Chicagoans picked their museum ready-made from the detritus of the fair. Haste was a virtue, as patrons had no intention of building their museum by patient accumulation. Indeed, instant prestige was the goal. Edward G. Mason of the Chicago Historical Society made this point in a stirring speech at the museum's opening-day ceremony. "Although so young," he exulted, "[the museum] already has a splendid past. . . . Other great museums have slowly grown. . . . This in an instant takes its place by their side."[2]

If instant prestige could indeed be had, then its purchase price would first have to be raised from the depths of America's worst recession to date. Edward E. Ayer, a Chicago lumber magnate who served as chair of the finance committee, was responsible for passing the hat. With persistent begging and pleading, he wrested a hefty contribution of $1 million from his merchant friend Marshall Field, the wealthiest man in Chicago. In gratitude, patrons named their new institution the Field Columbian Museum. Once Field committed to the cause, and the financial future of the institution seemed secure, reluctant Chicagoans like George Pullman,

FIGURE 1.1.

The Field Columbian Museum opened in June 1894 in the former Palace of Fine Arts, a sprawling brick-and-mortar building constructed for the World's Columbian Exposition. The building's plaster veneer would soon start decaying, endangering exhibits, staff, and visitors alike. Photograph courtesy of the Field Museum, CSGN21029.

Mary Sturges, and many others came forward to support the museum with generous donations.[3]

Field felt a personal interest in the welfare of his namesake institution. Although he visited frequently and offered abundant counsel, he nevertheless remained aloof from the formal administration of the museum. Ayer, on the other hand, who was chief instigator and who surrendered his sizable private collection of ethnographic objects to the museum, became president of its board of trustees. Martin A. Ryerson, a prominent Chicago lawyer and philanthropist, was first vice president. Harlow N. Higinbotham, an executive at Marshall Field's eponymous department store who'd served as president of the exposition, was chair of the board's executive committee.[4]

With financing secured, museum patrons began in earnest to acquire specimens from the fair. Early November witnessed the museum's first major purchases. On November 18, the Palace of Fine Arts, perhaps the most iconic building on the fairgrounds, was chosen to be the museum's temporary home. With its skeleton of brick and mortar, the palace was the exposition's most substantial building, although it wasn't designed or built to be a permanent structure. The building was also a sentimental favorite with fairgoers. "The greatest achievement since the Parthenon," sculptor Augustus Saint-Gaudens called it. On December 7, a group of museum organizers and their allies gathered at the palace to determine a preliminary arrangement for exhibits. At this meeting, Frederick J. V. Skiff assumed temporary charge of administering the museum. On January 26, 1894, the board of trustees appointed Skiff the museum's first permanent director.[5]

Skiff was a newspaper owner and manager from Denver, Colorado. He'd served two terms in the Colorado legislature. In 1889, the governor appointed him superintendent of the Bureau of Immigration and Statistics. In this capacity, he developed an exhibit on Colorado mining for the World's Columbian Exposition. So well organized and attractive was his exhibit that Skiff was tapped to serve the exposition as chief of the Department of Mines, Mining, and Metallurgy. He later ascended to deputy director general of the exposition under Higinbotham. Skiff was a dependable and trusted middle manager, amenable to the Chicago businessmen who would later become museum patrons. Yet he had no scientific training or any formal museum experience. Nor had he ever directed the work of research scientists.[6]

Director Skiff, who answered to the board of trustees, would be responsible for the day-to-day management of the museum. He took his responsibilities seriously and urged his staff to approach their work with similar gravity and dignity. Before entering the building each day, he doffed his hat "in token of the reverence which he felt for the activities being carried on there." Deeply loyal to the businessmen who gave him the opportunity to serve as director, Skiff felt it was his duty to manage the museum in such a way as to reflect well on his patrons, and to insulate them from criticism. He was particularly wary of the conduct of the museum's scientific staff. "The man of wealth whose investments extend into the public utilities," he wrote, "has a great responsibility in safe-guarding the public from the improper use of these utilities by the men employed to conduct them."[7]

Not everyone was comfortable with Skiff's qualifications. Some curators, for example, expressed misgivings about the new director's role, several insisting up front on maintaining complete control of their own departments as a condition of their employment. They were particularly concerned that Skiff's background might preclude him from managing the museum's scientific work. In fact, they took this idea for granted, one curator writing to Ayer that "Skiff has done wonders in arranging and running the Museum but of course he is not competent to criticize or arrange a Zoological Department." Skiff, who pitched himself as a facilitator, worked to allay the curators' fears. "I do not desire . . . to interfere in the slightest degree in the plans and the work of any [scientific] Department," he explained. "On the contrary, it is my desire, as it is my duty, to assist in carrying out the plans of the Departments, and to add whatever wisdom or energy I may have in supplementing the work so undertaken."[8]

Some of the harshest criticism of Skiff came from outsiders. The pugnacious Philadelphia paleontologist Edward D. Cope, for example, published a scathing editorial in the September 1895 issue of the *American Naturalist*. "Our hopes of the benefits to science to be derived from the Field Museum of Chicago have not been realized," he complained. He claimed erroneously that the museum's scientists had all quit "with expressions of dissatisfaction." He pinned the blame on upper management, singling out Field and Ayer for censure. He excoriated Skiff, calling him "impossible." Chicago's museum was beginning, he argued, "at the bottom of the ladder. . . . Perhaps Mr. Field . . . will some day come to the rescue, and insist that the director of the Museum shall be a scientific man of proved ability." Cope, an accomplished gentleman naturalist who'd squandered his family fortune through disastrous mining investments in the 1880s, now needed to earn his daily bread. No doubt he envisioned himself in the position of museum director.[9]

A FOCUS ON ZOOLOGY

Patrons intended their museum to be not only a place of public education, but also a research organization. Thus, a staff of scientific curators and technicians was necessary from the outset. These men—there were relatively few women in the masculine world of museum science in the late nineteenth century, and none at the Field Columbian Museum—were expected to assemble, care for, and exhibit their collections, and to publish original research as a part of their professional responsibilities. The museum's publication series, inaugurated in 1894, speaks to its mission to create as well as disseminate new knowledge. As Skiff explained, "a good museum, while devoted to the people, must also perform its function in the higher field of extending knowledge." Building collections and attracting a staff of first-rate scientists were imperative for fulfilling the museum's mandate for research.[10]

The museum made zoology one of its first and highest scientific priorities. In November 1893, at the close of the exposition, it purchased its first zoological collections from Ward's Natural Science Establishment. Large enough to fill thirty railroad cars, Ward's exhibit was a modest museum unto itself. According to a contract penned by owner Henry A. Ward, the purchase included three distinct zoological collections, namely: a collection of invertebrate animals, consisting of some 2,000 insects and 12,000 shells; a cabinet of stuffed and mounted animals with 305 mammals and 650 birds; and, finally, a comparative anatomy collection, which included 238 skeletons of birds and mammals, 47 fishes, and 64 reptiles and amphibians. Cases, labels, mountings, and fixtures of every kind were included in the sale. Also included was the delivery of these collections from the Anthropological Building—where they were exhibited during the fair—to designated rooms in the Palace of Fine Arts. The cost of this contract, the museum's first major purchase, was $100,000. The labor of dismantling, transporting, and remounting these specimens would be done by some of Ward's own staff. Among them was a young malacologist, Frank C. Baker, who—by virtue of being present, working with the collections, and organizing them into creditable exhibits during the first months of 1894—became the museum's first curator of zoology in March.[11]

Another blockbuster purchase, made early in 1894, was Charles B. Cory's private collection of birds. The asking price for this collection was $17,000. Cory negotiated a lifetime appointment as curator of ornithology at the new museum as a condition of the sale. Yet, he was ambivalent about the deal, writing: "I am both sorry and glad that the museum has decided to take my collection."[12]

Cory was born in Boston in 1857. An indulgent father, who'd made a fortune in luxury imports, encouraged his interest in natural history. At a young age, Cory began assembling his own collection of bird skins. In 1875, he made a grand tour of Europe. In Florence, Italy, while pitching pennies to beggars on the Arno River, Cory befriended another young American from Chicago—Martin Ryerson. Together they traveled to Egypt and hunted birds along the Nile. One year later, they were roommates at Harvard University. Cory worked at Harvard's Museum of Comparative Zoology under the de facto curator of birds and mammals,

Joel A. Allen. But, having little patience for rules and regulations, he never graduated.[13]

In 1877, Cory took an extended trip to Florida, thus beginning "a life of freedom and pleasure in the pursuit of natural history and sport which . . . might well be the envy of many a man." Over the next thirty years, he traveled extensively, collecting birds, hunting, fishing, and pursuing various outdoor sports and hobbies. He was a champion golfer and billiard player, a crack pistol shot, a raconteur, and a successful songwriter. He published a collection of "weird tales." With an ample allowance provided by his father, Cory "spent money freely and doubtless foresaw no future in which money would ever be a problem."[14]

When his father died in 1882, Cory inherited a sizable fortune. He built a summer retreat on a thousand-acre estate at Great Island, Massachusetts, where he established a private game preserve and bird sanctuary. He built a platform atop the abandoned Point Gammon Lighthouse for bird-watching. During this period, he authored books about his travels and articles describing new bird species in the *Auk* and other journals. Cory's friends and peers were other young men of means, including Chicago businessmen like Ryerson and Ayer.[15]

Cory's collection, which was strong in North American and West Indian birds, grew prodigiously, becoming more and more cumbersome. By 1892, it numbered some nineteen thousand specimens and filled three rooms in his Boston mansion. Cory began drawing up plans to build a small, private museum to house it. When Ayer learned of Cory's intention, he wrote a blunt and clumsy letter soliciting the collection for Chicago: "Of course you know we have started a great museum," he declared, "and it is the only place for your birds. What do you say?" Sensing an opportunity, Cory replied that he couldn't afford to give his collection away. As an alternative, he proposed to sell it for "a very reasonable price, lower still if I am made curator." By Cory's reckoning, the collection contained some fifteen thousand "selected" bird specimens, along with some eggs and nests and a few mammals. He had an excellent ornithology library, also, which "ought to go with it." The collection was in "splendid shape," according to Cory, and featured more than sixty of his own type specimens—the representative specimens he used to designate new species—as well as several "unique" and "some extinct species." Cory wanted the collection to remain in the United States; otherwise, he claimed, he "could have lately sold it abroad, but I should be like a carpenter without tools if I let it leave the country."[16]

Because he was still traveling and collecting extensively, Cory couldn't be tied down to the daily drudgery of curatorial work. Instead, he arranged to draw no salary from the museum and to serve in an honorary capacity. This agreement, which was inked in January 1894, was ideal for Cory's vagabond lifestyle. It gave him the prestige of an institutional affiliation and enabled him to continue building his collection without the responsibility of maintaining it. A succession of younger naturalists, including George K. Cherrie, William A. Bryan, and others worked as assistant curators in the department. These men did Cory's bidding and cared for the collection in his absence. The curator himself made one or two annual visits to the museum and offered general recommendations regarding his department to Director Skiff. As Cory explained, "I am at liberty to work as much or little as I please, but must direct the general government of the collection."[17]

SPENDING TO THE BEST ADVANTAGE

A flock of unsolicited advice accompanied Cory's birds. "Whether you take the collection or not I should like to talk to you about your new museum," Cory offered in a letter to Ayer. He knew his Chicago friends had ambitions to create something special right away. He encouraged their ambitions while also coaching patience: "If you go to work in the right way and spend your money to the best advantage," he advised, "you can have the finest museum in the world *in time*."[18]

On the other hand, Cory also stressed the urgency of making certain zoological collections quickly—before the clock of extinction had run down. Some of the world's rarest mammals, he noted, were not yet represented in museums. "Many of the beautiful antelopes of South Africa can now be bought for a reasonable price," he noted, "but like the Bison . . . are becoming extinct & soon cannot be bought at any price. . . . It is only a question of time when they will entirely disappear."[19] To build a zoology collection in a rapidly changing world was a matter of some urgency.

Cory was chiefly concerned with staffing. "If you want one or two good *men*," he advised, "I think I can help you to get the best . . . at reasonable prices." He assumed the museum would eventually want to publish its own journal, and knew quality scientists were necessary for sustaining a reputable publication. He knew whom to get and how to get them. "I know a splendid mammal and also a reptile man that I think could be had by going the right way about it," he insisted.[20]

The mammalogist Cory coveted was his former Harvard mentor, Joel A. Allen, who'd moved, in the meantime, to the American Museum of Natural History in New York. Cory offered to write a confidential letter to Allen, to see at what price the museum might retain his services. In June, he invited Allen to consider a position as curator of mammals and editor of the museum's scientific publications. He tried to sweeten the deal—for both of them—by suggesting that Allen could serve as acting curator of ornithology in Cory's absence and under his general supervision.[21]

Allen was flattered by the gesture, but not particularly tempted. He hinted that an increase in salary was appealing, but that "in every other way the change would be a great sacrifice." He balked at abandoning a zoological collection in New York that he'd worked so hard to develop, and that was finally yielding important returns. He was worried about the possible loss of autonomy in Chicago, insisting that as a condition of accepting the appointment, he would require absolute authority "to manage my own department as I see fit, subject only to the direction of the Trustees." And he was especially concerned about a potential clash with Cory over ornithology. "As regards birds," he explained, "I should not wish to interfere in any way with any plans of yours for literary work . . . but would not like to be cut off from original work in the line of ornithology." Despite his misgivings, Allen weighed the pros and cons of relocating to Chicago for five weeks. In the end, however, the American Museum made the decision easy by granting him a substantial raise. He wrote to Cory in July to thank him for his "kind intentions" and to decline the offer.[22]

This was a disappointment, of course, but it wasn't entirely unexpected. In any case, Cory had an alternative. In late June, when Allen was still considering the unofficial offer, Cory wrote to Ayer to warn that their prospective curator was "on the fence." Cory was by then all but convinced that Allen would never leave New York. He insisted that if they couldn't get Allen then they must get Daniel Giraud Elliot, instead. He closed the letter emphatically, urging: "Write me what I shall do about getting [E]lliot as I give you my word that the Museum cannot afford to lose" both. Cory suggested that Ayer should grant him authority to hire Elliot if Allen turned the museum's offer down.[23]

Two weeks later, when Allen's letter declining the position was forwarded to Cory's Great Island estate, Elliot was there for a visit. Cory offered him the position on the spot. But, like Allen, Elliot demurred. Cory sent Ayer a telegram to explain his reluctance: "Elliot will not accept unless in control of his depart[ment] only under trustees as regards scientific matters," he wrote, "will probably not accept less than 4000 may possibly accept 3500 no less what shall I do[?]" Ayer replied immediately: "No director would want to interfere with the scientific part of any department." But this was a clever dodge of Elliot's concern that promised nothing. Regarding a salary, Ayer stated bluntly: "We all think thirty five hundred ought to be enough." As if to emphasize the seriousness of the museum's offer, Skiff arrived for a hurried visit to Great Island the following day. Together, the three men hammered out a tentative arrangement whereby Elliot agreed to come to Chicago as curator on a trial basis for $3,500, plus $500 for personal expenses.[24]

It was probably the challenge of starting and running his own department—or of replicating or even besting what Allen had done in New York—that appealed to Elliot most. Rumors of the virtually limitless financial resources of the Chicago museum would've been a draw. Elliot accepted the challenge, taking—at fifty-nine years old—the first and only paid position he ever held. Cory, for his part, was pleased that his new zoology colleague was someone well known to him. Patrons of the Field Columbian Museum, too, felt fortunate for having secured Elliot's services, and especially for having lured him away from New York. As American Museum founder Albert S. Bickmore told a reporter for the *New York Times*, "Chicago would gain by the appointment of Mr. Elliot at the expense of New-York." The *Chicago Daily Tribune* likewise taunted the American Museum, stating that with the departure of Elliot for Chicago, the New York museum "will be a loser."[25]

Elliot was a good choice for the position. Given his unusual background and experience, there was arguably no man better suited in America to build a museum zoology department from practically nothing.

2

THE STRONGEST KIND
OF COMPETITION

Daniel Giraud Elliot was born on March 7, 1835, to a wealthy New York family. A child obsessed with natural history, he made and kept his own collection of bird skins. In Elliot's youth, Philadelphia was home to America's most active community of naturalists. He went there to learn ornithology under the wing of John Cassin at the Academy of Natural Sciences, where Elliot's feathered specimens jostled for table space with the reptiles of a young and precocious Edward D. Cope. Years later, Elliot recalled the loneliness and isolation he felt as a fledgling New York naturalist: "There was no one of my age . . . who sympathized with me in my pursuit; I was practically alone."[1]

Elliot had planned to enter Columbia College in New York in 1852, but frail health kept him grounded. Instead, he embarked on a series of trips to warmer climates for rejuvenation and self-improvement. At first, he confined his travels to the southern United States and the Caribbean. In 1857, he went to Brazil, where the strange fauna excited his interest in zoology. He sailed to Europe, then Egypt, everywhere adding to his growing collection of birds. From Cairo, he crossed the desert to Palestine on the back of a camel, touring Jerusalem and Damascus. He revisited Europe, establishing invaluable friendships with European naturalists. He also perused the storied natural history museums of London and Paris, which were filled with exotic animals plundered from the far reaches of their empires. He returned to New York a well-seasoned naturalist of independent means with a sizable private collection of birds. In 1858, he married Ann Eliza Henderson.[2]

Elliot published his first paper describing several new species of birds in 1859, only a month before Charles Darwin's *On the Origin of Species* appeared. Over the next decade, he published a series of lavishly illustrated bird monographs, painting most of the figures himself. These were the most elaborate bird books published in the United States.[3]

In the winter of 1868–69, when Albert S. Bickmore founded the American Museum in New York, he turned to Elliot for scientific advice. He also hoped to net the naturalist's collection of bird skins for the museum. Elliot, who planned to travel again indefinitely and had nowhere to keep the collection, agreed to donate it. Consisting of some thousand specimens, Elliot's collection was among the first material of any kind obtained by the new museum.[4]

Elliot then began a long period of living abroad, largely in London. He regularly attended meetings of the Zoological Society of London during this time, relishing the august company of eminent naturalists. "No such body of celebrated men . . . had ever before been assembled together, and we may believe it will be a long time before one equal to it will be again seen, for it was the height of zoological activity in the world," he remembered later. He spent countless hours in the society's incomparable research library, where he sometimes encountered luminaries like Darwin or Alfred R. Wallace. He continued to write illustrated bird monographs, including one on birds of paradise and another on hornbills. He also published his first book on mammals, *A Monograph of the Felidae*.[5]

While in Europe, Elliot examined and acquired several private natural history collections for the American Museum. The trustees, through Robert L. Stuart, the museum's second president, had commissioned him to make purchases of specimens that in his judgment would be advantageous to the museum. One day he happened by the window of a London taxidermist where he spied a specimen of the extinct great auk in winter plumage—a very rare bird, indeed—and bought it at a hefty price. When the trustees hesitated to honor the purchase, Stuart, who appreciated the great value of extinct species, agreed to pay for the specimen himself.[6]

Elliot, now forty-eight, returned from Europe in 1883 and settled on Staten Island. He continued to write descriptive papers and monographs on birds at the American Museum. When the museum's new president, Morris K. Jesup, organized his institution's first collecting expedition in 1887, Elliot was chosen to lead it. The expedition went to Montana to hunt bison in the same area where William T. Hornaday, taxidermist for the National Museum in Washington, DC, had—only the previous year—harvested dozens of specimens from one of the last remaining wild herds in America. When no living bison were found, the New York museum was forced to purchase several costly skins in order to move forward with a planned group display of these iconic mammals. The lesson was painfully obvious: collect specimens now, while they could still be taken in the wild, or pay a heavy price for specimens of the same species later. Joel A. Allen, in explaining the predicament of the bison expedition in the museum's annual report, clearly advocated for practicing salvage zoology: "In view of the fact that the larger mammals of North America are being rapidly exterminated . . . it seems highly desirable that the friends of the Museum should provide the means for securing groups of these interesting animals . . . before it becomes too late to obtain them."[7]

Elliot was among the most celebrated naturalists of his day. He had a reputation as a tireless, diligent worker, with great powers of concentration and a remarkable memory. His illustrated monographs had earned abundant accolades. He was an active member of many scientific societies, foreign and domestic. He was instrumental in establishing zoology at the American Museum. "With a practical knowledge of the requirements of a working museum, Dr. Elliot was also a man of affairs who could impress the trustees . . . with the soundness of his views," wrote American Museum ornithologist Frank M. Chapman.[8] In other words, he was a wealthy, well-bred man cut from the same cloth as the Field Columbian Museum's patrons. For this reason, they trusted his judgment to develop the scientific side of the museum in ways they might have been reluctant to do with a mere naturalist.

MAKING ROOM FOR ELLIOT

To make room for Elliot, the Field Columbian Museum had to dispense with Frank C. Baker, the sitting curator. Baker's tenure began to unravel when he and a coworker, Henry K. Coale, wrote a paper describing several new species from specimens in the museum's modest ornithology collection. (Cory's collection hadn't yet arrived.) The museum was soon to launch a fledgling scientific publication series, so Baker submitted the manuscript to Director Skiff for consideration. Knowing nothing of birds, Skiff passed the paper to Cory for his opinion. Cory read it and examined Baker's type specimens at the museum in June 1894. "I must say," he wrote to Skiff, "a more unscientific and badly written paper I have never seen." Later, in New York, he allegedly discussed the work with Allen, who, according to Cory, remarked "that if by any chance the Chicago museum had published it as it stands in its bulletin it would have been the laughing stock of the scientific world." The paper was so full of errors, in Cory's judgment, that it should be rejected rather than revised. "For heaven's sake don't print the thing," he pleaded, "let [the authors] send it to the 'Auk' . . . and they will hear a few remarks . . . which will allay their thirst for notoriety." Not content merely to kill the paper, Cory wrote to Ayer and urged him to "get rid of Baker." Cory argued that "by barring out incompetent hustlers, we will make that [zoological] department second to none, but it will take time, work and ability." There is no record of any action taken by the museum, yet something must've been done or said to Baker, who resigned on June 26, 1894.[9]

Baker's departure also spelled the end for his small staff, which created an opportunity to hire an assistant curator to serve under Elliot. In January, when Cory first started agitating to hire Allen, he'd also alluded to another potential curator. In August, after arrangements with Elliot had been finalized, Cory wrote to Skiff with specifics: Oliver Perry Hay, a specialist in fishes and reptiles, was a "very good man . . . if Elliot needs an assistant he would do well." Elliot met with Hay in October to offer him a subordinate position as assistant curator. "Dr Hay . . . has agreed to come as my assistant for six months at $90 per month, which I have stated to him was the outside figure at present," Elliot explained to Skiff. "I took him to the Librarian & authorized them to give Mr Hay the keys of the [exhibit] cases & have outlined sufficient work for the Dr to busy himself with."[10]

Cory, despite his long absences and more-or-less honorary position, soon showed himself to be prickly about his status in the museum and extremely territorial with respect to the birds. In June 1895, he wrote confidentially to Skiff demanding that in any official museum publication he be given the same title as his fellow curators. "I do not care for that sort joking in private," he explained, "but do care to stand on equal footing in all ways with my associates[,] especially as I have a right to the title."[11]

Baker's ornithological manuscript provoked a more serious crisis of authority. As Cory explained in an irate letter to Skiff, "Mr. Baker states he has received much assistance from his assistant in ornithology[,] Mr. Coale. [P]lease let us stop this farse. If my contract means anything I have [sole] control of my department." To ensure that his message was getting through to museum patrons, Cory went over Skiff's head with a second letter to Ayer highlighting many of the same complaints. He wrote: "Please see that a stop is put at once to any interference with my department. This was thoroughly understood as our contract, as you know, explicitly states I was to have entire charge of the department under no one but the Board of Trustees." He would never have accepted the position, he claimed, without such an agreement. Regarding Skiff, he wrote that he wouldn't allow an "unscientific man to dictate the . . . arrangement of the collection, nor anyone put in my place without being selected by me or with my permission."[12] Cory made it clear that not only was he opposed to Baker and Coale dabbling in ornithology, but he also resented Skiff's "interference."

Elliot would have to endure his share of conflicts with Cory. For example, Cory allowed that it was appropriate to exhibit the museum's bird skeletons together with other osteological specimens in a single hall. He wrote to Skiff, however, to suggest that "possibly it might be wise to have it understood that anything to do with birds belongs to my Department, and might prevent question[s] in the future when I might possibly desire to use them in my lectures or work." He was likewise possessive of the ornithological library he'd sold to the museum. Having heard a (false) rumor that Elliot wanted to relocate these books, he wrote a petulant letter to Skiff explaining that it was "understood between Mr. Ayer and myself that my library should be kept in *my room*. . . . I trust you will insist that this be done." Finally, Cory was adamant that Elliot be referred to officially as curator of the Department of Zoology except Ornithology. To do otherwise, Cory insisted, was an injustice to his position. "You remember you told Elliot he would be curator of zoology *except* Ornithology," Cory reminded Skiff. "Simply curator of zoology is not fair to me—this you told him but I do not think he understands it so—my department is of course entirely separate, by agreement."[13] Elliot objected to the title and was reluctant to use it, but he acquiesced in order to placate Cory.

The proper placement of bird skeletons and books, Elliot's title—these were minor squabbles, easily adjudicated. Years later, however, the overlapping research interests of Elliot and Cory would prove much more consequential.

BEEFING UP ZOOLOGY

To the public, the museum presented the appearance of comprehensiveness and completeness. The first edition of the *Guide to the Field Columbian Museum* crowed that the zoology collections were "very extensive, covering fully this field of science from *Protozoa* to man." Privately, however, the museum acknowledged that these collections "did not at the beginning rank with the other Departments." Consequently, for its first few years, the museum devoted the lion's share of its resources to beefing up zoology. Cory and especially Elliot benefited from the museum's liberality.[14]

In December 1894, Skiff was asked to estimate the cost of enlarging museum operations in the coming year. Museum patrons, after a long year of toil to get the museum ready for opening day, were dreaming of expansion. Skiff passed this mandate down the chain of command, asking each of the curators to convey their "wishes and hopes" for the futures of their respective departments.[15]

This was Elliot's first task at the museum. In reply, the new curator penned a thorough report concerning his department. He addressed the crucial issue of expanding its modest collections, enclosing several long lists of specimens from other museum zoologists and from various natural history specimen dealers. One of these lists was from London's Rowland Ward Limited, which included dozens of specimens from Africa. Of the specimens enumerated, Elliot noted, "with but few exceptions, there is not a single species named that is not most desirable to add to the Division of Mammalogy." Elliot emphasized the importance of acquiring specimens now, while they could still be found in the wild. African wildlife, he argued, was especially vulnerable: "The Antelopes of Africa are proceeding rapidly towards extinction & the day is not far distant when like our own Buffalo the majority of existing forms will disappear from the earth. The 'White

Rhinoceros' . . . is already practically extinct. There is not a specimen in this country, & but two or three in all the Museums of the World."[16]

Elliot stressed that the museum's newness placed it at a competitive disadvantage. "This is the youngest Museum in the World," he emphasized, "& it has entered the field at the eleventh hour. The time for acquiring large collections . . . , which . . . puts a Department at once . . . ahead of its rivals, has . . . passed." Regrettably, the world's great private collections had already been absorbed by older museums. "We have therefore to build our structure brick by brick, a slow and weary process." In Elliot's view, collecting was a zero-sum contest, whereby any specimen acquired by his department was one that rival institutions could no longer obtain. His gain was their loss. Curators in other museums viewed the competition in a similar light, or so Elliot believed. "From an experience of more than a quarter of a century in Museums," he wrote, "I know that we have . . . to enter into the strongest kind of competition, and that nothing that is rare, especially valuable or desirable will ever be permitted to enter the walls of this Museum, if rival kindred Institutions can prevent it." For the museum to succeed, he argued, it must "be prepared to take advantage of every opportunity for obtaining desirable material, and to reach a quick judicious decision that experience and familiarity with the subject will enable us to give."[17]

As part of their plans for expansion, the curators had been asked to submit a budget. Elliot, however, declined to recommend any particular sum. This, he wrote humbly, was "best left for the [Executive] Committee to decide." He would only say that "be it small or great, it will be used . . . to the best advantage of the Museum as my experience and judgment may guide me."[18] This was a thinly veiled plea for generosity and greater autonomy.

Skiff, impressed with the frankness of Elliot's report, forwarded it to the executive committee with his endorsement. Convinced that zoology at the museum needed to be completely overhauled, Skiff noted that "the most casual investigation of the real condition of the collections and individual specimens of this very important department, will convince any person that the Museum has made a very feeble and a very poor beginning in this field of science." He noted that the worst material acquired from Ward's Natural Science Establishment was the mammals, and that the installation done under Ward's contract was "neglectful and indifferent in the extreme." The material was "a poor lot to begin with and [Ward] injured it all he could in placing it in position." Apparently, these facts had already been "regretfully" discussed by the executive committee at a previous meeting.[19]

Skiff argued that zoology should be the museum's highest priority. "If the Museum is to expend any appreciable amount of money upon any of the departments," he wrote, "the expenditure should be made upon Zoology and Ornithology." Zeroing in on an exact figure, however, was difficult. "The department needs almost everything, and the amount that can be expended judiciously" was practically unlimited. He lobbied for "a liberal appropriation."[20]

Skiff had one additional concern about zoology. "The Curator," he noted, "does not give the importance I think should be given to the question of groups of familiar mammals showing their haunts and habits." The director was somewhat ahead of the curve in pushing his institution to favor a new style of lifelike group or habitat displays over more traditional taxonomic exhibits. (Most American museums only began to invest heavily in habitat displays in the 1910s and 1920s.) A capable taxidermist, Skiff suggested, using "easily procured" material, could greatly enhance the "interest and instructiveness" of the zoology exhibits with group displays. Thus, it would be necessary to add one or two skilled taxidermists to the zoology staff. He also advocated a large budget for specimen purchases and for fieldwork to acquire specimens locally. The total appropriation recommended for Elliot's department was a whopping $16,500.[21]

Cory, meanwhile, also penned a report to Skiff about the Department of Ornithology. Because he'd spent so little time at the museum, Cory's report lacked the details provided by other curators. Nevertheless, according to Skiff, "the Curator hits at once the weak point in this collection, native birds, and suggests a means for strengthening the department in this particular, by advocating . . . for field work." The bird curator also outlined a plan for developing new exhibits. Skiff was impressed, writing that Cory's "advocacy of the construction of impressive groups of birds, is in direct line with the policy of the Executive Committee as informally declared." To execute this work, however, another taxidermist was required. Lest the committee should see this request as an unnecessary duplication of personnel, Skiff explained that Cory insisted that such a position would be needed "exclusively for Ornithology, and that a man to work in both Zoology and Ornithology could not be satisfactorily arranged." The director conceded that considerable money had already been spent to acquire Cory's collection. "But it may be said that for the purposes of securing exhibition material, or improving or re-arranging the material with which the Museum started, nothing has so far been done since Mr. Ward placed the birds within the poor cases in which they . . . are installed." The department, Skiff noted further, was also understaffed. The assistant curator, George K. Cherrie, who was supposed to have immediate charge of ornithology, had been away doing fieldwork for most of the time since his appointment. Left unsaid was the fact that the department was run by an absentee curator. Skiff recommended a total appropriation of $5,000 for ornithology, including salary to hire a taxidermist, an appropriation for fieldwork, and funding to purchase new bird material and to cover the cost of creating group displays.[22]

CURTAILED SPENDING

The executive committee was slow to approve Skiff's recommendations. In fact, as winter turned to spring, and spring to summer, the very idea of growth in the scientific departments was abandoned in favor of an unofficial policy of maintaining the status quo. In the first few months of the museum's existence, spending had been profligate on new collections and on requisite upgrades to the building. However, the administration quickly adopted a new, more conservative approach to expenditures soon after the museum first opened to the public in June 1894. There were two reasons for caution. First, the founders had underestimated the staggering cost of operating a museum, and they were looking for ways to augment the institution's income or to build an endowment to maintain it. Second, they were reluctant to invest significantly to maintain the Palace of Fine Arts when it was obvious that a larger, permanent home would be needed for the museum in the long term. This conservative approach came directly from the museum's eponymous patron, Marshall Field. In a June 1894 letter, Field explained: "My judgment is that we should go slow in all expenditures from this time on, at least until we know definitely where the permanent home of the Museum is to be and where the money is to come from to maintain it."[23]

This new attitude curtailed spending for the remainder of 1894. The outlook for the following year seemed better, however, as the executive committee had solicited plans to expand the museum's scientific programs in 1895. Unfortunately, when cost estimates for expansion came in higher than expected, the committee hesitated. In fact, it debated the issue for months, withholding approval on departmental appropriations all year, and only approving expenses in a piecemeal, opportunistic fashion. This was a frustrating time for the curators, particularly for those—like Elliot—who'd come to Chicago because it seemed like such a promising opportunity to pursue science extravagantly, with few financial limitations.

For Elliot, the highest priority was to grow the museum's zoological collection. Ideally, he wanted to be granted authority to purchase specimens according to his own best judgment—just as he'd done for the American Museum—rather than submitting individual requests to the director for his scrutiny and approval. But the executive committee refused to make this arrangement. Elliot was especially anxious to acquire African mammals that were then being offered for sale by Rowland Ward. Africa was remote, exotic; its native mammals were large, showy, and strange. A collection of these specimens would be of great scientific value. At the same time, an exhibit of African mammals was likely to appeal to Chicago's museum patrons, who were keen to showcase the global reach of themselves and their city. "The . . . specimens are all desirable," Elliot emphasized to Skiff. There were twenty-five skins, and the price asked was "about $1350, an average of fifty-six dollars apiece, which I consider . . . a rather low figure." Most of the specimens were unmounted skins. Elliot thought this advantageous, however, as he preferred that "such important pieces should be mounted . . . under my supervision." On the other hand, he declined to recommend the purchase of an especially rare and costly white rhinoceros. He didn't think it was overpriced, and he regretted losing the opportunity to acquire an animal that was teetering on the brink of extinction. "Yet in view of the fact that the Department of Zoology . . . is . . . in need practically of everything . . . I cannot at the present time advise that so much money be expended for one specimen, no matter how great an acquisition it might be."[24] To acquire specimens by purchase, Elliot concluded, especially those that were rare or nearing extinction, was simply too costly. Finding a more economical means of acquisition became one of the curator's chief concerns.

Spending limits crimped scientific activity throughout the museum in 1895, yet Elliot was seldom idle. He completed an inventory of the collections, comparing the specimens present against an assortment of catalogs. He did preliminary research on the most interesting of the incoming specimens. In the spring, he gave a series of popular lectures at the museum. He agreed to give a zoology course at the University of Chicago in the fall.[25] And he supervised the work of the department in general. Busy though he was, he cannot have been happy with the relatively slow pace of activity in his department in 1895.

THE WINTER OF ELLIOT'S DISCONTENT

Elliot's first year at the Field Columbian Museum was a disappointment. He disliked Chicago and its pretensions as a city equal to his beloved New York. In a letter to Allen, asking him to send a publication he needed for work, he complained: "there is not a copy of this Journal in this highly cultured & only Metropolis." He was content with his Hyde Park neighborhood, he wrote, but largely because it was "so far removed . . . from the soot & general nastiness of the town itself." Even the city's tobacco stocks were unsatisfactory: "Since I have been practically compelled to smoke 'domestics' since my arrival in Chicago," he wrote in a note to Skiff enclosed with a box of cigars, "I take a kind of fiendish pleasure in causing another fellow to suffer likewise." Perhaps the best indication of Elliot's dislike for Chicago was his rampant absenteeism. He worked for several weeks in New York in April and early May. He took a family vacation of more than two months from early July through sometime in September.

And he spent several weeks attending a zoological meeting and working at the American Museum in November and early December.[26]

Elliot wasn't entirely happy with his circumstances at the museum, either. The paltry resources to pursue science and the needlessly cumbersome and time-consuming way of doing business at the Field Columbian Museum—as he saw it—were his most vexing problems. He'd established a good working relationship with Director Skiff, addressing him in personal letters as "my dear Skiff," "my friend," and "my trusty friend." Indeed, they spent so much time conferring with each other that they both began referring to the extra seat in Skiff's office as "Elliot's chair." Yet he also found the "ceremonious" requirements of their formal relationship tiresome.[27]

Elliot's relationship with Cory was more difficult. They'd known each other for many years and had even been friendly. In Chicago, however, they became rivals. Cory enjoyed privileged access to several museum founders, including Ryerson and Ayer, and he worked these channels constantly to defend his department from what he saw as Elliot's transgressions. Elliot knew about Cory's territorial behavior from all the complaints he had to field from Skiff: complaints about Elliot's title, and the name of his department; complaints about control of the bird skeletons in the osteological hall; complaints about the placement of the ornithology library. And Elliot still objected strongly to the seemingly arbitrary separation of ornithology from zoology, especially as he self-identified primarily as an ornithologist. Finally, it seems that Elliot didn't take Cory seriously as a naturalist. In 1895, Elliot published a popular book on North American shorebirds. Cory was upset, writing to Elliot that had he known about the book, he would've hesitated to publish his own work, which was coming out soon and apparently covered some of the same ground. Elliot was incredulous. "What is [Cory] going to astonish us with?" he asked Allen in a letter. "Have you heard?"[28]

In April, Cory had a near-fatal encounter while hunting in Florida. Elliot was blasé and even a little mocking in relating the news to a mutual friend. "Cory came near being killed by a panther," he wrote, "which . . . sprang at him, knocking the old man down, bit & scratched his face, shoulder & arms & would have killed him, had not the hunter with him shot it. . . . I suppose what is left of the wreck will be coming this way before long. . . . Mighty hunter!!"[29] Zoological fieldwork was often dangerous—even deadly—surely Elliot could be expected to show some compassion for a colleague mauled in the line of duty?

Elliot could also be fussy about his physical surroundings, and the museum's dilapidated building gave him fits of displeasure. He'd been perfectly comfortable at the American Museum, "a building," he wrote to Skiff, "with whatever defects it may possess, [that] is in my judgement the best and most excellent" purpose-built museum building in the world. The Field Columbian Museum's edifice, by comparison, left much to be desired. It was plagued by myriad problems: threadbare wiring; cracked, broken, and leaking skylights; falling plaster; dry rot; walls that oozed in humid weather; a mysterious odor of decay. The rats were so bad that Skiff deployed a dog, two cats, two ferrets, and no fewer than twenty-five traps to combat them. Worst of all, for Elliot, was a new steam heating system that was badly overmatched by Chicago's notorious winter weather. The curator complained frequently about the lack of heat in the museum. "I thought of you all hugging the radiators during the late blizzard," he wrote to a colleague. "Expect to arrive next week about Wednesday & begin to hug myself."[30]

When the one-year anniversary of his employment approached, Elliot began to grumble about leaving. What set him off was that he somehow got wind that the museum intended to reduce his compensation to $3,500. Elliot had declined to accept that sum the summer before at Cory's Great Island estate, so Skiff and Cory had conspired to offer him an extra $500. Apparently, they didn't make it clear that this was meant to apply to the first year only, to cover Elliot's moving expenses. One year later, the museum wanted to set the record straight on the discrepancy. Ryerson asked Cory to provide his best recollection of the negotiations with Elliot. "I certainly understood it to mean for the expenses of moving to Chicago and to apply to the first year only but it was not stated that it should be for one year, no[r] was it stated that it should be for more than one year," Cory recalled, unhelpfully. The idea, he explained, was to get Elliot to agree to accept the job on a trial basis and "see how he . . . liked the place and how the Museum authorities like him." Cory was certain that Skiff understood the arrangements exactly as he did and insisted that no business had been done after the director left.[31]

Skiff genuinely feared Elliot would leave. In fact, he was anxious to retain the services of other department employees, noting that "if there is to be a change in the head of the Department of Zoology, [they] will be very valuable to the Museum." Later, in a letter to Cory, Skiff once again addressed the ornithologist's repeated complaints about Elliot referring to himself as curator of zoology. The director assured Cory that in its official publications, the museum consistently maintained the distinction between the two departments. (This was not entirely true.) He then begged Cory to be patient, writing: "I apprehend that this and other matters pertaining to . . . Zoology . . . will have determinative action within a very short period."[32] In other words, Skiff expected Elliot to fly the coop.

The museum, however, was finally poised to implement its long-planned program of expansion. Members of the executive committee, who were more than satisfied with Elliot's job performance, were eager for the curator to remain. "We are much the weakest in Natural History," Ayer remarked in a December 1895 letter. "[We] are anxious to extend in that direction as fast as it may be found possible." Had Elliot decided to leave, it would've upset the executive committee's plans for rapid zoological expansion. Therefore, on December 13, Ayer and Higinbotham sent him a letter—now lost—outlining the terms of the museum's new offer. They didn't raise the curator's salary, but they did offer him something that compelled him to stay.[33] The obvious conclusion is that the museum assured Elliot that the trickle of financial resources for his department would begin to flow more abundantly in the new year. Elliot, it seems, was given the green light to begin planning something truly audacious.

3

WE LOOK FOR GREAT RESULTS

A fortnight later, in a hopeful letter to Skiff, Elliot made an extraordinary proposal. "Convinced as I am that the only satisfactory economical & proper method to be adopted for the development of the Department . . . is work in the field," he wrote portentously, "I wish to bring to your consideration . . . the advisability of an expedition" to collect animals in Africa.[1]

By the late nineteenth century, European imperial powers had effectively divided Africa among themselves. Their reasons for colonizing Africa were complex, yet the mission to spread Western civilization or Christianity was often their rationale.[2] The infrastructure they installed—ports, railroads, telegraph service, and more—helped make expeditionary science more practical in Africa by the 1890s. But, at the same time, their civilizing mission imperiled native wildlife by destroying natural habitats. Western naturalists assumed the role of colonizing Africa's flora and fauna, taking specimens in prolific numbers for their museum collections, and naming and describing plants and animals as they saw fit.

Elliot argued that African mammals were quickly disappearing, which leant an urgency to his proposed collecting mission. "Many species which formerly like our Buffalo were seen in herds of thousands are already extinct, & it is but a question of a very short time when most of those remaining will be gone forever," he pleaded. The curator raised the specter of extinction not to tug at Skiff's heartstrings or to advance some kind of conservation agenda. Rather, Elliot's motive was salvage. He urged Skiff to support an expedition now, while African mammals could still be had for the taking. Wait too long, he warned ominously, and the window of opportunity would be closed for good. Swift and decisive action would give the Chicago institution a leg up on its American rivals. "There is no collection of these animals in this country," Elliot emphasized, "& if we can secure an adequate representation of them, the Field Columbian Museum will in that respect surpass all kindred institutions on this Continent."[3]

On Elliot's recommendation, the museum had recently purchased a variety of African antelopes from Rowland Ward. Mounted for display at the museum less than a week earlier, they were already buzzing with appreciative crowds. Elliot wanted to expand this collection, in part, due to its obvious popular appeal. But acquiring specimens one at a time by purchase was slow and cost prohibitive. Larger and rarer game, like elephants and rhinos, was especially pricey.

Fieldwork was Elliot's solution. By hunting and trapping animals in the field, the curator could cut out the middleman. He could collect exactly what he most wanted, in ideal condition (e.g., largest antlers or tusks, most colorful coat). In an animal-rich environment, he could afford to be choosy about which specimens to keep, and which to discard. "We can get enough fine specimens," Elliot explained later in his journal, "& need not encumber ourselves with poor ones."[4] Moreover, by collecting en masse, he would enjoy a proportionate savings in cost. Surplus specimens could be exchanged with other museums, keeping the best specimens for his own collection. Better still, he could make direct observations of the animals' ecology and behavior while hunting. He could also acquire authentic accessory objects for building habitat groups, including rocks, soil, plants, and more. Finally, Elliot was an avid sportsman, and fieldwork involved the curator's direct intervention in nature. Success in the field required masculine virtues like courage and perseverance. To collect animals for oneself was to master them, and to earn and to deserve them in a way that couldn't be conveyed by purchase.[5]

Elliot proposed to do a single season's collecting, leaving Chicago in March and returning in November. He estimated his expenses to be between $8,000 and $10,000. This sum wouldn't have to be spent at once, he emphasized, but could be paid out gradually, as needed, over the course of the trip. He could make the expedition cost effective by accumulating specimens worth between $25,000 and $30,000. "I feel very strongly in this matter for I *know* its importance," he argued, "and although it will be no holiday trip, but . . . a journey of hard work, anxiety and great discomfort, yet its . . . importance makes me willing . . . to lead such an expedition which will . . . be of the greatest benefit to the Museum." It was time-sensitive work, he stressed. "A delay of even one year may be fatal to the enterprise."[6]

Elliot's proposal was "considered and discussed at length" at the next meeting of the museum's executive committee, on New Year's Eve, but a final decision was deferred until the following more sober meeting in three or more weeks.[7] The curator would have to be patient.

Meanwhile, the new year brought good tidings. First, Skiff informed Elliot that $6,000 had been appropriated to reinstall nearly every exhibit in his department. This would require dozens of new, higher quality, custom-built, mahogany and plate glass display cases. Skiff emphasized that he wanted this important work to begin at "the earliest possible moment."[8]

More importantly, Carl Ethan Akeley accepted a permanent position in Elliot's department as chief taxidermist.

ENTER CARL AKELEY

Akeley was born on May 19, 1864, in Clarendon, New York, and grew up on a small family farm. He loathed school and had limited formal education. But he loved the outdoors and its wild animal life. Awed by a visit to a free exhibit of stuffed and mounted animals, Akeley taught himself the rudiments of taxidermy with the help of a how-to book. He soon had business cards printed proclaiming that he did "artistic taxidermy in all its branches." Eventually, he determined to seek a position at Ward's Natural Science Establishment, in Rochester.[9]

Years later, Akeley still remembered the crippling anxiety he felt when arriving at Ward's. There was an ominous-looking gate made from the toothy jaws of a sperm whale. A sign was posted there, warning: "This is not a museum but a working establishment, where all are very busy." This would seem to bode well for a hard worker like Akeley, but the sign made him feel unwelcome. He was admitted to see the proprietor, Henry A. Ward, a "very busy, very brusque, and very fierce man." Akeley recalled that he'd never had "a worse moment than when this little man snapped out 'What do you want?'"[10] Akeley sheepishly handed Ward his business card and was hired on the spot.

Ward's seemed a promising fit for young Akeley. A booming commercial outfit that supplied thousands of natural science specimens of all kinds to American colleges, museums, and private cabinets, Ward's was a place to hone his skills, learn the latest and best techniques, and earn a respectable living. But Akeley quickly grew disillusioned with Ward's factorylike methods, which privileged quantity over quality. One regrettable expedient was to sew up a skin like a pillow, stuff it with straw or excelsior, and then pull it in with needle and thread to ape the shape of the animal's body. This technique, which Akeley referred to disparagingly as the "old straw-rag-and-bone method," produced disappointingly unlifelike results. Akeley experimented with new methods of skinning and mounting animals on his own time, with some success but little or no encouragement. After four relatively frustrating years, he decided to move on.[11]

In 1886, Akeley relocated to Wisconsin's Milwaukee Public Museum at the invitation of a close friend and former Ward's colleague, William M. Wheeler. He worked first on a contract basis, later succeeding to full-time taxidermist in 1889. He worked in his own studio and was free to experiment with new techniques, including mounting animal skins on lifelike plaster mannequins. The pinnacle of his work there was his now-famous muskrat diorama—one of the first of its kind in America—completed in 1890. He resigned his position in March 1892 to concentrate on his private taxidermy business in Milwaukee, where he mounted deer heads and made bird panels and "grotesque groups" of anthropomorphized animals for home decor. His most important commission during this period was a trio of mustangs he mounted for the Smithsonian Institution's exhibit at the World's Columbian Exposition in Chicago.[12]

Alas, the market for artistic taxidermy was stiff and lifeless. By 1895, the Carl E. Akeley Company was failing. Akeley was working obsessively on a revolutionary diorama of Virginia deer that he hoped to debut—and sell—at another world's fair in Paris. But this speculative, costly, and labor-intensive venture was far from finished. In the meantime, to make ends meet, he was reduced to refurbishing fur coats. Meanwhile, Wheeler—now at the University of Chicago—convinced William H. Flower of the British Museum to offer Akeley a position as taxidermist. With no better prospects in the offing, Akeley accepted. On his way to New York, however, he stopped to visit his friend in Chicago and made a fateful visit to the Field Columbian Museum. On a tour—allegedly with Elliot—he gaped at the museum's grand interior spaces and its menagerie of shabby-looking Ward's mounts. The museum must have impressed the cash-strapped taxidermist as a golden opportunity to make his mark. He decided to stay in Chicago.[13]

Akeley worked first on mammal skins Elliot had purchased. These included monkeys and orangutans from Borneo, a Philippine deer, and a series of antelopes from Africa. At least one of the antelopes was spoiled and failed in the mounting. The seller, Rowland Ward, substituted the skin of a slightly more expensive white-tailed gnu, sending a bill for the difference. Elliot was pleased with the exchange, explaining to Skiff that "as this animal is now *extinct* & the probability of our getting a specimen in the future not very good, I do not [resent] the small advance in price." Elliot earmarked the gnu for his new taxidermist, whose superlative skills he immediately recognized. In fact, he was ecstatic about Akeley's first mounted mammals, which were seamless and eerily lifelike, a monumental improvement on the wooden-looking mounts purchased from Ward's. To a colleague he boasted: "I got a string of Antelope into the cases on Saturday . . . , & they drew crowds yesterday. As specimens they are not to be beaten anywhere."[14]

The work was done at a new warehouse and workshop space at the corner of Fifty-Sixth Street and Jefferson Avenue, a short walk from the museum. The museum purchased this Hyde Park lot with a house and barn in 1895 to provide much-needed room for storage. Another critical need was to find a space apart from the main building for some of the more noxious and messy museum functions, including carpentry, plasterwork, and, especially, taxidermy, "three kinds of labor which could not be permitted within the Museum building."[15]

By October, the appropriation for Akeley's contract work was dwindling.

Elliot brought the matter to Skiff's attention: "As you are aware I have engaged now for some months . . . C. Akeley, to mount the Antelope skins purchased early in the winter," he wrote. "The work that he has accomplished and that which I have laid out for him to do, will exhaust the appropriation." With his funding drying up, Akeley approached Elliot to inquire about his long-term prospects with the museum, hinting that he would soon be forced to make other arrangements. Elliot was loathe to see that happen. "Mr. Akeley's work is equal to the best I have ever seen in any land, and it is rare to find a man that can do nearly as well. . . . It would be a misfortune . . . to lose his services," he wrote with studied understatement.[16]

Elliot then raised the issue directly with Ayer, who asked the curator to ascertain how much his taxidermist expected in salary. Akeley asked for $2,500, which Elliot considered "fair" remuneration. "There are not a half dozen men in the United States, perhaps not three, able to equal his work," Elliot explained. "There is a great deal for him to do, much bad work to do over when possible, beside the new now on hand, & that I hope to acquire by new material." A taxidermist, he pointed out, "is a necessary adjunct to . . . Zoology, as there is always something to be done, and no method is so expensive as having the work done by the piece."[17]

Discounting Elliot's advice, the executive committee offered Akeley a full-time position for $2,000 per year, on a month-to-month engagement. Akeley reluctantly accepted the lower salary, but he strenuously objected to the lack of job security. Skiff, concerned that Akeley wouldn't accept, wrote to Higinbotham to plead Akeley's case. "I feel that we are very fortunate in securing Mr. Akeley's services at the price stated," Skiff wrote. "He is a superior workman. . . . He feels that he should receive a higher salary than he has concluded to accept, and only does so because he is interested in the Institution." When the executive committee adjusted its offer in his favor, Akeley accepted, beginning his new appointment on January 1, 1896.[18]

One week later, Skiff threatened to fire him. The director had a litany of complaints that he enumerated in an indignant letter to Elliot. First, Akeley had been entirely too careless about mitigating fire hazards in the taxidermy workshop. The building superintendent had tried and failed to impress on him the gravity of his responsibility. Skiff urged Elliot to try anew. Second, Skiff had been informed that Akeley had "left his work without leave of absence from the Director." Finally, Skiff complained that the taxidermy workshop had been occupied late into the night, and that gas had been seen burning in the building as late as nine o'clock.[19]

Exasperated by what he regarded as needless oversight, Elliot replied to the various charges in a letter written the following day. Akeley, Elliot explained, had gone to Milwaukee at the curator's direction to secure an assistant taxidermist. "He left on the 6 P. M. train & returned the next morning." Gas had indeed been burned late at night, Elliot admitted, "for I have been there myself as late as 9:30 . . . overseeing Museum work." The curator explained that it was often necessary "when an especial piece of work is in progress to continue until it is finished, no matter what the time may be, or run the risk of spoiling the specimen." As for the fire hazards in the taxidermy workshop, Elliot conceded that there would always be some risk of fire, but insisted that Akeley, as a matter of professional routine, reduced those risks to a minimum. "Mr. Akeley is not a child," Elliot protested, "but an expert in his business, and is fully alive to the risks from fire, and better able to appreciate them than all his critics who have

FIGURE 3.1.

This orangutan exhibit was among the first habitat groups that Akeley completed at the Field Columbian Museum. It was given pride of place in the museum's West Court. Elliot and Akeley wanted to use this mode of display for the mammal specimens salvaged in Africa. Photograph courtesy of the Field Museum, CSZ6235.

reported to you on these various points." The curator then suggested that Akeley's detractors didn't understand museum zoology, and at the same time implied—consciously or not—that Skiff didn't understand it either. Their opinion, he wrote, "I cannot help but know has been founded upon their great ignorance of the necessities of Taxidermy, and of the work that is being performed for my Department."[20] Skiff cannot have been happy with the condescending tone of Elliot's letter, yet he let the matter drop.

Akeley had been remarkably productive during his brief tenure at the museum. He'd mounted eighteen large mammals, mostly African antelopes. These were all done in the traditional way, as individual specimens meant to serve as representatives of their species. But Elliot wanted to augment these typological displays with more elaborate, more dramatic habitat groups featuring multiple specimens in active, lifelike poses and naturalistic-looking settings. Early in 1896, Akeley completed his first three habitat groups for the museum, one of proboscis monkeys, one of orangutans, and one of musk oxen. Staff working under Akeley's direction made the accessories for these groups, including artificial tree trunks, branches, leaves, fruit, and more. As soon as they were completed, Elliot gave the new groups pride of place in the center of the museum's West Court, a soaring, skylighted gallery, and one of the grandest exhibit spaces in the entire building. The beautiful artistry and naturalism of these displays showcased Akeley's obvious fitness to lead the museum's taxidermy initiative.[21]

FRENZIED PLANNING AND PREPARATION

The museum's executive committee reconvened at the Chicago Club on January 18. There, Higinbotham and a quorum of committee members debated the merits of Elliot's proposed African expedition. The high cost of the proposal, naturally, gave them pause. Elliot had penciled a hasty memorandum, itemizing expenses, including train and steamer fare, supplies of all kinds, hotel bills, and incidentals, estimating the cost at about $8,000. A second major bone of contention involved Akeley's participation, a point on which Elliot insisted. Ayer was strongly opposed to this plan. Skiff, too, was opposed, at first. Both felt that improving the state of the current zoology exhibits should be the museum's highest priority, and that the chief taxidermist should be left behind to lead this crucial effort. Elliot, however, convinced the director that if Akeley could see and observe Africa's animals in their native habitats for himself, he "would bring to the art of mounting the material" secured by the expedition a "precision of dramatic effect" and the "realism of still life" painting. The time lost during Akeley's proposed absence, Elliot argued, would be more than compensated by the high fidelity of his taxidermy. Skiff consented, then helped convince the committee of the wisdom of Elliot's plan.[22]

The committee agreed to finance the expedition. Skiff wrote to inform Elliot formally "that the Executive Committee has authorized an expedition to Africa under your direction, substantially as outlined in your communication . . . and in accordance with your memorandum budget." Elliot had deliberately kept his cost estimates low in order to win the committee's approval, but this strategy now came back to bite him. Skiff emphasized that "the Committee made . . . an appropriation of $8,000 for . . . said expedition with the proviso that this sum shall cover all expenses, and with the hope that you may be able to return a portion of it to the Treasury." The committee had other stipulations, as well. First, the museum would retain ownership of anything that Elliot might prepare for publication, whether "a scientific bulletin, memoir or monograph, or as a book of travel, experiences, life in Africa, etc." Second, the committee insisted that Akeley sign a contract to last at least three years after his return, agreeing to mount the animals acquired, and remaining at his current salary. Finally, and most curiously, Skiff made it clear that Elliot was not to "render any service of a professional character" to any other institution "either as a courtesy or for compensation" while in Africa without the approval of the executive committee.[23]

Skiff, for his part, was "delighted." He'd long recognized the museum's deficiency in natural history, and he viewed the expedition as a "welcome solution" to a "hard problem." In an interview with the *Chicago Times-Herald*, he opined that "nothing more important for the completeness of the museum could be undertaken than the enterprise in which Prof. Elliot will engage. . . . His undertaking is hazardous, but it has been worked out in plans carefully, and we look for great results."[24]

Frenzied planning and preparation began at once. Elliot wanted to depart Chicago on March 1. This afforded about five weeks for the adjustment of his personal and professional affairs. He arranged for his salary to be sent to his wife, Ann Henderson Elliot, during his long absence. She, together with their unmarried daughter, Margaret, would be staying with friends and family—first in New York and later in Montreal—for the duration of the expedition. Elliot and Skiff conspired to keep the curator's family misinformed about the true nature of the expedition by feeding Ann a steady diet of carefully edited news. "If you write to Mrs. Elliot say nothing about the mutinies & disagreeable things I tell you," Elliot later wrote in a letter to Skiff. "She is naturally anxious & I only give her the bright side of affairs."[25]

Elliot was so busy that he didn't have time to attend to all the minutiae of the expedition personally. So, he called on friends and colleagues for specialized help. He wrote to Frank Chapman, for example, "the great 'mouse hunter,'" for information about traps of all kinds. "I shall require some small traps like those you took away with you on your various trips,"

Elliot explained. "I do not know either how many I should have or what the sizes are, but I want enough & the right kind." Because he'd be in New York so briefly before sailing, he asked his colleague to order on his behalf whatever kind and whatever number of traps that Chapman, himself, would take were he going "on such a trip as I propose." Elliot particularly wanted traps that packed very small.[26]

As the date of departure crept closer, Elliot began to have misgivings about his budget. Funding was the key variable on which the success of the expedition depended, but that was entirely out of his control. He thought it prudent to write something to Skiff before leaving. In his memorandum budget, he explained, $8,000 was cited as the "very lowest amount which should be allotted to an Expedition of this character." He'd arrived at that estimate based on the reports of "several different parties sent out by the Royal Geographical Society of London and by different gentlemen who have visited Africa from time to time for the pleasure of hunting." Now Elliot feared that the amount placed at his disposal by the museum wouldn't suffice for an extended trip, nor permit him to remain as long in the game country as was necessary to ensure the success of his expedition. In short, he needed more money.[27]

Elliot assured Skiff that he had no intention of spending recklessly. He would keep the director informed of all expenditures, even promising to mail vouchers back to the museum whenever possible. He pledged to buy "nothing except what is really necessary, luxuries on a trip of this kind not being required or expected, but what I do purchase will be of the best, my long experience . . . proving . . . that the best is always the cheapest." He hoped the executive committee would be willing to send a supplement if his funds ran low. "In regard to this," Elliot pointed out, "there is one point well worthy of consideration, that after having incurred a considerable expense in outfitting and transportation, whether it would be wise to withhold additional funds after the expedition is in the field and compel it to return with its work incomplete." He warned, though, that it would be necessary to know of any additional sum before he set off into the wilderness—where communication would be sketchy—otherwise the increase would be of no use. He concluded optimistically: "I presume above all else it is the desire of the Ex[ecutive] Committee as it is my own, that this Expedition should be a success. . . . I realize fully the responsibility placed upon me, and shall do all that man can to bring credit upon the Institution." He vowed to return to Chicago with "ample material that will enrich the Museum & increase constantly in value."[28]

NEW YORK BOUND

Elliot and his retinue left Chicago on the 8:00 p.m. train on February 25, arriving in New York the following day. They lodged at the Murray Hill

Hotel near Grand Central Station. Elliot made this his headquarters while chasing around Manhattan, "trying to get the few things together that I consider advisable to take from here." The first order of business was to arm his party with guns from a Broadway outlet of the Winchester Repeating Arms Company. But there was a problem. The store had "not a rifle in stock such as I wanted," Elliot regretted. However, the manager phoned the factory in New Haven, Connecticut, where the company's vice president—recognizing the potential public relations value of arming a scientific expedition to Africa—"promised to deliver my weapons [in two days], if they had to let other work slide or even *steal* from other orders." Next, the curator arranged transatlantic passage for himself and Akeley, wanting something suitably comfortable for a man of his station, yet hoping to avoid the exorbitant cost of first-class passage. He called on James A. Wright Jr., manager of passenger business for the American Line, explaining his mission for science. Wright, Elliot was glad to report, was "very polite, and gave me one of the best rooms on the ship, but put Akeley in with me, promising, however, if the boat was not crowded to give Akeley another room by himself." Finally, Elliot acquired a veritable pharmacy of tropical medicines from a New York apothecary.[29]

Elliot went to Philadelphia to confer with Arthur Donaldson Smith, an American doctor who'd just returned from a two-year exploration of British Somaliland. Elliot, who was anxious to get whatever information he could of that region, plied his new colleague for recent intelligence over lunch. He inquired about expenses, "as at present that phase of the subject is giving me most concern." Elliot suspected that "the general run of outlays" in Somaliland would give him a good sense of what to expect anywhere on the east coast of Africa.[30]

It was Smith who broke the news that Elliot's labor costs would be much higher than anticipated. The headman, for example, "upon whom depends so much for the success of an expedition," would be very expensive, Elliot relayed to Skiff. Smith had paid his headman $120 per month. A good headman, the curator emphasized, was crucial. He "has charge under the Leader of the caravan, looks out for the goods and stores, manages the native carriers, and acts as interpreter." As for the working men of the caravan, Smith reported hiring 75 men and paying them each $9 per month. He armed these men with Snider rifles at $5 each. Smith bought 150 camels and 30 to 40 donkeys for $12 and $5 each, respectively. "Add this all up," Elliot suggested, including the cost of five months' worth of supplies for the caravan, bribes and presents for local officials, transportation, and a "thousand and one incidentals." Obviously, with his present funding only, Elliot's sojourn to the interior would have to be brief. "I cannot . . . stay longer than I know I have enough money on balance to pay all my indebtedness and get back . . . to London."[31]

Assuming that Skiff was in "full sympathy" with him, Elliot suggested that the matter would eventually have to be brought before the executive committee. "I know," he wrote, that "if it lay in your power, you would give me unlimited means to command success." But what would the committee do? "It would be a lasting disgrace," the curator argued, "if this expedition failed for lack of means. Do not let us go on the penny wise and pound foolish plan, and spend a lot of money and then fail because we would not spend a little more."[32]

Elliot wrote with customary frankness and earnestness. But, concerned that his tone was too alarmist, he tried at the same time to be reassuring. "Above all," he wrote, "I do not write to worry you, for I know and feel that you are the one strong friend and prop in all this undertaking that I have. But I want to do *you* credit as well as the Ex[ecutive] Com[mittee]." The key was funding: "I somehow feel that I can succeed if I have the necessary financial aid," he insisted, "but the whole affair will be a 'botch' otherwise."[33]

Nor was all of Elliot's news bad. Thanks to his sterling social connections, he'd managed to acquire an impressive set of letters of introduction, a critically important asset for a nineteenth-century scientific traveler. He had one to the legal advisor of the sultan of Zanzibar, and another to its consul general. He was yet hopeful of obtaining a letter from Joseph Chamberlain, the English secretary of state for the colonies, "so if I go to any British possession, I shall probably get all the assistance the Crown can give." Elliot was friendly with Thomas F. Bayard, US ambassador to Great Britain, and he was confident the ambassador would "do all he can to further my wishes." Other influential letters were forthcoming. "I shall leave no stone unturned on my side to achieve success," he wrote, "and will command it if I only have the funds."[34]

4

A REAL PARROT AND MONKEY TIME

Elliot and Akeley sailed for England on the morning of March 4 on the elegant and speedy steamship *New York*. Elliot's extended family assembled at Pier 14, in Lower Manhattan, to see them off. The curator was "very glad when the sailing day arrived," his wife wrote to Skiff, "for he was completely worn out. A week in New York proved quite too short for all the business arrangements he had to attend to & the very flattering desire of his friends & relatives to see him." She reported that Elliot and Akeley had good rooms on the ship—with hot and cold running water and electric light and ventilation—and seats at the captain's table under a great dome of glass. She was delighted with Akeley, to whom she gave "many instructions as to the great care he was to take of our treasure." So thorough were her suggestions that she feared she was chipping away at her husband's reputation for ruggedness. Akeley, she worried, suspected that "he had a very precious bit of fragile porcelain intrusted to him."[1]

From the pilot boat Elliot sent Skiff a long letter in which he wrote "very hopefully and confidently" about the expedition's outlook. He was obviously pleased to be underway. A thoughtful night telegram from Skiff read: "Your associates in Museum wish you pleasant and prosperous voyage."[2]

They arrived at Southampton on March 11 and traveled directly to London, where Elliot lodged with a friend in South Kensington, near the British Museum (Natural History). Akeley, meanwhile, boarded in a trio of London hotels, each one cheaper than the last. Elliot soon began to inquire "into African matters" of his friends and colleagues in London, then wrote to Skiff with an account of his results. He boasted foremost of his "exceptional advantages" for getting information. "My long residence in England," he explained, "and my membership in the various scientific societies, has made me acquainted . . . with many persons of influence, as well as travelers[,] all of whom are anxious to help." He assured the director that he could accomplish more than a stranger could in twice the time. He added that the expedition "is regarded both with astonishment and admiration, that the Executive Committee . . . have had the energy and pluck to send it forth, for no other scientific Institution has attempted a similar one, and the result of my efforts will be watched with the liveliest interest." The British Colonial Office had already written Elliot expressing its willingness to aid him as much as possible. Its officials merely wanted to know where the expedition would be going.[3]

Officers of the Field Columbian Museum were likewise keen to know Elliot's destination, for the curator hadn't yet reached a decision. He'd initially thought to go to Mashonaland (present-day Zimbabwe). Several interviews convinced him to look elsewhere. For starters, there was no recent news available about the region's game. Worse, costs there had risen dramatically in recent months, and Elliot feared that no bank there would honor his letter of credit. So, Mashonaland "not seeming promising," he began to inquire about other destinations. He considered the Maasai country (present-day Kenya and Tanzania), but quickly decided against it. "There is war going on," he explained. Besides, he'd learned from someone who'd lately come from there that a deadly disease originating with imported cattle had spread to the game animals, "and those had become nearly exterminated. One more . . . method to cause the extinction of . . . wild creatures."[4] Clearly, European colonial expansion was already wreaking environmental havoc. Time was running out for Africa's game, Elliot hinted.

This left British Somaliland as Elliot's "last hope." Britain's Somaliland Protectorate (as it was officially known) occupied most of the Gulf of Aden coastline on the Horn of Africa. Though never one of Britain's more prosperous colonies, Somaliland was, nevertheless, of vital strategic importance. After Napoleon invaded Egypt in 1798, British colonial officers in India insisted that a permanent naval presence on the Red Sea was essential to protect the long communication and supply lines linking India to Great Britain. Aden, on the southern shore of the Arabian Peninsula, with its deep and roomy harbor, was chosen as an ideal naval base. Early in 1839, the British East India Company landed a force of marines there to occupy the port and to protect British shipping to its Asian colonies. Aden quickly assumed the role of Britain's most important stronghold in the western Indian Ocean.[5]

Certain British officers next began to covet the Somali coast, across from Aden, to further consolidate their strategic position. In 1827, using the pretext of exacting retribution for the looting of a British vessel stranded on the Somali coast—and the massacre of its crew—a Royal Navy ship landed a party of marines to force local Somali tribes into signing "a treaty of mutual recognition." A second treaty, signed in 1840, required the Somalis to prevent the French from establishing a settlement anywhere along their coast. Aden, meanwhile, had become dependent on Somali supplies of fresh meat and ghee, which was a potential strategic vulnerability. To keep these supplies coming, Aden officials sometimes involved themselves in intertribal affairs in Somaliland. Even so, the British government declined to authorize any deeper or more permanent commitments in Somaliland itself.[6]

Once the Suez Canal opened in 1869, British warships and merchantmen began to use the new and much shorter canal route to get to India. In 1870, an undersea telegraph cable was laid between Aden and Suez, providing a direct link from London to Bombay. Extensions to Australia and South Africa were soon added, "making the waterway between the Mediterranean and the Indian Ocean one of the most important communication choke points in the British Empire."[7]

Despite the strength of their position at Aden, British colonial officials worried that other European powers (especially France) might try to gain a strategic foothold on or near this crucial waterway. Therefore, the few deep-water harbors along the Red Sea and the Gulf of Aden became important desiderata. Britain's colonial secretary, Sir Michael Hicks Beach, insisted that all such ports must be held "either by England, or a Power under the influence of England." This explains why, in 1875, when Egypt's ruler, Khedive Isma'il Pasha, declared the entire Somali coast as a part of his domain, the British government promptly recognized the claim. A Somaliland in the nominal control of an allied power was infinitely preferable to one controlled by the French. So long as Egypt occupied Somaliland, "Britain would remain the dominant naval power in the Gulf of Aden."[8] Unfortunately, Egypt's hold was not to last.

Elliot had learned what he called "authentic and recent information" about conditions in Somaliland. A copy of Harold G. C. Swayne's 1895 book *Seventeen Trips through Somáliland* served as an invaluable reference. Swayne was a noted British soldier and big-game hunter with ample experience in East Africa. Frank L. James's *The Unknown Horn of Africa* (in its second edition in 1890) was another important source. In Philadelphia, Elliot had conferred with Arthur Donaldson Smith, who'd been in Somaliland the previous winter. Better still, through his connections at the British Museum (Natural History), he met Edward Dodson, a British hunter and taxidermist who'd accompanied Smith's expedition. Dodson, who was familiar with the game, the landscape, and the people of Somaliland, was willing to join Elliot's expedition for a modest salary. Dodson had gone farther into the country than any of Elliot's more seasoned friends, "but not so far as I would go if I had the means, for no doubt there are many new and rare things to be obtained in Somalia if one could go far enough and remain as long as necessary." From these sources, Elliot received "encouraging reports" of the region and its fauna. "The species are scattered," he explained, "which will make it necessary for me to make long journeys to reach them, but they are there, unless some sudden visitation has carried them off in the last few months." With summer approaching, it was not the ideal season for hunting in Somaliland, but Elliot dismissed this objection as trivial. He would deal with the heat, "if I can only get what I am after and make a success of the expedition."[9]

LONDON DISBURSEMENTS

Having settled on Somaliland as their destination, Elliot and Akeley went to work at once to lay in supplies and equipment, "all of which must be taken from here, even down to such material as flour, for the entire time I shall be in the country." Elliot explained to Skiff that he intended to take "a pretty big stock, in hopes that . . . the Executive Committee in view of my representations . . . are willing to increase my appropriation . . . to enable me to remain in the country as long as I wish."[10]

Elliot suggested that the museum should send him $5,000 more, but he promised not to spend it unless he absolutely must. He then made a spirited plea for support, writing, "having put our hand to the plough, it will never do, for the credit of all concerned, for us to look back. Let us make this a memorable expedition in every way, and one for similar institutions to copy if they can." He wanted to stay in the field for eight months, unless the trip was so "successful as to warrant me to leave before that time has elapsed." Elliot urged Skiff "to lay the matter before the Committee." He hoped to depart England by steamer on March 27 and asked the director to telegraph him "care of Cowasjie & Dinshaw," a trading company based in Aden. "If you telegraph *Move*," Elliot instructed, "I shall understand that if I stay longer in the country than my present means warrant me in doing, that I shall find another letter of Cr[edit] awaiting me at Aden, but if you wire *Stop*, then I shall know the Committee will give no more, even [if] it insures the failure of the Expedition."[11]

The curator's letters traveled at the speed of steamships and trains. Replies in kind could take weeks in transit. Elliot often repeated himself because he had to assume that one or more of his letters might miscarry. When they didn't, his frequent pleas for money read as harping. In his latest letter, he explained, he hadn't "relied solely upon my own experience" in outfitting his expedition. Instead, he'd sought the advice of British army officers who'd been on the Somali coast, as well as sportsmen who'd ventured into the interior for trophies. "Many things I would have taken if this had been a private affair, have been left out, & in every way I have tried to keep down the expense," he noted.[12]

Elliot decided, for example, that he couldn't justify the purchase of a camera "such as would be required for the Expedition [to] do first-class work." Yet he recognized the importance of having a camera and said as much to Akeley, whereupon the taxidermist decided to buy one, spending his own money on a pricey photographic outfit. Elliot offered to furnish the glass plates required on the condition that "all negatives made on the journey should be the property of the Museum," and Akeley agreed.[13]

Elliot explained that to get giraffe, zebra, and other large and desirable game, one "must cross the Shebeyli River into Galla," far in the interior. This, he estimated, was a month's journey from the coast without halting. However, "rapid travelling will not do for me, as it would frustrate the object of my trip, as one must hunt for game, and wherever I find a good place for collecting I should . . . remain . . . until I have exhausted the neighborhood." He estimated that the expedition would cost about $1,000 per month. "I shall be happy if I can reduce this some," he wrote, "& shall do my utmost to bring about such a result. I do not want to spend money, quite the contrary, but now that we are all embarked in the enterprise I want it to be a success, something that everyone can be proud of, and that will be of great advantage to the Museum." His present funding would sustain the expedition for only two and a half months, "not enough to enable me to get to the Shebeyli and back." He could make a collection in this time, he conceded, but it'd be an inadequate representation of the region's fauna, and this simply wouldn't do. "The country I propose to penetrate," he explained, "has the most animals of any in Africa . . . and I want to bring a full series back. Very few of the species of large mammals have been brought to Europe in the entire skins, and fine examples are not seen in the Museums."[14]

Skiff hadn't yet responded to any of Elliot's pleas for money. Still in dire need, but worried that he was asking too much, Elliot mentioned a smaller sum. "Need $4,000 more," he wired. If the museum was "willing to see the Expedition through & forward the above sum," Elliot wanted Skiff to wire him that fact at the Imperial Ottoman Bank in Aden, then forward a new letter of credit to the same address "so that I shall receive it on my return from the interior." He needed to be sure the money would be waiting in Aden when he returned from the hunt, as his plan was to "time my stay away from the coast by the amount of money at my disposal on starting." He didn't mean to suggest that he'd linger in the bush until the money was gone, but only that he would remain so long as he found it "advisable & profitable to the Museum to do so." But, he warned, with his present means he wouldn't be able to carry out his plans "nor achieve the result the Ex[ecutive] Com[mittee], and I may say the Scientific World expect, and which I myself hoped to accomplish."[15]

The expedition would pay for itself, Elliot argued. The skins of some animals, he estimated, were worth as much as £20 ($100) each, and the largest skins were worth £100 ($500). Elliot intended to collect skins by the hundreds. Therefore, $12,000 wasn't "an excessive price to pay for such an Expedition as this, nor for the collection of rare forms it ought to secure." The matter, he wrote, was entirely in the museum's hands. "I will do all I can with my present means, & if I am obliged to return prematurely, I shall be mortified naturally, but I shall feel the result as beyond my control." He hoped the committee would approve and pledged to "do all in the power of man to make the Chairman & members satisfied."[16]

Elliot sent a personal letter to Skiff outlining his plans in greater detail, while also fishing for sympathy. "I am pretty well used up with the incessant & annoying work of the past two weeks," he griped to the director, "but everything is now aboard ship & I sail tomorrow." The passage would take about eighteen days on the steamship *Britannia*, "a good boat but fearfully crowded." He was looking forward to an extended period of "peace & quiet."[17]

They were destined first for Aden. There, Elliot would seek to hire a headman. From Aden they would cross over to Berbera on the Somali coast, where camels could be bought, and men hired. "I suppose I shall have a real parrot and monkey time getting things together and in shape, and I shall not feel contented until civilization is left behind & my face is headed towards the unknown interior," he wrote hopefully. With letters of introduction to the governor of Aden and the chief British official at Berbera, Elliot felt he'd be arriving "pretty well recommended." The finances of the expedition remained Elliot's main concern. He regretted having started "on the small sum provided." He remained optimistic, however, that he'd "come out all right." After all, it'd be very uncharacteristic of Chicago men, he flattered, "if once they put their hand to the plough, they stop & look back."[18]

GIVING THEMSELVES OVER TO NEPTUNE

They shipped on March 27, as scheduled, and enjoyed an uneventful cruise through the English Channel. The following morning, however, in the Bay of Biscay, they had "a gale from the North West & a heavy sea abeam, & the way we rolled was a caution even to sea dogs." No one could stay put, save by "lying on his back." Furniture slid back and forth across the deck of the ship. "It is a wonder no bones were broken," Elliot remarked. "I distinguished myself by turning a somersault in the cabin which would have done credit to a circus rider." A nasty gash to his temple and a humorous caricature drawn by a talented eyewitness were Elliot's souvenirs.[19]

Dodson was in his element. "I have had no mal de mer," he boasted to his parents. Elliot was impressed with his plucky new hire, "a good sailor, & a bright smart young fellow & I can see he is going to be of the greatest service to me," the curator wrote. "His knowledge of the people & country . . . will be invaluable." Akeley, by contrast, was miserable, as Elliot described: "[He] is a new man or ought to be, for there is not much left of him but his outside skin. Neptune has the rest."[20]

The curator was anxious to disembark and begin the real work of the expedition. "All this seems like wasting time," he grumbled impatiently about the voyage. Yet the shipboard hours they spent together were good for cultivating the party's esprit de corps. Elliot, their leader, was difficult to get to know. He could seem rigidly formal—almost unapproachable. "To strangers he appeared reserved," a colleague and friend later wrote, "but this natural dignity . . . was merely the shield of a gentle, kindly, sympathetic nature, behind which one found a charming and congenial companion [who] was as fun-loving as a boy." Familiarity broke the barriers of formality down. "The more I see of Elliot," Dodson wrote, "the better I like him and we bid fair to be as merry as crickets."[21]

Akeley, too, was a complex character. Hard-working, driven, and ambitious, he had a self-confidence that was often seen as arrogance. His outspokenness sometimes bred resentment. He was quick to anger. He tended to be solitary, yet he could be funny and charming company when inclined. He was fiercely loyal to his friends, but a terror to anyone who'd fallen out of favor. He had a thin, wiry body that masked his toughness, and a broad skull and high forehead crowned with a thinning tangle of wavy hair. His handsome face was strong but kindly. In Africa, he grew a beard for the first time, possibly to mark this trip as a decisive break from his peripatetic past.[22]

Regarding Dodson, little is known. Born in 1872, he was the oldest son of a successful London carriage maker. He worked as a taxidermist at the British Museum (Natural History). As a young man, he had a yen for travel and adventure, hiring himself out to several foreign expeditions as a hunter and skinner. He was fit and very fair haired. Though he was the youngest member of the expedition, he'd spent the most time in Africa. His experience would prove of the greatest utility to Elliot.[23]

The junior men of the expedition were mostly content to pass their time afloat attending social activities. A gifted tinkerer, Akeley also experimented with his new camera. Elliot, however, spent most of his time afloat fantasizing about his game bag and mapping out the route he proposed to take. He hoped to find new species unknown to science. "Who can tell what I shall find there," he wrote, "to enrich the Museum and surprise Naturalists? We shall see." He thought greedily of sweeping up all the available game: "I want to take my own time, go slow, and leave little behind for any who may follow me." He hoped to get "into the unknown country," beyond the Juba River. "This country has not been disturbed, and is uninhabited, being a sort of no man's land between rival tribes. In such a place many animals are found, not unlikely of new species."[24]

To discover new animals, to describe and name them, was a godlike act of creation. This was the ultimate goal of every death-defying naturalist since Carl Linnaeus and the birth of modern taxonomy in the mid-eighteenth century. The scientific success of the expedition would be judged, in part, on the discovery of new species. The remote, "unknown country," where new species were likeliest to be found, had the highest potential for important scientific results. Elliot expected to find animals new to the museum's collection literally everywhere. Animals new to science, on the other hand, would be more elusive. For this reason, it was important to stay longer, to travel farther.

Their ship made port in Gibraltar, Malta, and Brindisi. Once through the Suez Canal, they sailed southeast through the Red Sea to the port of Aden, arriving on the morning of April 14. A spacious natural harbor in the crater of an extinct volcano, Aden had been a way station for mariners since ancient times. In the mid-nineteenth century, it was a busy free trade center, handling spirits, wine, opium, coffee, hides, ivory, feathers, and aromatic gums. By the 1890s, Aden was controlled by the British. Because of its location near the mouth of the Red Sea, Aden quickly became a key transit port and coaling station, especially after the Suez Canal opened in 1869. Despite its strategic importance, Aden had a poor reputation. British explorer Sir Richard Burton derided it as "a hot bed of scurvy and ulcer."[25]

Before he'd even landed, Elliot—at last!—was handed a telegram from Chicago advising him that the museum's executive committee had appropriated an additional $4,000 for his expedition. This news "gave me most welcome relief," Elliot revealed to Skiff, "for I have been much troubled about the financial question." The curator now had "no fear but that I can carry out my part of the undertaking and send to the Museum a splendid collection of the rapidly disappearing big animals of Africa." But this would take money, more money than Elliot, "carefully as I had made my calculations, had expected, and every year the expense is going to be much greater with the chance of success much lessened." In Somaliland, "there is probably more game today than in any other section of Africa but it will be gone in a few years, & I begin to think that if this Expedition had delayed even a year to start, it would have been too late."[26]

Aden was the expedition's final port of call before embarking for Africa. Here they completed the details of their outfit and finalized arrangements with colonial officials. Elliot was gratified to report that he'd been courteously received by British military authorities and every assistance had been offered to the expedition. In fact, he'd requested and received permission to collect zoological specimens even in the so-called game reservation, a tract of land about eighty miles wide on the coast of British Somaliland that the military maintained as a hunting preserve exclusively for British officers. Brigadier general Charles A. Cuningham, the local governor at Aden, had offered to let Elliot stay with him at his official residency. This honor, however, Elliot politely declined, as he was intent to remain with

the rest of his party at the Hotel de l'Europe and see to it that the expedition was quickly and efficiently outfitted.[27] He hoped to cross the Gulf
of Aden for Africa in a week or less.

The expedition was "regarded with much favor" by colonial officials.
The reason, according to Elliot, was the curator's status as a scientist, and
the stark distinction he drew between himself and his collecting rivals. "I
am not a butcher," he elaborated, "killing for the love of taking life, as are
too many so-called sportsmen." In fact, Aden authorities hoped that Elliot would make a scientific report of the probable extinction of animals
in Somaliland, "which will enable them to enact stringent [game] laws,
which will be enforced, wherever British authority in Africa extends. . . .
I should be very glad indeed to give all the assistance I could for such a
laudable object," he wrote.[28] This is an especially interesting episode, as
it's the only hint in Elliot's records of a conservation rationale for his expedition. It must be noted, however, that there is no evidence that Elliot
ever wrote the report. Moreover, the curator's willingness and ability to
make such a report was predicated on the opportunity first to take his fill
of specimens. If, as a result of his report, further collecting in Somaliland
should be restricted, so much the better—no rival museum could repli-
cate his results.

One unforeseen problem at Aden was the shockingly high price asked
for camels: "I regret to say camels are very dear, double the usual price."
Unfortunately, Elliot needed many camels to complete his caravan. He
urged Skiff to send the new letter of credit "as quickly as possible, as I
must not get out of funds *now*."[29]

Days later, on the eve of their departure for Africa, Elliot was again
feeling "slightly discouraged" with the financial outlook. With the funds
available, even including the additional $4,000 that Skiff promised to send,
Elliot was concerned that the duration of his trip into the interior would be
too brief to guarantee success. "The fact is," he explained testily to Skiff,
"no man living can name an actual sum which will be sufficient to carry
such an expedition as this through its proposed route, when most of it is a
practically unknown land." Elliot boasted that he had "as much experience
in traveling in foreign lands away from the routes generally followed, as
most men." But his experience, thus far, had been to no avail. He'd made
himself miserable with worry over the financial health of the expedition.
He pledged that "not only would I be unwilling ever again to lead an expedition for any Institution, but no man, a year from this time, even *with
my accounts to guide him*, would be able to name an amount sufficient for
a similar trip." Prices were constantly changing, he marveled, and money
appeared to diminish in value almost continuously. He wrote this "in no
spirit of complaint," he said, "for the Ex[ecutive] Com[mittee] have come
forward to my aid as I requested." He had no intention of asking for more

money. If further sums were to be sent, he suggested, it would have to be on the museum's initiative, and "not at my request."[30]

Yet, the museum, he hinted further, had to think about posterity. The country where he intended to collect was "fairly teeming with game from Elephants to mice, and I *can* make a splendid collection for the Museum which will increase in value year by year until the time comes, which is near at hand, when no living representatives of the species remain."[31] But he needed time and money to accomplish great results.

Elliot had one more card to play in his gambit for more funding; he planned to monetize a part of his collection. The elephant country, he explained, was a journey of six to eight weeks from Berbera. A pair of elephant tusks averaged about seventy-five pounds. Ivory was then worth about $3–4 a pound, or about $250–300 for a pair of tusks. "I have always expected to pay a portion of the expense of the Expedition from the sale of ivory," he wrote. He'd avoided mentioning this plan previously because of the uncertainty. He refused "to have any money advanced by the Ex[ecutive] Com[mittee] on the supposition . . . that they were going to have it refunded." Either the refund would come "as an agreeable surprise or not at all." But now that the prospect of reaching elephant country was growing dimmer, he confided to Skiff by way of leveraging a larger budget.[32]

Some money would also be recouped by selling his camels at the end of the expedition. But, Elliot warned, this would probably not amount to much. Many animals would die en route, and these would have to be replaced. Those that survived would be returning exhausted and emaciated and would be worth a third or less of what Elliot was paying for them up front.

"Now I hate to write in this strain," Elliot continued, "but it is proper I should do so and place you in full possession of the *facts*." The expedition's finances had caused him nearly constant anxiety. "I am much older than when I left you," he complained, "simply from worrying over the state of affairs which I had no power to alter." He knew that as his expenses mounted, the expectation of greater results rose accordingly, yet he remained confident that the museum would get "full value" for the expedition's cost. In fact, it was "the best investment a scientific Institution could [make], & one constantly increasing in value." He knew it would be hard to get the executive committee to understand, however. "*I know* what I am about," he insisted, "& fully appreciate the inestimable value to the Museum the specimens I shall get will be, even now, not regarding at all the time when obtaining such a collection will be impossible." He reiterated that he was making "no formal request" for money. "But," he concluded, "if I receive no telegram by the beginning of June at least, I shall return to the coast in August, for the simple reason that I can stay out no longer."[33] The onus was on Skiff to decide what, if anything, to do about Elliot's predicament. To approach the executive committee with this new intelligence and ask for even more funding would be his responsibility and not the curator's.

COMPLETING HIS OUTFIT

The weather in Aden was appallingly hot, which didn't help expedite the work. Neither did their busy social schedule. They all dined one night with the helpful and accommodating American consul, William W. Masterson, a "right good fellow," according to Elliot. Despite the stifling heat, they wore their formal attire. "Fancy the horror of a dress suit in this climate," Dodson wrote.[34] The curator was fearful that the really intense heat was yet to come, and he hoped to arrive in the higher elevations of the Somali interior before it commenced in earnest.

From Masterson, Elliot learned that it'd be impossible to draw on his letter of credit from a Chicago bank in Africa. He was forced to withdraw the remaining balance, amounting to £715 ($3,575). He carried part of this total in cash and the rest in checks drawn on a banking house well known on the coast of Somaliland, Cowasjee Dinshaw & Brothers, which would honor Elliot's checks in Berbera. In case this sum should be insufficient for the expedition's needs, Elliot personally advanced an additional £100 ($500) of his own money. This amount he would refund to himself once the new letter of credit from the museum arrived. The fact that the Aden branch of the Ottoman Bank was closed for business forced Elliot to make other arrangements for obtaining cash, yet another unexpected source of anxiety.[35]

During their weeklong stay in Aden, the expedition procured its first scientific specimens. "We filled a small barrel half full of fish of various kinds," Elliot wrote. Some of these specimens they caught using Dodson's fishing tackle, others they purchased from local fishermen. The barrel ultimately proved to contain some 114 specimens, representing 39 species of marine fishes.[36]

Elliot worked hard putting his outfit together. In addition to pack animals and men, he had to stock sufficient supplies of cloth, beads, and other items "required for trading with natives," as well as supplies for a large caravan "which could not be procured in London to equal advantage." He bought three horses and three mules for riding, though not as many as he thought he needed: "without such animals not a day's march into the interior could be made."[37]

Dodson, who'd done business in Aden before, did most of the bartering. He was also helpful when hiring local men to accompany the expedition, many of whom had previously worked for Smith. He regretted that some of the men he'd worked closely with were now unavailable. Mahomet, his former personal attendant, was dying. One of Smith's gun bearers and bug collectors, Yusif, now had only one arm, the other having been mauled by a crocodile. Smith's other gun bearer, Karsha, was willing and able to join the expedition. "I shall have Karschar," Dodson wrote to his parents, "the sole survivor."[38]

Isman, a young shepherd who'd been on Smith's expedition, called to see Elliot at his hotel. The young man had received "two dreadful wounds in the back" while guarding Smith's flock. Smith described this incident vividly in his book: "I was startled by a boy's piercing screams. . . . I grabbed a few medical appliances, and rushed to the place where it was now reported that one of my boys had been wounded." There he found Isman, badly lacerated by spears. Smith hastened to stop the bleeding, "and while thus engaged I was shouted to that one of my men, Elmi, had been killed. Poor Elmi lay dead behind a bush, horribly mutilated,—his intestines scattered about, and a part of him carried away as a trophy by his murderers." Once he'd revived from his shock, Isman explained that he'd been ambushed by three Borana.[39] The young man ultimately recovered, and now he gamely wished to accompany Elliot's expedition. The curator admired the shepherd's grit, but, having no sheep or goats in his possession, he kindly declined.

Elliot had hoped to speed away from sweltering Aden as soon as his outfit was complete. Unfortunately, the decrepit steamer he'd hired, the *Tuna*, wouldn't sail until the evening of Monday, April 20. He expected the passage to Berbera to take as many as three long days, as there were several planned ports of call. He dreaded it. Frank L. James, a British adventurer who'd made the same voyage only a few years earlier, described it as "one ghastly thirty-eight hours' nightmare." His traveling companion called it "simply *hell*."[40]

5

RATHER A BAD BEGINNING

April 20–30

The *Tuna* was a small, steam-powered cargo vessel built in England. Bound for New Zealand, she was wrecked near Aden on her delivery voyage in August 1885. The star-crossed steamer was then salvaged and sold to Cowasjee Dinshaw & Brothers, which used her to communicate with its commercial outposts on the east coast of Africa. She was scheduled to leave Aden for Berbera on April 20, but her time of departure had been left indefinite. She was supposed to weigh anchor in late afternoon, load up the expedition's ammunition in the harbor—a legal requirement—and then steam for Africa around sundown. The ship's owners told Elliot that he could take a tender out to the *Tuna* once the ammunition was stowed. With the temperature above ninety degrees Fahrenheit, and no apparent reason to rush, Elliot and his party dawdled at their hotel. They ordered dinner, a bottle of Saint-Julien wine, four sodas, and a chilling twelve pounds of ice. They "were taking things as easily as possible" when a breathless courier arrived with an urgent message that they must get on board the steamer immediately.[1]

Elliot grumbled bitterly about the slapdash arrangements of the "wretched boats" running between Aden and Berbera and along the coast. "One is lucky," he griped, "if everything turns up at the end of the voyage."[2] Any credit for success was due to the frantic efforts of the passengers, not to any officials. Following their urgent summons, Elliot and his companions crammed their clothing and personal effects into their trunks, which were then loaded onto a handcart and wheeled expeditiously to the wharf. From there, Elliot could see the *Tuna* cruising in slow, lazy circles in the harbor. A dilapidated launch then carried the party out to the steamer. From the deck they could see their personal baggage, as well as some of the equipment for their caravan, coming along slowly in rowboats. Elliot was concerned that he could recognize only a few of the men he'd hired for his expedition. After much trouble, the rowboats finally pulled alongside

the steamer and the expedition's equipment—along with fourteen of its men—was transferred aboard.

More troubling still, the headman, Dualla Idris, was nowhere to be found. A well-respected figure in African exploration, Dualla commanded extraordinarily high fees as a headman and guide. A Somali man of about thirty-three years, he'd been to the United States as a boy. In 1887–88, he led the Count Sámuel Teleki expedition across what is now Kenya. For six years he'd been one of Henry Morton Stanley's most trusted followers in the Congo, then he traveled with Stanley in Europe. He spoke English, Arabic, Swahili, Somali, and Hindi. He'd been Arthur Donaldson Smith's headman and came recommended by him. Dualla had initially declined to go with Elliot. But, after visiting with Elliot for a short while at his Aden hotel, taking the curator's measure, he changed his mind. He insisted—prophetically, as it turned out—that this would be his last expedition.[3]

Apparently, Dualla had boarded the steamer in the afternoon, bringing a load of goods, and then went back to shore for unknown reasons. Elliot prevailed on the *Tuna*'s captain to delay, hoping the headman would soon appear. But "after knocking about for half an hour," they started without him. Fortunately, the second headman, Selon Mohammed, was aboard, and he took charge in Dualla's absence. Selon had been second headman on Smith's expedition, also, and was likewise well regarded. He was "by far the best man I had in the caravan," Smith wrote. Nevertheless, Elliot was vexed by Dualla's absence. "Rather a bad beginning to commence a long journey without the Head man who is responsible for the Caravan," Elliot lamented.[4] But there was nothing to be done about it now.

Their ship was crowded with adventurers and their bulky outfits. A pair of young, English sportsmen were also heading to Somaliland for a hunting trip. Their outfit—a headman and several camel men, with their baggage and equipment—was comingled with Elliot's outfit all over the boat. A solitary German traveler, and an agent for the Italian army, were likewise taking passage on their steamer. There were no private accommodations on the boat, so the American and European passengers sat up all night in chairs on the captain's bridge. Though cramped, exceedingly uncomfortable, and skittering with monstrous cockroaches, it was still preferable to the teeming open deck. Elliot and the Englishmen had agreed to pool their resources, and, by means of a bribe, they'd hired the boat to go directly across the Gulf of Aden to Berbera, rather than cruising along the coast and stopping at the ports of Zeila and Bulhar, as was her normal route. By this means, they spared themselves "two days and one night of misery."[5]

Viewed from the bridge, where Elliot and his mates "were all packed with less comfort than herrings in a barrel," their ship presented "an extraordinary sight." The *Tuna* was "a wretched little thing" of about eighty-five tons. Elliot called her "as vile a craft as floats and yet calls itself a

FIGURE 5.2.

Nearly every inch of deck space on the *Tuna* was crowded with men, animals, and equipment destined for Berbera. Photograph courtesy of the Field Museum, CSZ5936.

steamer." She was so overloaded and riding so low in the water that even a gentle swell would likely have swamped her. The horses and mules were tied along the rail near the stern, "where they had a fine view of the sea."[6] In fact, they could almost dip their muzzles in the water. Every inch of deck and every hatch was occupied by Somalis or Hindus. Shouting and laughing among themselves, they raised a din that was louder than the ship's limp and mournful steam whistle.

They'd gone only a short distance from the harbor when the engine sputtered to a stop and all headway ceased. The boat began to swing around helplessly in the current. Elliot wasn't surprised—a mechanic had been tinkering with the *Tuna*'s grimy, rusting engine for several days prior to departure. An overheated rod, he learned, was responsible for the stoppage. The boat drifted aimlessly for some time before "the wheezy old thing was induced to get to work again & we once more headed for the African Coast, not any particular part of it, though our destination was Berbera, for the prow rarely pointed to any one point of the compass for any length of time."[7] A meandering wake stretched behind the boat in great, drunken arcs that reminded the zoologist of a slithering snake.

Their supper, served in "a little hole called a cabin" close to the ship's boiler, was, in Elliot's memorable phrase, "an impossible repast." The oppressive heat and the reek from the galley ruined his appetite. He and his companions "all sought the cooler though hardly less satisfactory accommodation of the bridge." With nightfall, the Africans prostrated themselves pell-mell on the deck, swaddled in their tobes—versatile, white cotton robes—looking like so many bales of goods. The moon was bright, and the night, which was still with a placid sea, wore on interminably. Elliot enjoyed "occasional snatches of sleep" as he "happened for a moment to find a less uneasy position in his chair after many trials." Dawn broke at long last with no wind and a dead-calm sea. The trip from Aden to Berbera could normally be done in twelve hours. The *Tuna* needed fully nineteen, using "all the struggling and groaning of which she was capable, not enabling her to make much over six knots."[8]

ARRIVING AT BERBERA

By ten o'clock in the morning, a line of tawny-colored mountains shimmered indistinctly above the horizon. An hour later, the low African coast came into view, showing a long stretch of sand dotted with small brush—the heated air dancing above it in waves—and a line of white foam describing the beach. Berbera could soon be seen from the bridge.

With its deep and sheltered harbor, Berbera was the best natural port on the African side of the Gulf of Aden. The maritime plain enveloping Berbera, on the other hand, was a hot, dry wasteland. In the summer months, the region was almost uninhabitable. Traditionally, Somalis used the port only as a seasonal marketplace. Egypt, however, having opened the Suez Canal in 1869, took an interest in developing the Somali coastline. Egyptian forces occupied Berbera in the winter of 1873–74, sending a permanent garrison of soldiers in 1875. They built a lighthouse, a pier, blockhouses, barracks, and administrative and residential buildings. To ensure a steady supply of drinking water, they installed a six-inch iron pipe from Berbera to the base of Dubar Mountain—seven miles away—where a hot, brackish flow oozed from multiple fissures in the ground. Such amenities improved the lives of the few Somalis who fished the gulf or who conveyed passengers or cargo across to Aden. Moreover, local pastoralists now had a reliable market for their livestock at the new port. Their steadiest customer was the hungry British garrison in Aden. With the Egyptian improvements, Berbera became a more-or-less permanent center of commercial activity.[9]

Uneasy with the Egyptian occupation of Somaliland, a British consular official in Aden, Frederick Mercer Hunter, took it on himself to form a special paramilitary unit of forty armed policemen, backed by a small artillery piece, which was kept on permanent standby for speedy deployment to Berbera in the event of an emergency. This was apparently done without the approval—perhaps even without the knowledge—of the British government in London.[10]

In Cairo, in 1881, Khedive Isma'il was overthrown. At the same time, an Islamist revolt erupted, led by a charismatic cleric who called himself the Mahdi. Armed Mahdist rebels quickly overran most of Egyptian-controlled Sudan. Meanwhile, Egypt's war of conquest against Abyssinia was going very poorly. Bankrupt, and staggered by war and rebellion, Egypt saw its grip on the Somali coast weaken by the day. In August 1884, Hunter sent his deputy, Langton Prendergast Walsh, to Berbera, together with his improvised paramilitary force, which acted as Walsh's personal security. Hunter next forged a letter from the Egyptian government in Cairo to Berbera's Egyptian deputy governor, Boorhan Effendi, ordering him to return immediately to Suez. Walsh then quietly and bloodlessly assumed the role of acting governor of Berbera, reporting to Hunter. Tacitly acknowledging the new reality on the ground, Egypt evacuated the last of its soldiers from Berbera in late summer 1884, relinquishing all claims to the Somali coast. Fearful of losing Berbera to a rival European power, the British government assumed responsibility for Somaliland. Between 1884 and 1886, Britain made treaties with several local Somali tribes. The British formally established a protectorate on July 20, 1887, with Berbera as its administrative capital.[11]

Lying approximately 160 miles due south of Aden, Berbera commanded most of the commercial ship traffic of the western portion of the Gulf of Aden. Richard Burton, sensing its potential, called it "the true key of the Red Sea." Akeley biographer Jay Kirk, on the other hand, called it an "impoverished blister of a town."[12]

The *Tuna* rounded the coral spit of Point Tamar and puttered slowly into the harbor. Berbera appeared to Elliot to be divided into three distinct districts. There was a newly constructed government quarter, close to the water, where the British resident official lived, along with several British army officers and a garrison of Indian troops. The one-story, flat-topped houses there were built of whitewashed rubble masonry. There was a town some distance inland, consisting of low, white houses, inhabited by "Hindus, Jews, &c." And, finally, three-quarters of a mile to the east were the neat, domed huts of the Somalis huddled together, "looking from a distance like a collection of black mushrooms."[13]

The harbor was clear and deep—even near shore—the water still and intensely blue. The ship's anchor plunged prodigiously before finally touching bottom. Long boats with sharp, beak-like prows soon paddled alongside. These were rowed vigorously by nearly naked crews, using long poles fixed with wooden discs. The rowers joined enthusiastically in the incessant chatter of the Somalis on the steamer. Many appeared to be friends or relatives. Together, they raised such a racket that it seemed impossible to make any sense of the conversations.

A long, iron pier jutted from the beach. Permission received from an official onshore, the *Tuna* raised her anchor, lurched shoreward, then tied up to the end of the pier. Elliot was allowed to land with all his equipment, without the formality of getting cleared by the local customs house. A platoon of men clad in loincloths then unloaded the boat quickly and noisily. The commotion was bewildering, "as bodies of men . . . glistening like bronze lifted the boxes from the hold" and shunted them onto the pier.[14] With his belongings safely unloaded, Elliot marched to the government treasury—in the customs house, immediately in front of the pier—behind three men carrying the bulging bags of cash he'd brought from Aden. The men then gave the money over to the care of a British officer.

When their steamer left Aden, Isman—the shepherd who'd survived the brutal spear attack—was found hiding on board. As the ship approached Berbera, the captain took stock of his passengers. Isman, who had no money for his passage, was seized. No one from his tribe was on board, so other Somalis took no interest in him. He was brought to the bridge to be surrendered to British authorities in Berbera, who would place him in a chain gang and make him work off his fare with hard labor. Elliot, however, realized he would eventually have need of a shepherd, as it was necessary when on the march to take sheep and goats for sustenance. Isman had proved himself a "useful and very plucky" member of Smith's party and would be a valuable addition to Elliot's caravan. The curator, therefore, let him remain in confinement until it was almost time to turn him over to the authorities. Then he called him over and, "after lecturing him upon the impropriety of his action . . . I gave him the three rupees he needed." Isman repaid Elliot's kindness many times over "by doing the work of two men and making himself useful in every way."[15]

Elliot called on Captain Herbert D. Merewether, the British resident official. Merewether had served as assistant resident at Aden before assuming the same position in Berbera, in 1891. He had ascended to resident only a few months before Elliot's arrival, when his predecessor, Captain Henry M. Abud, was recalled. Good-looking, quiet, and shy, Merewether would prove to be an able ally of the expedition. He received Elliot and his small party politely and introduced them to his young wife, Kathleen. Elliot was aghast at the thought of the apparent loneliness and isolation of their lives. "A miserable time they would have of it," he assumed, "with no one but Hindus, & savages to speak to."[16] The captain's residence—the old harem of the former Egyptian governor—was a one-story, whitewashed building with a courtyard open on the side facing the sea. The courtyard was planted with palm trees and a magnificent acacia that soared like a green parasol above the roof, casting a cool and pleasant shade. Merewether's house was full, he apologized, and thus he was unable to offer Elliot a room.

As an alternative, the captain led Elliot's party to the former governor's house and gave them permission to reside there as long as they needed.[17]

FIGURE 5.3.

The former Egyptian governor's residence served as the expedition's commodious headquarters in Berbera. The man standing in the middle archway could be Captain Merewether. Photograph courtesy of the Field Museum, CSZ5876.

The house was a sprawling one-story masonry building fronted by an arched arcade. It enclosed an interior courtyard—a gurgling fountain at its center—where they let their horses and mules loose. The rooms, which opened into the courtyard, were large and well ventilated. Elliot was delighted to discover a bathroom with a commodious tub. They set up their camp furniture in the house, including tables, chairs, lanterns, and folding beds. They laid down green canvas flooring from their tents. The walls of Elliot's room were festooned with colorful flags, including the Stars and Stripes, a flag for the British territories, and a proud, custom-made banner representing the Field Columbian Museum African expedition. Elliot, Akeley, and Dodson slept on their camp cots. The other men spread their tobes on the floor or in the courtyard and slept wherever they could find room. Some slept in tents pitched around the house. Despite the suffocating heat, the house remained fairly cool and comfortable during the day. Elliot's lone complaint was the fountain, which bred clouds of blood-thirsty mosquitos.

Merewether invited Elliot to dinner that first evening, and the curator gratefully accepted. The two English sportsmen joined them. Henry F. G. Barclay, Elliot learned, was a wealthy, Cambridge-educated big-game hunter and adventurer, and a member of the Royal Geographical Society. He later donated several rare skins and skeletons to the British Museum. Rowland Ward's *Records of Big Game* notes several record mammals shot by Barclay in Somaliland. Barclay's companion, Lionel C. G. Sartoris, was a first cousin to Algernon "Algie" Sartoris, the late husband of Ellen "Nellie" Grant, President Ulysses S. Grant's beloved only daughter. Together, they made a merry company. There was "not a breath of wind," according to Elliot, and the heat was "considerable." Nevertheless, they "passed a very pleasant evening."[18]

FOR WANT OF A CAMEL

The next day was exceedingly hot. Elliot spent much of it seated at a makeshift desk, reconciling his accounts and writing letters, a blue coil of smoke rising intermittently from the bowl of his pipe. He was scribbling thoughtfully in his journal when Selon, the second headman, brought bad news: there were then only five camels in all of Berbera, and their owners were asking eighty to ninety rupees ($39–43) apiece, more than double the usual price. Elliot refused to pay this. Selon assured him that more animals were expected shortly, which should drive prices down. He returned later, however, to report that camels remained scarce and very pricey, the agent from the *Tuna* having begun stockpiling the animals for Italian soldiers stationed at Massawa.

In the second half of the nineteenth century, the Horn of Africa became the setting for a colonial tug-of-war between Great Britain, France,

Elliot writes in his journal in his room at the expedition's headquarters in Berbera. He often wore pajamas around the house in an effort to stay cool. Photograph courtesy of the Field Museum, CSZ5910.

and Italy. In the early 1880s, Britain and France maneuvered to establish a foothold on the Somali coast in order to protect their interests in India and Southeast Asia. France declared a protectorate in 1885. Britain followed suit two years later. Meanwhile, the Ethiopian ruler, Menelik II, was expanding his influence into the Somali interior in the Haud and Ogaden plateaus, perhaps as a check against European incursions inland. Massawa, an island off the Red Sea coast, together with the nearby mainland port of Arkiko, handled most of the trade of northern Ethiopia. Several interested parties had been vying for control of Massawa for centuries, including Egypt and the Ottoman Empire. Ethiopia's interest was in maintaining access to the sea. Massawa was first occupied by Italian troops in 1885, ostensibly to aid the British in policing the Sudan, but in reality to establish a foothold for an African empire. In 1896, the Italians were fighting a losing campaign against the Ethiopian forces of Menelik II. They'd been soundly defeated at the Battle of Adwa on March 1. Camels were needed for a possible punitive expedition.[19]

Captain Merewether heard that the Italian agent paid an extraordinary 114 rupees ($55) each for two camels. Various rumors were circulating as to the number of camels needed to satisfy the Italians, some making it as high as two thousand. A tent boy—a kind of personal servant—told Elliot that he'd heard they wanted only a hundred, and that once this number was acquired the prices would plummet. "I sincerely hope so," the curator wrote, "as it is no fun staying here."[20]

Word spread quickly that Elliot was a buyer. One morning he was visited by a boisterous crowd of Somalis bringing five scrawny, young camels to sell. The owner of the best camel of the lot demanded $65, or about 135 rupees. "The entire crowd had as much to say about the sale as the owner himself," according to Elliot, "and one would think each individual expected to get the . . . money into the folds of his own tobe." After much haggling and many protestations, Elliot finally offered 85 rupees ($41) for the finest camel. Insulted, the Somalis left in a huff. Desperate though he was to acquire pack animals, Elliot made no effort to detain them. He worried, naturally, that if he showed a willingness to pay exorbitant prices right away, he would never get a camel at what he considered a reasonable amount thereafter. Selon continued to assure him that more camels would be brought to town soon. Elliot, too, believed that at the prices then being paid, "all the camels in the country would be brought down to the coast."[21] Another group of Somalis visited Elliot with a proposition to go into the interior and gather a hundred camels for 100 rupees ($48) each, bringing them back to Berbera in a month. Elliot declined, though he found the offer encouraging. It showed, at any rate, that there were still plenty of camels in the interior and that they must be selling much cheaper there, for no doubt the Somalis expected to make a handsome profit.

While Elliot awaited word about camels, he sent Akeley and Dodson out to collect. They returned with a spoonbill, a heron, and six other birds, which were made into skins that same evening, as Elliot feared that the weather was too warm to leave them overnight. These were the first African specimens obtained by the expedition. "All birds at Berbera," Elliot noted approvingly, "are exceedingly tame, no one molesting them. . . . Crows and Vultures will hardly get out of the way of the pedestrian." Gulls loitered on the beach, "unmindful of the children playing."[22]

Akeley and Dodson tried to make the best use of their enforced idleness. One morning they went out early to test the expedition's weapons, Captain Merewether having arranged for a target to be set up on the beach near their quarters. Their powerful Lee-Metford rifle, they found, shot "splendidly." The Lee-Metford had a detachable ten-round magazine and a quick-action bolt. It fired a .303-inch lead bullet and had an effective range of eight hundred to a thousand yards. They later discovered that the nickel-pointed, solid bullets they used to hunt game made only a very small entry wound in an animal's hide. On another occasion, Dodson went after crows he heard cawing near the house. He soon returned with a pair, male and female. Another morning they both started out around six o'clock for birds and returned with two vultures. Akeley captured a pair of eight-inch-long brown skinks, later identified as a new species, *Lygosoma akeleyi*. At the beach, they collected some one hundred marine shells representing at least thirty species. On still another day, they went fishing and "caught some wonderful fish striped like zebras." One brilliantly colored lionfish was later identified and described from the lot as a new species, *Pterois ellioti*.[23]

By this point, Elliot had only hired a score of men, who, he noted approvingly, "are very attentive to us and try to do everything they can." Abdi Kareem, who spoke some English, served as tent boy and translator for the curator. Abdi often wore a shirt that appeared to be made from a coarse fishnet with a short, brightly colored jacket. He kept his hair shorn close and daubed with red clay, a local fashion. On more formal occasions, he wore a white tobe, "gracefully folded & draped about him." He was a helpful and accommodating attendant, yet Elliot suspected him of an inflated sense of self-importance, "as the servant of the Leader of the Expedition." The cook, Musa Mohammed, "a rather pleasant faced fellow with a wonderful arrangement of hair that stands out in spiral from all parts of his crown, is an expert in his calling, & provides us with excellent meals." According to Dodson, the "splendid cook" served up "a regular French dinner every night." Elliot was pleased with his hired help. "Once in your service," he wrote, "they are . . . faithful and will fight to the death in defense of the caravan."[24]

Young Somali men with no livestock or other capital were dependent on the modest cash income that came from working for safaris. This helps explain why so many were willing to submit to the orders, the expectations, and the small (and large) indignities imposed by foreigners like Elliot. Those Somalis who worked the hardest and were the most compliant gained a reputation for usefulness. Some even received a written reference attesting to their effort and loyalty. This improved their prospects for future employment. Since so many safarists hailed from Great Britain, many Somalis learned to speak and understand rudimentary English—an invaluable skill in the new safari economy. Indeed, a French Roman Catholic mission in Berbera taught English to young Somalis.[25]

Elliot rose at half past four most mornings to enjoy the cooler weather.

"It was very pleasant on the plain" before dawn, he recorded, but the temperature climbed quickly with the sun. A hot, dry, southern monsoon—the *karif*—would often rise late in the morning and blow with great, gritty ferocity until it died away around eight at night. Moonlit nights were "magnificent beyond description," he wrote. The buildings of the government quarter "stand out clear and sharp in the bright light as though formed in spotless marble, the plain stretching away . . . to the distant rocky hills looks like a beautiful buff colored carpet, the . . . thorn bushes scattered over its surface forming a pattern of deepest green." To the north "lies the tranquil sea shining like molten silver . . . while overall was the unfathomable sky thronged with most brilliant clusters of stars . . . with the southern cross scintillating . . . in opposition to the north star and great Bear showing above the northern horizon." From the plain, at intervals, the "shrill cry" of a fox or the "derisive laugh of some skulking hyena" could be heard. These sounds filled Elliot with loneliness and impressed on him the fact that the expedition had finally "entered the borders of the Dark Continent." Despite the revelry, Elliot was feeling restless: "It is dull work waiting here & we are all anxious to be away," he wrote.[26]

DUALLA ARRIVES

Dualla appeared belatedly on the afternoon of April 23. He'd sailed from Aden in a small lateen-rigged sailboat—a dhow—the passage taking three nights. "He had evidently had a very miserable trip in an open boat with very little wind, exposed all day to the heat," according to Elliot. Yet, the headman got down to business right away. He went first to the camel market and reported to Elliot that he hadn't seen a single good-looking beast of burden. He had, however, bought "one very fine animal which had been fattened to eat."[27] Such a beast would have to be broken and trained to carry a load.

Later, Elliot—mounted on an Afghan mare—together with Dualla, rode to the Somali quarter to confer with a trader on whom he had a letter of credit. This was probably Mahomed Hindi, a Hindustani merchant living in Berbera, who was a partner of Cowasjee Dinshaw & Brothers. The dignity of their errand was reflected in Dualla's outfit, a "gold embroidered turban & vest, with an Arab sword in an inlaid scabbard." Although it was still early in the morning, the sun was "too warm to be pleasant." Despite the stifling heat, Elliot was impressed by the "cool and clean" look of the "low white walls of the town." Crowds of men and children filled the streets. Over the wall, the Somali huts were packed tightly together, covering an immense area. Beyond the huts, the camel market was in full swing, with more men, however, than camels. Apparently, news that the Italians were buying hadn't yet reached the interior, and few animals had been brought to town. Dualla purchased two more fat camels. Elliot joked in his journal that the expedition would "not starve . . . even if we do not get away on our journey."[28]

Elliot heard whispers that a plague had broken out at Bulhar, about forty-five miles away, and that camels there were dropping like flies. In the 1890s, a devastating rinderpest epidemic swept through Africa, killing millions of domesticated animals as well as wild populations of buffalo, giraffe, and wildebeest. It was probably introduced into Eritrea in 1887 by Indian cattle brought by the Italians for their military campaigns. It then spread through the Horn of Africa.[29] Elliot feared that the dreadful disease might reach Berbera.

Long days passed with little or no progress. Elliot was exasperated by the lack of camels, and he regretted the lost time. "The outlook is not promising, and it is most unsatisfactory being obliged to remain here on the verge of the promised land," he griped in his journal.[30] He detailed Akeley and Dodson to collect whatever they could locally. But this pattern didn't content them for long. Unfortunately, mammal life on the barren Somali coast was conspicuously absent. Wary animals were occasionally seen on the plain, but it was difficult to get near enough to shoot them.

Meanwhile, breaking their camels to carry loads proved more challenging than expected. Hearing a camel bellow in protest as it was brought into camp, Elliot went out to watch. "An unbroken beast, it refused to do anything . . . & as the man who had it in charge was trying to put a rope around its fore legs it reared up & struck him over the head with its foot & knocked him sprawling into the dust." Elliot feared momentarily for the Somali's safety, "but the fellow got up and rubbed his woolly pate." Once a rope had been passed around the camel's lower jaw, which was done with the aid of several men, it was led to the vicinity of the tents. "It is a good looking animal," Elliot remarked, "but it is very evident there is going to be lots of fun . . . before it consents to carry a load."[31]

Later, they spent a raucous afternoon training the camels. At first, the obstinate animals refused to cooperate. But, by persistent work, they were made to kneel down and rise at command, which was a promising beginning. To make them follow in a train, each camel was tied in a row, the head of one fixed to the tail of the one ahead. A Somali trainer then took a rope tied about the head of the lead camel and pulled mightily, while other Somalis pushed, pounded, and beat the beasts from behind. "Sometimes the[y] got along very well for a short distance," Elliot wrote, "& then the procession came to a halt, & camels and men were badly mixed up & the line had to be straightened out again." Next, the camels were trained to carry a load. Each animal was forced to kneel, then a Somali would jump on its hindquarters, clinging precariously to the hair of its hump. The camel was then made to walk around with a man hanging on behind. "Throughout the performance the growling was terrific," the

FIGURE 5.5.

Breaking the
camels and
training them
to carry loads
near expedition
headquarters
in Berbera.
Photograph
courtesy of the
Field Museum,
CSZ6045.

curator remarked, "the beasts never ceasing to express their disapproval of the entire proceedings." Elliot called it a "great circus."[32]

Having traveled by caravan before, Elliot had strong opinions about camels. "A camel . . . is an abominable beast," he recalled. "He can kick worse than a Virginian mule, with any or all of his four feet. . . . He can bite as hard as a horse, and buck equal to a bronco." When on a galloping camel, a rider suffered a sensation like "an earthquake" as he was violently bumped and jostled seven feet off the ground. One final "endearing trait" of a camel was its ability to twist its "long snake-like neck around . . . poking his ugly nose into his rider's face and growling out his displeasure. . . . The rush of his fetid breath is enough to shoot one into the middle of the next week." Yet the camel was the only beast of burden capable of enduring the desert climate of Africa, he conceded, and without it "all intercourse with various parts of the continent would cease."[33]

When another day arrived with "but few camels," and with the asking prices still high, Elliot reluctantly decided to send two men to Bulhar to buy camels there, despite his fear of plague. The Somalis evidently favored the market at Bulhar over Berbera, as the former featured more convenient pasturage. Elliot's emissaries started from Berbera in a dhow at noon and were expected to reach their destination that same night. If his men should succeed in getting ten camels, Elliot's plan was to make a short excursion of a few weeks to hunt Somali wild asses and other mammals near the coast. The curator hoped that Dualla—who was staying behind—might pick up additional camels in his absence, or that the asking prices might fall, there being then no other European buyers in town. He was irked by the prodigal Italians, who were inflating the prices for camels by "foolishly giving four times what the animals are worth."[34]

One morning, Elliot's three fat camels were loaded with cartridge boxes—weighing fifty pounds each—and marched onto the plain for practice. Though there was "much growling & swearing on the animals part & they made a tremendous row before they got away," Selon reported that he was satisfied with the trial. The camels behaved remarkably well, and he expected they'd be completely broken in two days. The men sent to Bulhar were expected to return at any moment. "I should like to see them coming in leading a string of good animals," Elliot wrote hopefully.[35]

Meanwhile, a curious legal entanglement triggered some excitement in camp. Selon had hired a blacksmith to hammer fifty iron tent pins. According to the second headman, the price agreed was two annas each. But, when they were finished, the blacksmith demanded three. When Selon refused to pay this price, he was summoned to court. Elliot described the summons with great curiosity: "The maker of the pins went to a policeman and stated his grievance. The policeman then picked up a stone and gave it to one of my men and told him to give it to Selon. The latter on receiving it came to me & showed me the authoritative rock and asked permission to . . . appear at court." Elliot marveled "at the simplicity of the proceedings." A man who'd received such a stone, the curator mused, "would never dream of disobeying." Hearing the evidence, the judge rendered a decision. Selon would have to pay three annas for thirty of the pins and two each for the rest. Elliot was miffed by the added expense. "Verily a wise Judge who endeavored to please both by cutting the child into two pieces," he grumbled.[36]

BAGGING THEIR FIRST MAMMALS

One sultry night, while Elliot was paying another visit to the Merewethers, Akeley and Dodson went out on the moonlit plain with a rifle and shot repeatedly at a hyena. Though badly wounded, the hyena managed to escape when the ammunition gave out, Dodson having taken only three rounds. Akeley, meanwhile, succeeded in capturing two live hedgehogs, *Erinaceus diadematus*, in his hat. These were the expedition's first mammals. The hedgehogs offered no resistance, but "immediately rolled themselves into a ball, only attempting, by a sudden muscular movement, to drive the spines into the hand of anyone who touched them." Akeley brought them back to the house, where he let them roam free. They scurried about, examining their new confines, keeping mostly to the shadowy corners. They only assumed the protective ball shape when handled.[37]

The next afternoon, Dodson suffered a chill followed by a mild fever. The cause, Elliot speculated, was getting into a great heat chasing after the wounded hyena and then "foolishly standing in the wind and getting a check of perspiration. One cannot do such things with impunity in Africa," he scolded.[38]

Encouraged by the previous night's results, Akeley went back onto the plain alone with his rifle. The moon was full, and the landscape was bathed in a silvery light. Hardly a breath of wind was stirring. Prowling about in the stifling air, he was soon soaked in sweat. He concealed himself behind a makeshift screen, having taken part of a sheep's carcass, tied a rope to it, and dragged it through the dust leading away from his position. Some jackals, *Canis anthus*, scented the bait and "followed it straight to the ambush where they were warmly received."[39] Akeley managed to take four and wound a fifth.

Skinning a large mammal in the field was a meticulous, messy, and malodorous task. Akeley, who was bothered whenever stiches in a mounted animal were plainly visible, had experimented extensively to develop his own method for skinning an animal that required the fewest possible incisions. He began by opening the animal's belly with a clean, straight cut. In making this initial cut, it was best to insert the point of the knife under the skin, blunt edge up, to avoid cutting any of the specimen's hair. He then disjointed the limbs from the body through the opening. Next, he made a cut above each hoof, then rolled the skin of the limbs back as one would roll up a stocking. Then he removed the limbs. The complete skin of the animal's body was thus left intact. The only cuts he made were along the belly, above each hoof, and along the mane. Once the skin had been removed and thoroughly cleaned, he sewed any tears or holes from the inside.[40]

Heads were tricky. On an animal with horns or antlers, it was necessary to make an opening at the back of the neck in the shape of a Y. He then cut completely around each horn at the base. Next, he skinned the head by carefully working downward over the forehead and cheeks. The ears he severed close to the head. The skin he turned inside out over the head and peeled back to the eyes. Working slowly and deliberately with a fine knife, he cut close to the edge of the bony orbit, until he saw, through a thin membrane, the iris and pupil. This membrane he then cut through to expose the eye, carefully preserving the eyelids. He skinned down to the end of the nose, cutting through the cartilage close to the bone, and cutting down to where the upper lip joined the gum. He cut both lips away from the skull, close to the bone, all around the mouth. Thick and fleshy lips had to be split open from the inside and flattened, so that the flesh in them could be carefully pared off. When paring the flesh from the inside of the face, he had to be careful not to shave off the roots of the whiskers, or they might fall out. The membrane inside the eyelids was then pared away. The ear was turned inside out at the base, in order to remove the flesh around it. If the ears had hair, they had to be skinned from the inside and turned inside out all the way to the tip. This separated the outer layer of skin, which held the hair, from the cartilage of the ear. He removed the skull through the Y-shaped cut. Finally, he withdrew the cranium through the occipital opening at the skull's base using a specialized tool called a brain scoop. The skull was then cleaned carefully and smeared with a thin arsenic soap, tagged, and allowed to dry thoroughly.[41]

In general, it was easier to remove the skin clean and free from flesh as he cut it from the carcass, stretching it tight with his left hand in order to trim the flesh off cleanly. Excess flesh left on the skin added to a specimen's weight, required a longer time in curing, and often resulted in a waste of preservatives. A clean, thin skin was more easily and quickly cured and carried. Ideally, he cleaned the skin so thoroughly in taking it off that no paring down was necessary before curing it. It was also important, as much as possible, to keep blood from staining the specimen's hair—especially with fair-haired animals. Blood had to be washed out immediately, even when water was scarce. If a wound bled profusely, one trick of the trade was to heap dry dirt or sand on it to absorb the blood and keep it from spreading. The blood-clotted dirt could then be knocked off.[42]

Once removed from the carcass, the skin had to be preserved. It was far easier to mount a pliable specimen than an old, dry skin with no elasticity. Work on a soft skin was easier, quicker, and produced vastly better results. Wet preservation, or preserving skins in a briny bath, was best for specimens intended for exhibit. The formula for the brine could vary, and Elliot's recipe is now lost. Generally, for every gallon of water, Akeley added about a pint of alum powder and a quart of salt, then brought the mixture to a boil and stirred occasionally until the alum and salt dissolved. This solution was then poured into a wooden barrel. Once cool, it was ready to use. Salinity was key. Too much salt, and the skins would stiffen. Too little, and they were liable to shed their hair. Salt could always be added to the bath, but it couldn't be removed. Skins were simply immersed in the brine. Large skins were treated individually, giving them ample space in which to be stirred around in the solution, thoroughly soaking every nook and cranny. If the skin was folded or clumped, the preservative might fail to penetrate, and certain spots were then liable to lose hair. A small or thin skin was finished very quickly. A large skin often had to be stirred and soaked over a period of several days. As fresh skins were added to the bath, it gradually lost its potency. It was also likely to become, over time, very dirty with blood and grease. Especially greasy skins had to be treated in isolation.[43]

Once treated, skins could be packed together in a brine-filled barrel with other specimens. When the barrel was filled, the lid had to be tightly closed, as the liquid could leak or evaporate. Even when the best and most exacting methods were used, however, field conditions often made it so difficult to care for specimens properly that "only a very small percentage ever reach the taxidermy shop in perfect condition."[44]

GETTING UNDERWAY

One evening, the party saddled three mules, and, as soon as the moon was up, Elliot, Akeley, and Dodson rode onto the plain, together with some of their men, to hunt hedgehogs. This was hardly the stuff from which to build great museum dioramas, but the exercise was good for morale. They rode about five miles but saw no animals save a solitary, speckled lizard, *Eremias brenneri*, which one of the men captured. They also spied a venomous snake, brown and white striped, which another man pummeled with a volley of stones. Unfortunately, the snake was "battered out of all shape," and was thus worthless as a specimen.[45] The expedition returned to camp with their luckless lizard.

Akeley tried to shoot some sandgrouse, which drank regularly at a well near their quarters. He found them very wild and succeeded in obtaining only one. This bird, *Pteroclurus exustus*, was seen periodically in small flocks near the water at dawn and dusk. Some frequented a small ditch on the beach, where water overflowed from a large cistern.[46]

Elliot's new Winchester paradox gun arrived from London. A combination weapon that fired a cone of pellets like a shotgun and bullets like a rifle, the paradox was especially useful for hunters who might encounter birds or big game on any given day and needed a single weapon that could handle either scenario. Harold G. C. Swayne recommended a spendy, eight-bore paradox gun manufactured by Holland & Holland. Elliot couldn't afford this weapon. The Winchester was his reasonably priced substitute, and he was eager to test it. Toward evening, he went out to try for more sandgrouse. When the grouse didn't cooperate, he managed to shoot a white vulture and several wagtails. He was surprised to find that he'd obtained three different species, including *Motacilla campestris*, *M. borealis*, and *M. cinereicapilla*, with a single shot. The different species had been comingled in a large flock. Elliot assumed there was only one species present until he sorted through the carnage.[47]

At long last, on the morning of Wednesday, April 29, the men returned from Bulhar leading seven sturdy camels. They made the forty-five-mile return march, a lengthy journey even for unladen beasts, in one night. They'd bought four "& had an excellent outlook for getting the ten I sent for," Elliot recorded, "when some fellow who went over with them . . . told the natives that they were buying for Europeans."[48] The asking price was

FIGURE 5.6.

Akeley's attendants. The man in the front is his camera bearer. Photograph courtesy of the Field Museum, CSZ6136.

then immediately raised. They paid ninety rupees ($43) each for the seven acquired, which was cheaper than the current prices in Berbera. Elliot now had thirteen camels and planned to start the next afternoon on a short excursion for wild asses. Dualla would stay behind to buy more camels. Elliot was certain that once he was seen riding away with his small caravan, prices would drop. The rest of the day they spent preparing for departure.

On the last day of April, Elliot kept busy all morning. He arranged with Captain Merewether to furnish Dualla with funds to buy camels, instructing the headman to purchase twenty-seven more, if possible. With forty camels—half the number originally proposed—Elliot intended to start for the far interior. Perhaps there, where the prices were presumably lower, it might be possible to increase his caravan. Akeley and Dodson, meanwhile, dismantled the encampment and packed everything necessary for a short excursion.

Elliot hired new men, advancing them a month's wages. A well-run, nineteenth-century safari required many personnel. The headman, Dualla, who answered directly to Elliot, oversaw the locally hired labor. He was assisted by a second and a third headman as the party expanded. Shikaris served as hunting guides, one or two for each hunter. Gun bearers carried and cared for weapons on the march and during the hunt. Akeley also had a single camera bearer. Tent boys—often the youngest members of the safari—served as personal attendants. They cared for the tents, clothing, and personal belongings of each hunter. They also served the hunters at meals. Syces took care of the horses and mules and their riding equipment. The most numerous group was the camel men, or porters, whose duties were manifold: loading and unloading the camels, gathering water and firewood, bringing game back to camp, beating for lions, setting up and striking tents, and handling other miscellaneous duties. Cooks and their assistants prepared the meals. Finally, askaris were armed and served as guards. There was, on average, one askari for every ten men in the safari.[49]

The white hunters were utterly dependent on their African hired hands, and not just for the manual labor they performed. Local guides provided essential knowledge, without which a successful expedition would have been nearly impossible. Their guides knew the routes and distances between water sources, for example. They knew where good grazing was likely to be found. Their shikaris, meanwhile, had uncanny skills at tracking and following the spoor of big game that the white hunters couldn't hope to match. Even the least skilled of their hired men possessed useful, local knowledge. They knew, for example, which snakes were most dangerous, or where to find certain game. They also knew the local languages and customs, without which the foreign visitors would have been helpless.[50]

Elliot came to appreciate the vital role of his hirelings. He especially respected their superlative hunting and tracking skills. He also felt deep sympathy for the sick and injured among them, often going to extraordinary lengths to help alleviate their suffering. He expressed great admiration for many of them individually, yet he often had disparaging things to say about them collectively, calling them boys or savages (or worse), criticizing their beliefs and customs, and deriding their religion. Many of their habits—their loquaciousness, for example—made him uncomfortable. In general, he seemed to feel a strong sense of fatherly superiority. He was arguably no more racist than the garden variety European or American traveler in Africa in the late nineteenth century, and a good deal less racist than the worst.[51]

In midafternoon, the entire company of twenty-one hired men, led by Dualla, marched to Captain Merewether's residence. There, the captain read aloud an agreement that Elliot had written. He read the document first in its original English. He then translated it into Hindi for his interpreter, who then rendered that into Somali. However mangled Elliot's conditions were translated, his men assented. The curator watched the curious proceedings with fascination. "They were a queer looking lot," he wrote, "as they stood in line with much less clothes per man than Fallstaff's ragged bravados possessed, but hardy & of fine physique, all capable of standing the journey before them."[52]

With the formalities finished, the men hiked back to headquarters, where the chaos of loading began. Men pointed and shouted instructions. Camels roared in protest. "The space behind the house was an ever shifting, kaleidoscopic scene . . . running hither & thither carrying boxes and packages of all sizes and shapes and putting them on the backs of the camels which greeted each addition with a fresh roar." The men had difficulty with the fat camels, which were "particularly obstreperous." Yet in an hour the loading was done. The caravan, led by "a steady reliable beast," then plodded slowly away from Berbera across the plain, bearing southeast.[53]

The hunters and their attendees waited a leisurely hour before following, Elliot riding a speedy mule and leading his horse. The late afternoon sun had lost much of its potency, and a stiff breeze was blowing. They rode pleasantly over some of the same ground they'd covered a few nights previously when they hunted in vain for hedgehogs. As they marched, the Somali men probably chewed khat leaves—a mild stimulant that rendered the exertion more tolerable—and talked or sang snatches of traditional songs.[54] The plain surrounding Berbera was barren and covered with a crust of white pebbles. A mile or more from town, the ground became sandy, with a profusion of waist-high, flat-crowned acacia shrubs and a scattering of thorn bushes, some reaching twelve feet in height. They headed for some low hills called the Kaffir Range—apparently not far away, these hills seemed to recede from them as they advanced, until it seemed as if they would never reach the landmark. The hunting guides, or shikaris, along with the gun bearers, walked in front, keeping a sharp watch ahead for any game.

After the party had marched for an hour, four gazelles were seen lying prone on a distant ridge. Elliot tried to approach them with his shikari, but the skittish animals soon leaped up and fled. He took a long shot, which "went harmless," and the animals quickly disappeared among the low bushes.[55] Shortly thereafter, another distant gazelle appeared. Elliot sent Akeley to see what he could do, but the taxidermist was unable to get near enough to attempt a shot. The game near Berbera, Elliot suspected, was hunted so regularly that the animals were very wary.

The sun set red and hazy while they were still crossing the plain, and, "as is always the case in the tropics, darkness fell upon us at once." The waning moon would be rising later, so the hunters were obliged to stumble forward as best they could. Around half past seven, they finally reached the limestone hills, where they entered a wide, sandy gorge. In the dim light, Elliot could make out what appeared to be the banks of a dry riverbed on either side of their path. "Nothing but sand rock & thorn bushes were about us, and we kept on going a little east of south." By nine o'clock, "the men began to halloo, in hopes of hearing an answering cry."[56] Soon they spotted two campfires. Reaching a village, they found a scattering of huts surrounded by a stockade of thorn bushes called a zareba, which helped repel predators. The villagers informed them that a caravan had passed by previously. The fatigued curator followed in the direction of their gestures. Shortly thereafter they heard a friendly shout. Approaching another blazing fire, they soon found themselves in the midst of their own camp.

The camels, which had already been unloaded, were stretched about the camp helter-skelter, resting. They'd ceased their growling and "were either quietly chewing the cud, or steadily gazing out of their great eyes over the sands." Musa was busy making dinner over the glowing coals of a small cooking fire. It was half past nine. The moon, which had just risen, "was shedding a pale light over the scene." It reminded Elliot of similar nights he'd spent long before in the Arabian Desert, and "although the localities were widely separated, the same attractive charm was over both. It was a scene of peace and rest."[57] Many of the men, exhausted from the march, had dropped to the ground and were already fast asleep. The tent boys, meanwhile, were busy arranging bedding for Elliot, Akeley, and Dodson. It was so late, and so warm, that no tents were pitched. Instead, their cots were assembled together on one side of camp, where they would sleep in the open under an incomparable blanket of stars.

Supper over, camp grew quiet. Guards were set. The hired men wrapped themselves in their tobes and stretched out on camel mats on the ground. Elliot "sat up for a long time smoking, enjoying the beauty of the night with its countless stars." He listened thoughtfully to the "mocking laugh of a hyena."[58] At long last he went to bed with a feeling of deep contentment. The expedition was away, at last. Success was now a matter of hard work and luck. He slept fitfully.

During the night, a curious camel ambled over and gazed down dumbly on the sleeping hunters.

6

NEVER A FINER DRINK

May 1–5

Elliot didn't sleep long. The sounds of voices, shuffling feet, and low, discontented grunts woke him with a start. Peering into the darkness, he could see the men were beginning to load the camels. Dim lamps swayed in the gloom like will-o'-the-wisps. A cooking fire burned brightly. Immediately, he sprang from his cot. He wanted to make an early start in order to reach the next watering hole before the worst heat of the day.

By four o'clock that morning, they were in motion, the hunters leaving ahead of the caravan. They kept to the same sandy plain, rising steadily in elevation as they made their way inland by the dim, gray twilight. As dawn broke, they came to a rough, weathered country, "broken up into great fissures, and fantastic shaped hills." Elliot recalled the badlands of western North America, where he'd once hunted unsuccessfully for bison, hoping to salvage some specimens for the American Museum before the species disappeared. A lone gazelle grazed at the edge of this country—it was soon chased off by the crack of Akeley's rifle. Later, two diminutive dik-diks, among the smallest antelopes, bounded from a bush and "scampered away like rabbits."[1]

They descended a ravine by a steep, rocky path. By six o'clock, they reached a marshy expanse with beautiful, bright green shoots of new grass and two small, spring-fed pools. Muddy hoofprints were everywhere. They unsaddled their mules and turned them loose with the horses, which clipped the fresh grass greedily. Half an hour later, the caravan arrived, and the camels were quickly unloaded. They pitched their tents a short distance from the water. Elliot unpacked his aneroid barometer and measured 1,050 feet of elevation. Akeley went searching for gazelles, while Dodson blasted a sunbird and two warblers. By late morning the heat was oppressive.

They loitered in camp long after lunch, as Elliot judged that it'd be useless to hunt in the heat. The late afternoon temperature was more tolerable,

FIGURE 6.1.

Tents and camel
mats at a field
camp in Guban.
Musa's kitchen
is under the
shrub on the
left. Photograph
courtesy of the
Field Museum,
CSZ6063.

though, so the eager hunters saddled their mules and set out in different directions. Elliot went south through the badlands, where sandstone hills were carved into spires, pyramids, and domes, red in color, with broad, green bands and thin veins of milky quartz. "The whole country was sand interspersed with places covered thickly with broken quartz [and] flints," he wrote, "with thorn bushes growing along the banks of dried up streams. It was a most desolate and forbidding country." He crisscrossed the landscape until nightfall. Except for a few birds, he didn't see a living thing.[2]

Arriving back at camp, he found Akeley and Dodson already there. Each had succeeded in killing a Pelzeln's gazelle, *Gazella pelzelni*, one a young buck and the other an old doe. "Beautiful little creatures they are," Elliot wrote, "so delicately formed, with long slender legs and delicately shaped heads." Both sexes bore modest horns; the female's were slimmer and straighter than the male's. Later, after more specimens had been collected, the curator determined that there was "considerable variation in the coloring of individuals. . . . The typical style has a broad conspicuous chestnut band running lengthwise along the body just above the white of the belly." Some individuals, on the other hand, evidently of equal age and killed at almost the same time and with the same condition of coat, "were entirely without this distinguishing mark." It was after midnight before the skins were removed, preserved, and stowed away.[3]

In the afternoon, a villager had wandered into camp, asking after the expedition's purpose. At an audience with Elliot, he claimed he could guide the hunters to a spot where he'd often found droves of wild donkeys. He'd guided hunting parties before, he said. Elliot offered a reward if his report proved true. The enterprising villager promised to return the following day at first light.

In certain parts of Somaliland, especially in the desert strip along the coast, the Somali wild donkey, *Equus asinus somalicus*, roamed. It was "a very handsome animal," wrote Elliot, with a large, well-shaped head. It wore a "bluegray coat, relieved with the white nose and belly, and the striped whitish legs." The creature was "exceedingly wary . . . , always on the alert," and it was "no easy matter to approach within even long shooting distance of a single animal." They were generally solitary, sometimes congregating in sets of two or three. They lived in "sterile, rocky districts, the ground covered either with sand or broken stones." Elliot wondered what these animals lived on, grass being exceedingly scarce. They seldom wandered far from reliable water sources. Their tracks were always found on the muddy margins of spring-fed pools, showing that they came regularly to drink. Despite their "alertness, swiftness and other game qualities," hunting them wasn't very sporting. Elliot likened it to "slaughtering horses." Had they not been hunting "for scientific purposes, none of my party . . . would have molested them," he insisted. The Somali donkey wasn't plentiful. The curator suspected that continued persecution would "speedily extinguish the race."[4] Yet, this was the inoffensive beast that the expedition had set its collective sights on, hoping to bag several representative specimens before the species disappeared.

Elliot and his hunters were ready early the next morning, waiting with growing impatience until nearly dawn. The eastern horizon began to glow pale blue, the stars winking out one by one. At last, a woman came from the village to say that a relative had died during the night and so their guide couldn't come. Annoyed but undaunted, the curator sent his hunters out in different directions to see whether they could find game for themselves.

The hunters rode in tandem past the muddy spring near camp, then parted. Soon after splitting from the rest of the party, Elliot saw two gazelles. Dismounting, he'd gone only a short distance when he spied two donkeys about eight hundred yards away, a female and a half-grown juvenile. Elliot and his gun bearer began to cautiously stalk the more valuable donkeys, when they realized that the animals were slowly, carelessly grazing in their direction. The hunters were downwind, so they had merely to sit tight and the animals would come. They crouched behind a thorn bush and waited. The donkeys had approached within two hundred yards when the shikari—thinking they'd been detected—urged the curator to shoot. Elliot fired at the female, aiming for the shoulder. At the sound of the shot, the animal wheeled quickly and bolted at a full run, followed by the juvenile. "Missed," the shikari whispered.[5] They rose as one and ran toward the place where the donkeys had been grazing. There they found a modest splatter of blood and shredded tissue, which became more plentiful as they followed the telltale trail on foot. In a mile, they found where the donkey had stopped and coughed up a puddle of frothy blood, showing that Elliot had shot it through the lungs. The wounded animal's trail then led up a steep slope. The heat had become very intense, and Elliot was spent. He sent his shikari to follow while he went back for his mule and rode to camp to rest in the shade of his tent. If he should find the animal, the shikari was to send for a camel to bring it to camp.

On arriving, Elliot was delighted to learn that Akeley had killed a donkey. A camel was then bringing it into camp. It proved to be "a splendid old mare, beautiful in color & in fine condition." Akeley photographed it when it lay dead, and again after it'd been loaded onto the camel. The carcass weighed over seven hundred pounds, which made a heavy burden. Dodson soon arrived, and he and Akeley spent the rest of the morning preparing the skin. Several choice cuts from the haunches and the saddle went to the cook to make a soup and a roast, for Elliot was curious to know how wild donkey tasted. "The flesh of these animals is very good," he later recorded, "almost the best we ate in Somaliland, being more tender and

FIGURE 6.2.

Akeley's attendants with the expedition's first Somali wild donkey loaded awkwardly on a camel. Photograph courtesy of the Field Museum, CSZ5990.

having much more flavor than any of the antelopes."[6] A dozen Somalis harnessed themselves to the flayed carcass with long ropes and—while singing a song of victory—dragged it a short distance from camp. Elliot then had several traps buried in the sand around it, hoping to catch any scavengers that might pay a visit overnight.

The shikari returned late in the afternoon. He'd spotted Elliot's wounded donkey but couldn't approach. He then lost its track on difficult, stony ground where it'd taken refuge. Elliot was surprisingly blasé, scratching "Better luck next time" in his journal. Akeley and Dodson went out again in the evening. Dodson brought back a Pelzeln's gazelle buck with "fine horns," nearly eleven inches long.[7] Akeley saw nothing. They passed the late evening smoking and swapping stories of the hunt while laboriously skinning Dodson's gazelle.

At sundown, a group of men aligned themselves in front of Elliot's tent with their faces turned toward Mecca. Their spiritual leader chanted a prayer while the others joined in response. There was, "much genuflexion and prostration." Once prayers were over, the camels were brought into camp and made to lie down near the tents for the night. They'd been browsing contentedly on shrubs all day, and the grinding of their jaws as they chewed the cud was heard through the night. Elliot marveled at any animal that could feed from these plants, which bristled with needle-sharp

FIGURE 6.3.

A shaded kitchen setup at a field camp in Guban, with one of the cooks. Photograph courtesy of the Field Museum, CSZ5911.

thorns. A thistle "that would pierce the hand and inflict painful wounds" was "a delicious morsel" to a camel, which "rolls it under and over his tongue" with apparent delight.[8]

Elliot later wrote pointedly about thorns, which abounded on nearly every tree and shrub in Somaliland. "Marvels of ingenuity," he called them. "No matter how one may encounter a bush, the thorn of exactly the right shape & position is there to hold him & every movement made for release is met by others whose shapes & positions exactly fit them to aid their fellows." The thorns came in every size and shape, "from straight lance-like business implements of two inches or more in length to tiny ones just standing above the bark, but these last go in pairs to make up I suppose for their diminutiveness." Others, "shaped like fish hooks," penetrated deeper into the flesh the more one tugged at them.[9] A man caught fast in thorns would often lose his temper, and simply wrench himself free, leaving behind bloody shreds of clothing. The cuts and scratches inflicted were painful, often taking days to heal.

Akeley and Dodson, thoroughly tired from a hard day's work, went to investigate the decaying donkey carcass before going to bed. They found a hyena as big as a lion caught in one of Elliot's traps. The hyena was "in a very ugly mood."[10] As soon as they approached it lurched away in fear, dragging the trap and yanking the bush to which it was anchored out of the ground. Though it was already dark, they fired two aimless shots in the hyena's direction. Too tired to pursue, they let it go. Later that night, from the comfort of his cot, Elliot heard the painful yip of a jackal and supposed it'd gotten caught in another trap.

The next morning—Sunday, May 3—Akeley and Dodson started out around four o'clock. Elliot stayed in camp for a sabbatical, a pattern he followed for the remainder of the expedition. The jackal he'd heard was found in a trap, killed, and brought to camp. He sent two camel men to find the hyena, but they soon returned, saying they'd lost its trail in the rocky hills. Elliot sent a shikari after it instead, hoping to recover the trap, at least, if possible. But the hyena managed to get away despite being encumbered with a trap clamped to its wounded leg. Elliot gave it up as lost. There was "not a breath of air stirring," and the heat was withering. Elliot, thinking of his absent hunters, judged that it was "really dangerous to remain long exposed to the sun." For the previous two afternoons there'd been storms raging in the hills to the south. He could see the lightning flashes distinctly and hear the rumble of distant thunder. "A shower," he admitted, "would be refreshing."[11]

Late in the morning, Dodson returned, bringing four more Pelzeln's gazelles and one baboon, *Papio hamadryas*, having stumbled across a colony of the latter in the nearby hills. He saw no donkeys. Soon, Akeley came in "thoroughly used up with the heat." He'd seen two donkeys and had taken a long shot at one with no result. Elliot now realized the difficulty of obtaining specimens to build a group of these rare and wary animals. Yet he was determined to have them for his collection. If he couldn't salvage sufficient specimens now, he would probably never have another chance. On the other hand, the expedition had already procured enough specimens to build a group of Pelzeln's gazelle, "& these graceful, beautiful creatures will prove a great attraction . . . in the Museum."[12]

Buzzards, which had discovered the decaying donkey, spiraled overhead in growing numbers. The bearded vulture, Elliot remarked, was a "magnificent bird when on the wing, his great pinions bearing him aloft with scarcely a perceptible motion, & all his movements in the air are performed with a grace and easy power." He watched the birds intently through his field glasses. At times, they would swoop down to the carcass and "fight and scramble for choice bits, chasing each other over the ground." He tried to capture one in a trap but failed. The vultures made short work of the carcass. "It looks as if the beasts of the night would have poor pickings left for them," he wrote.[13]

As they were sitting by their tents in the afternoon, Akeley and Dodson working on the skins brought in that morning, one of the men ran up, shouting that there was a hyena caught in a trap. Dodson ran immediately for his gun, but before he could reach it a report was heard, and then a shout, and Elliot knew the hyena had been killed. It'd abandoned the donkey carcass and was escaping along the dry bed of a stream as quickly as it could manage, dragging the trap behind it. One askari noticed and shot it with his Snider carbine. Four men then carried the dead animal to camp, followed by several others who were happily singing a song of praise. The Somalis detested hyenas, for they were known to carry off sheep and goats, "and even young children from the huts," according to Elliot. The men crowded around the shooter, who was "the proudest man in camp" that day. The carcass, which was laid down carefully by Elliot's chair, proved to be "a splendid specimen of the striped hyena," *Hyena striata*, a relatively scarce and, therefore, very desirable creature.[14]

The afternoon had been exceptionally hot, "the sun fairly burning everything it shone upon." The evening, by comparison, was "most refreshing."[15] They sat up late into the night, smoking, stoking the coals of their ebbing campfire, and swapping tall tales about the day's hunt. They dreaded to enter the oppressive heat of their tents. With so many skins to care for in camp, Elliot decided that Akeley and Dodson wouldn't hunt the next morning. He was determined to find an ass himself.

Elliot rose at four o'clock, ate a light breakfast, and then set out with his attendants on what ultimately proved to be a long and unproductive meander. He went southeast, through rugged country, but saw only two wary gazelles. There were no donkeys, although he found plentiful tracks, including some recently made. The early morning air was pleasant. It'd rained all night in the nearby mountains, and the sky at dawn was mercifully overcast. Once the sun ascended, however, the clouds evaporated and the heat soared. Elliot returned to camp at midday. He found that Dodson had bagged three new birds and Akeley had attended to all the skins. A horde of vultures was feeding on the hyena carcass. Elliot shot one with his .22 rifle.

Late in the afternoon, Elliot set out again, taking a long route through the hills to look for baboons, but seeing no mammals at all. He returned with one weaver bird, and a four-foot-long rock monitor, *Varanus albigularis*, that he shot as it was sheltering under a shrub. Ahmet, a gun bearer, said it was "the child of the crocodile and very good luck to kill it." The curator also saw a flock of ground-dwelling francolins but couldn't get close enough to shoot one. The elusive birds "ran with great speed & the bushes were many & thick."[16]

On his way back to camp, Elliot encountered a wandering Midgan. A minority tribe concentrated in northern Somalia, Midgans were regarded by other Somalis as an inferior caste. From this wanderer he purchased a bow and quiver with eight small, iron-tipped arrows, two of them poisoned. A small knife in a sheath was attached to the quiver. The leather strap on the quiver was "ornamented in crude patterns in black with bits of red worsted sewed in various places." The quiver was made of hide in the shape of an hourglass and had a cover to keep the arrows secure. The knife was a thin, narrow blade with a wood handle. The bow was about five feet long, altogether out of proportion to the foot-long arrows, which were heavy and coarsely made. With these weapons, Elliot marveled, the Midgans "do not hesitate to attack the largest animals."[17] Their poison was potent.

Each afternoon of late there'd been storms in the mountains to the south. Rolling thunder and flashes of lightning were commonplace, but the rain held off in camp. But for a cool wind that sprang up in the late morning, the heat would've been intolerable. "Everything in & about the tent becomes red hot & the gun barrels cannot be touched." Their work in camp finished, Akeley and Dodson would crawl under the bushes and either nap or play cards in the paltry shade during the hottest part of the day. The Africans apparently didn't mind the heat: "The way they stand & walk about under the full glare of the sun with bare heads some of them shaven is extraordinary," Elliot wrote.[18]

While the men were at rest, an exuberant Somali visited, insisting that he could lead them to a place, some hours away, where donkeys were plentiful. "If we did not get an ass's head," Elliot recorded, "he would give us his."[19] The curator decided that such a claim must be investigated. If there really was an abundance of donkeys, he would move camp. Elliot was eager to secure enough donkey skins to make a museum group. This accomplished, he could make for the mountains to the south and the promise of rain and cooler weather.

That evening, some of the men organized a dance that lasted late into the night. The dancers formed a line, alternately advancing and retreating, sometimes pivoting on one foot. They clapped their hands to the rhythm of a caller's song, the whole party joining in a throaty chorus. "The light of the fires glanced upon them," Elliot wrote, "and the swell of the deep toned chorus as it rose and fell in rather melodious unison sounded finely echoing from the surrounding hills."[20] They were still so near Berbera that a zareba was unnecessary. Guards were posted, however, and their calls and the tramp of their sandaled feet as they walked their rounds, gun set on shoulder, could be heard through the night.

I SHOULD NEVER BE A SPORTSMAN

Early the next morning, Akeley and Dodson marched off with their cock-sure guide. Elliot stayed put. The hopeful hunters brought several camels to carry specimens back to camp. Unfortunately, no one thought to bring extra drinking water. This was a day when poor planning, together with Akeley's fiery temper, nearly brought the expedition to ruin.

Hours later, they were traversing a dry, sandy plain studded sparingly with dwarf shrubs and tufts of wiry grass. Suddenly, a gun bearer pointed to a dark, hazy silhouette on the horizon. That, he emphatically insisted, was a wild donkey. They advanced cautiously. There was no cover, and therefore no possibility of a stealthy approach. The likelihood of getting a shot at close range seemed remote, "for we had found in our previous experience that the wild ass is extremely shy and when once alarmed travels rapidly and for long distances." But this time was different. They managed to approach within two hundred yards and had begun to suspect that this must be a domesticated donkey. Suddenly, the animal became uneasy and bolted. But curiosity brought it about for a last, fatal look at the hunters, who took advantage of the opportunity and fired. The donkey was "hard hit," but recovered and stood facing its tormentors. They stole even closer. Not wanting to take chances, someone fired again at close range. The donkey merely walked around a bit at random, making no attempt to escape. They continued to carefully approach the animal, which showed no sign of fear. At last, Akeley put a hand on the wounded donkey's withers and simply pushed it over.[21]

This was slaughter, not sport. There was no thrill at all! No chase. No fair play. "I began to feel," Akeley later wrote, "that if this was sport I should never be a sportsman." They skinned the donkey on the spot, sending the partially prepared hide back to camp on the back of a camel. Then they went on.[22]

The hunters soon discovered that their supply of water was exhausted. The day was punishingly hot, and although they wanted to prolong the hunt, they decided that to carry on without water was too risky. Their guide had assured them there would be plenty of water on their route, but he was mistaken. They knew it was at least five hours back to camp, and that riding without water in the midday heat would be torture. "It is said," Akeley wrote, "that in that region thirty hours without water means death to the native and twelve hours is the white man's limit." The guide insisted that there was a reliable source of water an hour distant, so they went on. Their trail led them "under a pitiless noon-day sun along a narrow valley shut in on either side by steep, rocky hills." They rode head-on into "a strong, hot wind that drove the burning sand into our faces and hands."[23] After what seemed like several more hours of hard marching, however, they arrived at an empty well. They clawed at the dry bottom of the hole, quickly realizing that there was no hope of finding water there.

The guide recommended another well about an hour distant. He volunteered to lead the gun bearers there to collect some water and bring it back. They set out, while Akeley and Dodson unsaddled their mules and hunkered down for rest under the feeble shade of a leafless tree. Hours went by. Their guide and gun bearers finally returned at four o'clock in the afternoon, with no water. Akeley and Dodson, stupefied with thirst, heard the news, gazed at each other resignedly, and cowered beneath the shelter of their saddle blankets. Death by exposure was only a dream away.

Later, Akeley was roused by a gun bearer, who'd spotted a pair of men leading some camels. The thirsty taxidermist summoned the energy to look. Each man, he noticed, carried a bloated goatskin slung over his shoulder. Assuming the goatskins were filled with camel's milk, he was relieved to think that their suffering would soon be over. He instructed the gun bearer to bargain for a portion of the milk, then covered his head again to escape the blowing sand.[24]

Akeley and Dodson were both by then in a "semi-comatose state," and the former paid no attention to the proceedings until he was urgently prodded again by the gun bearer, who was now greatly agitated. The gun bearer gestured at the receding camels, explaining that the men had refused to part with any of their milk. "The white men might die for all they cared," Akeley later wrote bitterly. Shaken out of his stupor by the gravity of their predicament, Akeley became desperately angry. "There seemed to be only one solution," he wrote. He seized his rifle and aimed at one of the departing men and called on Dodson to draw a bead on the other. He waited an awkward moment while Dodson came to his senses. Just when his partner was ready, and Akeley was about to give the word to fire, the men halted, gesticulating wildly. The gun bearer "told us not to shoot, that the milk would come, and it did," Akeley wrote. "Milk! Originally milked into a dung-lined smoked chattie, soured and carried in a filthy old goatskin for hours in the hot sun. But it was good. I have never had a finer drink."[25]

It's tempting to read something about racial attitudes into this ugly episode. Yet, the expedition wasn't in the habit of taking supplies from villagers by force. And, although he never addressed this fact in his memoir, Akeley almost certainly insisted on paying for the privilege of drinking the milk he'd seized at gunpoint. This singular episode was driven by desperation. Akeley reflected on the incident later, and the lesson he drew from it was that one must take responsibility for oneself while on safari, or the consequences could be fatal.[26]

Meanwhile, back in camp, just as Elliot was sitting down to dinner that evening, a jackal yelped painfully near the donkey carcass. Some of his men dashed off, returning with a struggling jackal, still stuck fast to a trap. Elliot shot it with his .22 rifle, but not before it'd caught one of his tent ropes in its snarling, yellow teeth, yanking it loose. The jackal was "a mangy little beast & the skin not worth preserving." Elliot discarded it. Damaged, diseased, runtish, or otherwise imperfect specimens weren't wanted for the museum's collection. An unknown number of these were shot, examined, and then left for the vultures. Curiously, when the jackal first began to howl, a group of the men were praying in front of Elliot's tent. With the excitement, however, the prayer meeting collapsed, all the penitents joining in the revelry of the others. "The leader tried to get them to go again but with poor success," Elliot reported. Yet he later heard them praying after dark, "the leader evidently having got some remnant of his flock again together."[27]

An hour before sundown, Akeley and Dodson started back to camp. At dusk, the shadowy forms of five donkeys darted across their path not fifty yards ahead. They fired, hearing a bullet strike home. One donkey fell behind as the others bolted away. The wounded animal soon stopped, and the hunters approached cautiously, their rifles raised. "As we got near," Akeley recalled, the donkey "turned and faced us with great, gentle eyes. Without the least sign of fear or anger he seemed to wonder why we had harmed him." The donkey's only wound was a minor one that would have caused no serious trouble had the animal continued with the herd. The hunters encircled the wounded animal, approaching within six feet. "It would have been child's play," Akeley claimed, "to have thrown a rope over his head."[28] They processed the specimen quickly and set out again.

Elliot, who knew nothing yet of their brush with catastrophe, had expected his companions to return in the afternoon. Around midnight, just as he was beginning to genuinely fear for their welfare, they finally stumbled into camp exhausted. Akeley announced that if any more donkeys were needed for the collection, then "someone else would have to shoot them." He'd had "quite enough." They brought another donkey skin and a tall, tan and white Soemmerring's gazelle, *Gazella soemmerringii,* on the second camel. A buck with a fine head, the gazelle would make "a splendid specimen" for exhibition. The locality they visited was about a seven-hour ride from camp, too far to go and return in a single day. Altogether, though, they'd seen some twenty donkeys. Akeley worried that the skin of the donkey brought in at midday might spoil, so he and Dodson stayed up half the night preserving it. Elliot noted ungratefully that "the heat had been so great that I much fear it will not turn out first class."[29]

7

A CROW'S IDEA OF TENDERNESS

May 6–20

Elliot spent the next morning merrily examining his treasures. The accumulated skins were in excellent condition, except one donkey, which had lost a precious patch of hair. Akeley thought it could probably be saved. The cautious curator decided to send their first specimen barrel, now at capacity, back to Berbera for safekeeping. It would be folly to carry the cumbersome barrel any farther than was necessary. It would also be risky—why risk losing specimens?

The material success of the expedition weighed heavily on Elliot. He'd used the promise of a bonanza of rare African specimens—some nearing extinction—to justify the exorbitant cost of his expedition in the first place. Much was riding on this promise: the possibility of future fieldwork, for example; the growth of his department, its collection, and exhibits; Akeley's position as taxidermist; maybe even his own position as curator. Everything depended on specimens. For the curator, who'd staked his professional reputation on the expedition's success, skins stowed securely in briny barrels back at Berbera were like money in the bank.

Akeley left in the afternoon, under a welcome cover of low-hanging clouds. He expected to reach Berbera by midnight and return the following evening. Elliot sent two camels with him and let him have his mare to ride. Dodson gave him letters to mail.

Elliot had his first serious labor issue when several camel men—who'd refused to clean some bones he'd given them—attempted to organize. He quelled the dispute by having the leaders brought to his tent, where he informed them "that they must obey my orders no matter what. . . . That I would stand no disobedience . . . & if they refused to do as they were bidden I would send them into Berbera . . . charged with insubordination & they should never go with any party again." This threat evidently succeeded, as the ringleader approached the curator afterward and pledged to do anything he was asked. "I think they were trying to see what kind

FIGURE 7.1.

Akeley poses with a horse. The golf cap he wears gave rise to the myth that Akeley—a novice—didn't know how to dress for a safari, yet he also sometimes wore a pith helmet. Photograph courtesy of the Field Museum, CSZ6064.

of a man I was, & I hope they are satisfied with their discovery," Elliot wrote smugly.[1]

The next day, with Akeley away, Elliot and Dodson decided to hunt baboons. But the threat of rain delayed their departure while the men—anticipating a downpour—dug trenches to divert runoff away from the tents. Setting out from camp, at last, at daybreak, they were immediately caught in a tremendous thunderstorm. Stubbornly, they rode for five hours, pelted by rain and with lightning flashing all around. They saw only a pair of distant gazelles and returned to camp soaked and empty handed.

The temperature dropped precipitously as the rain fell, which was "most welcome." Elliot wanted to relocate in the cooler weather but needed to await Akeley's return. Instead, the remains of the day were passed quietly in camp, with brief showers coming and going at intervals. The men were delighted with the rain, romping and splashing until they were "as wet as drowned rats."[2] A nearby dry river flashed suddenly into a fearsome, three-foot-deep torrent. It flowed brown with sediment and debris for several hours, carrying everything before it, then disappeared again into the desert.

Akeley returned early that evening. He'd stowed the expedition's lone barrel of pickled skins at the house in Berbera. It was "a great satisfaction" to Elliot to know they were safe. The taxidermist brought three additional camels bought by Dualla that same morning. These were "fine animals capable of carrying heavy loads."[3] He reported that the plain was full of gazelles and dik-diks, drawn by the abundant rain and fresh shoots of green grass.

Dismounting Elliot's mare, Akeley winced, clutching painfully at his side and walking awkwardly. He felt a terrible, stabbing pain in his kidneys. Elliot looked him over. Akeley had attempted to do too much before becoming fully acclimated to Africa, the curator concluded. The long ride, together with the hard work of the previous several days in the extreme heat, had taken a toll. Elliot sent him to bed at once, slapping a stinging mustard plaster on his lower back and plying him with Dover's powder, a potent opiate. Fortunately, he had no fever. In a short time, the curator predicted, Akeley would be himself again. The patient stayed in bed the next day, felt better but awfully stiff the following day, and was finally back on his feet on the third day after his return.

Meanwhile, their camp had run out of meat in this game-poor area. "We cannot depend upon our rifles to get anything to eat here," Elliot lamented. Compounding their troubles, the local villagers were so poor that the expedition had difficulty finding sheep to buy. Worse still, the men refused to eat the meat from any of the game killed by the expedition's hunters because the curator wouldn't permit the throats to be cut, as it compromised the skins. "Unless an animal's throat is cut before it dies," Elliot explained, "no good Mohammedan will eat it."[4]

Elliot expected "the meat question" would be only one of a multitude of cultural differences to contend with on the expedition. "What with their frequent prayers, sometimes occurring at most inopportune times, and their prejudices in regard to food," the Somalis were "not an easy lot." A pig they wouldn't touch, not even the bones after they'd been scraped clean. Elliot intended to collect warthogs for a group display at the museum. But, since the men wouldn't even handle a sack containing the prepared skin and bones of a hog, he expected to have to manage those specimens himself. "Whatever it may be necessary for us to do will have to be done, for I mean to have the pigs," he vowed.[5]

Itinerant flocks of sheep and goats passed by the curator's tent periodically. These were often guarded by a Somali armed with spear and shield. The latter was "a small round affair made of Oryx hide," which was carried on the left arm by an inside handle. Elliot sent Selon to buy an animal from one of these shepherds. Selon brought a man to Elliot's tent who had two sheep and a plump goat for sale. After prolonged bargaining, Elliot acquired all three for twenty-one rupees ($10). "Berbera prices," the frugal curator scoffed. "We cannot expect to get them cheaper until we reach the interior and away from the influence of too much money."[6]

The sheep of Somaliland were what Elliot called "the fat tailed variety." Their fatty rumps weighed two or three pounds and were considered "a great delicacy." Musa, the cook, asked for these to be sent directly to the mess to be used for cooking. "Fat," Elliot explained, "is esteemed as much among the Somalis as among the Esquimaux." The only fat the Somalis enjoyed in their diet came from sheep and goats, as the local game animals were typically very lean. The men had "a great feast" with the sheep they were given.[7] Half a dozen fires blazed throughout the camp, each surrounded by a cluster of contented Somalis. Elliot heard loud talk and laughter long after he'd retired to his tent.

Despite the trials of the previous few days, the curator's mood was buoyant. "The sunset this evening was gorgeous," he marveled, "producing a wondrous effect [on] the vast masses of black clouds that shrouded the rest of the heavens, and the peaks and outlines of the hills stood out sharp and clear in lines of black against the golden background."[8] A barrelful of donkeys and antelopes stowed safely in Berbera added to his joy.

A NEW SPECIES?

One morning, Dodson went out early to hunt, returning with what appeared to be another Pelzeln's gazelle. Elliot thought the specimen looked unusual. Could it be a new variety, or even a new, yet-to-be-described species? "There is no chestnut band along the side," he wrote of Dodson's unusual specimen, "that part being very pale, while the nose is decidedly inclined to a ridge like that of Speke's." It was definitely not that species, however. The horns were flatter and heavier, he thought, and there appeared to be differences in the skull. These animals were very wild, he noted. They were also possessed of great curiosity, which sometimes caused them to stop and look back when fleeing. "They know well the range of the ordinary rifle" and usually kept out of harm's way. Their bodies were long legged but small, presenting a difficult target. When partly hidden by stones or scrub vegetation, they were difficult to see. The male and female both carried "long slender annulated horns, curving backwards & inwards at the tips." The odd appearance of Dodson's gazelle was possibly a case of individual variation, Elliot speculated, "but this can only [be determined] by comparison with other specimens." Having just sent his skins back to Berbera, however, he had nothing to compare. Elliot's curiosity was piqued, yet final determination of this specimen would have to await a museum visit.[9]

New species were the much-desired currency of nineteenth-century zoology. To identify and name a new species was to make a real contribution to an ever-growing body of scientific knowledge. A new species would forever bear Elliot's surname as its original describer. Permanent acknowledgment in the scientific literature or on museum labels conveyed a kind of immortality that naturalists coveted. Vladimir Nabokov, the Russian author who moonlighted as a lepidopterist, insisted that the everlasting glory of naming a new butterfly exceeded literary acclaim. To discover new species, the English poet and novelist Charles Kingsley wrote, "brings with it the temptation to look on the thing found as your own possession, all but your own creation; to pride yourself on it . . . even to squabble jealously for the right of having it named after you, and of being recorded . . . as its first discoverer." The pleasure of discovering a new species, he noted, was "too great."[10] Elliot hoped to find new species to pad his reputation and to augment his collection with type specimens.

While Elliot was minutely examining his new prize, a hard rain commenced, unleashing incessant thunder and lightning. This continued for most of the afternoon. The men enjoyed the rain, stripping down to the skin and amusing themselves with all variety of "jumping, running, throwing the spear etc." The men somehow gave their spears "a tremendous rotary motion, which keeps [them] straight," and made them appear to quiver in flight. The spears were of two distinct kinds. One was long, with a broad, powerful blade, used for thrusting and stabbing. The other, shorter spear was for throwing. An expert could hit a modestly sized target at thirty yards. The gun bearer Ahmet was particularly "adept at this business."[11]

Intermittent thunderstorms continued the following day, so Elliot stayed in camp. He passed the time observing the omnipresent vultures and crows. "Impudent things," he called the latter. They strutted brazenly into camp, climbing among the camel blankets and scrutinizing everything in sight. They would frequently steal into the mess tent and make off with a scrap of meat. "A crow's idea of a tender attention," Elliot observed, "is perhaps natural but hardly acceptable to a non-feathered biped." He was observing a crow tear at a piece of meat it had found when another crow alighted nearby. The first crow brought a morsel and "walking up to his friend proceeded to place it into his open bill, the recipient acknowledging the favor with quivering wings. It was the first time that I had witnessed . . . carrion received or given as a mark of . . . consideration."[12]

Camp was also frequented by a brown kite of the genus *Milvus*. This bird, too, was very bold, swooping down and seizing bits of meat from their very midst, then speeding away like a thief. Its movements were graceful, and it was "a pretty sight to see it feeding as it flies along, the talons brought forward holding the meat, which it tears with the beak, evidently as much at ease as if upon the ground."[13]

LETTERS FROM HOME

A messenger arrived from Berbera with mail. Elliot had a letter from his wife, dated April 14. She'd attended a lecture in New York given by Arthur Donaldson Smith, the Somaliland explorer, who had been especially complimentary of Dodson, calling him "most useful." He congratulated Elliot in absentia for securing Dodson's services.[14] Dualla had also sent word that he'd purchased nineteen additional camels, bringing the expedition's total to thirty-two. With luck, there would soon be enough to leave for the far interior. Elliot gave orders to move camp again, then retired to his tent for the evening.

The camels began to growl by three in the morning. Soon, the entire camp was in an uproar. Immediately after breakfast, Dodson and Elliot rode away to hunt, leaving Akeley, who was still achy, to come along slowly with the main caravan. They soon spotted baboons scampering among the rocky hills. These were wary and wouldn't permit the hunters to get closer than three hundred or four hundred yards. Yet they stood conspicuously on prominent outcrops and howled "in the most frightful manner." The hunters aimed a few shots from their powerful Lee-Metford rifles, striking perilously close and agitating the monkeys into fits of rage. "Missed a shot by a shave," Dodson wrote. Eventually, the baboons fled over the hills, still calling obscenely. "They have a peculiar hoarse bark," Elliot remarked, "which can be heard for a long distance."[15]

Shortly thereafter, they came to Durban, a plain stretching away to the mountains. It was here that Akeley and Dodson had seen so many donkeys the previous week. By now, however, numerous Somalis had brought their flocks to graze and were pitching huts and constructing a zareba. The hunters assumed that the timid donkeys had probably moved on. As it was getting late in the morning and the sun was already hot, Elliot decided to go no farther. They unsaddled their horses and sat in the shade of a tree. They rested for an hour or more, watching the Somalis unpack. Each camel, Elliot observed, was led by a woman. It could carry on its back the entire contents of a hut, including all the household furniture and utensils. "It certainly does not take a Somali long to change his locality and set up his mansions," Elliot remarked, "& he goes through life very much as the Patriarchs did." They bought a quantity of camel's milk, which Elliot found "very refreshing." One of them captured a small, brown skink with pale lateral stripes, *Mabuia varia*.[16]

When the caravan arrived, camp was quickly established near a reliable well called Better-An. Large droves of camels were pastured in the area. A great many were milling about the well, creating an interesting spectacle. Nearly every adult female camel was accompanied by a juvenile. "Queer comical looking creatures," Elliot called them, "with legs too long for the body[,] a diminutive hump, grave face, and the sedate solemn walk of the adult." The growl of the camel was "very like that of a lion," Elliot observed, "while the cry of the young is a magnified cross between the bleat of a sheep and a goat, & when about a hundred are trying . . . the concert that follows is not captivating."[17]

Dodson killed a Phillip's dik-dik, *Madoqua phillipsi*, with his shotgun. This was one of the smallest of the African antelopes, no larger than a rabbit. Elliot described the specimen as "a beautiful creature with grey hair and a reddish crest on top of the head. The eyes are very large, black with a dark orange border." The heat of the day was tremendous, until clouds formed in the late afternoon, bringing welcome shade. A still hobbling Akeley took a shotgun and hunted small birds that were flitting about the bushes near camp. He returned with a beautiful sunbird. He went back out with a larger shotgun and bagged a male Phillip's dik-dik, a single no. 4 shot striking it fatally in the head. It was "a most attractive little animal in its beautiful grey and fawn coloring & would make a splendid specimen." The skin was soon removed and deposited in a barrel of formalin in camp. Reptiles were also taken at this camp, including a pair of false sand lizards, *Eremias brenneri*.[18]

Early the next morning they moved camp again. In the afternoon, Akeley and Dodson went out to hunt large game, while Elliot took a shotgun and chased after birds. He bagged a "beautiful sun bird with a long tail, blue & purple head & breast & orange yellow belly"—probably the Nile Valley sunbird, *Hedydipna metallica*. He also collected weavers and warblers. The industrious weavers had been busy making nests, great numbers of which were swinging freely from tree branches. The nests "exactly resemble one half of an old fashioned purse well filled, and are composed of fibres and grasses closely woven together," he noted.[19] Dodson brought in a Soemmerring's gazelle buck in beautiful condition. Akeley stayed up late making a plaster cast of its head and taking careful measurements of the body. He was likewise busy the next morning photographing the gazelle in various positions and taking notes that would be useful in mounting it for exhibit. These casts and photographs, Elliot noted, were among the most valuable results of the expedition.

CAMEL STAMPEDE

The next days were unspeakably hot. But for an occasional breeze in the afternoon, it would have been intolerable. Tents provided the only shade to be had during the day at their desolate campsite. Yet, they radiated heat back at their restless occupants through much of the night. There was no tree under which to shelter; nothing but low bushes dotted the sandy plain. The surrounding hills were a jumble of bare, weathered, sunblasted rock.

Elliot went hunting for donkeys but saw nothing. Dodson spotted three oryx, but he ignored them, as Elliot had promised British authorities not to kill any of these animals within the reservation. He did, however, take another Phillip's dik-dik with a shotgun near camp. With no donkeys in sight, Elliot wanted to move. Their guide warned that there was no water at Boholo, the next well on their line of march. Probably there were no donkeys, either.

Elliot watched from a distance as Dodson ran down another Pelzeln's gazelle. "It led him a fearful journey in the sun," Elliot wrote, "but after a long chase he finally got a chance & killed it."[20] Arriving in camp, he learned that Akeley had killed an animal of the same species. Akeley's specimen was a female with a swollen udder, suggesting that a helpless juvenile was hidden somewhere in the brush nearby. Elliot longed to find it for his habitat group.

Meanwhile, a Somali told Elliot that he'd seen several donkeys—some with foals—at a nearby pool of rainwater. Elliot wanted these juveniles for a group display too. He sent some men with a telescope to the crest of a hill to reconnoiter the land. They returned and reported no game in sight. Elliot and Dodson then went out to spy the pool of water in the evening, hoping donkeys might appear at dusk to slake their thirst. When none came, they returned to camp defeated. Late that night, after Elliot had gone to bed, he heard a donkey braying a taunt from the direction of the water.

Hunting resumed the next day. Elliot killed a Pelzeln's gazelle with a single shot from his Lee-Metford, but then found that he was out of ammunition. Certain that he'd filled his ten-round magazine that morning, he was never able to solve the mystery of what had happened to the bullets. His other gun, a .50-100 Winchester, was too heavy for the small game he was seeing, so he ended his hunt early and returned to camp.

Earlier that morning, not far from camp, one of the men spotted an enormous scorpion sheltering under a shrub. He had no way to capture it, so he built a cairn of stones to prevent its escape. Later, Elliot sent Ahmet back out with a pair of long forceps and a bottle. Ahmet soon returned with the specimen. It had two monstrous pincers shaped like lobster claws, eight pale legs, and a long, ringed tail tipped with a yellow poison sac that warned of a formidable sting. The fearless scorpion was spoiling for a fight. It struggled for a long while before finally succumbing in a 12 percent solution of formalin.

As they were sitting outside their tents late in the evening, the camels, which had all been reposing peacefully, suddenly stampeded onto the plain. One huge beast tried to pass directly through Elliot's tent, which stood open at both ends to catch any breezes, but it became hopelessly entangled in a web of guylines. Much shouting and rushing of men in the darkness ensued, in which all the animals were quickly recovered. Soon after they resettled, however, they broke loose again, this time fleeing in a different direction. They were captured once more and made to sit in the campsite. A ring of bonfires was lit around them, and they remained still for the rest of the night. Camels, Elliot observed, "certainly have as little sense as anything can have." It is "an unmitigated horrid beast. The only qualities it is has . . . are its ability to carry a load a reasonable distance & to go a few days without drinking."[21] When a note from Dualla arrived advising that forty of these animals had now been purchased, Elliot replied with instructions to buy six more.

With game scarce, Elliot decided to start the next morning for the Golis Range, where he hoped to find kudu, as well as cooler temperatures. Akeley's lower back continued to give him occasional trouble, when he was "almost incapacitated from doing any work." Elliot began to feel anxious about his health. "He certainly will be able to do little or nothing in the difficult Koodoo country," Elliot wrote, fearing that he and Dodson would have to compensate.[22]

The men, meanwhile, went to a dance at a nearby village. They stayed out most of the night. The next morning, they were all miserable. Even the tent boys were incapacitated. They spent most of the day napping. Elliot, stuck in place for another day, would be reluctant to give the men permission to go carousing in the future.

Two women, each with a baby on her hip, and another naked youngster of about two years old toddling alongside, came to camp to sell camel's milk. It was difficult to judge how old the milk was, so Elliot declined to buy it. The women had a peculiar walk, the curator noticed. Their feet were thrown outward with each step—just as a lion moves its hind legs. This was caused, he speculated, by the habit of carrying their children astride the hip. With the weight concentrated on one side of the body, the leg was obliged to swing in a semicircle to support the burden. Akeley tried to get the women to pose for a photograph, even offering them a rupee for their trouble, but they refused. He managed to catch them unawares, however, as they passed in front of his tent. They were, as a rule, exceedingly shy of the foreign men in Elliot's entourage.

Elliot sent lookouts back into the hills with his spyglass. They reported seeing no game. Frustrated, the curator determined again to abandon this locality. It was about a ten-hour march to Haili, a well in the Golis Range, which Elliot planned to make in two marches, leaving in the small hours of the morning to avoid the worst of the heat.

FIGURE 7.2.

Somalis fetching water at a well. The availability of water often dictated the expedition's movements. Photograph courtesy of the Field Museum, CSZ5986.

INTO THE HILLS

On Saturday, May 16, the men awoke at three in the morning and departed an hour later. They headed south across the plain, which bristled with low bushes and a few stunted trees. The early morning air was fresh and pleasant, reminding Elliot of June at home. The sound of birdsong trilled from every green clump. They saw abundant game, including one Soemmerring's gazelle, four gerenuks, a Waller's gazelle—their first—and many dik-diks. These were all extremely wild and wouldn't permit the hunters to approach. Meanwhile, Elliot felled one bustard, *Lophotis gindiana*, as it soared gracefully overhead.[23]

After marching four hours, they arrived at a hole in the dry bed of a river. Several Somalis surrounded it, and one young boy was in it, scooping damp sand from the bottom and flinging it out with his hands. There was at the very bottom only a small quantity of muddy water, which the Somalis carefully dipped and decanted into the large bottles called chatties that the women routinely carried on their backs. In this thirsty land, Elliot observed, every drop of water was precious, and people would travel great distances to reach a well like this one. Elliot found their habits at the wells off-putting, however. "The first thing a Somali does," he wrote, "is to wash himself in the well & then drink the water. It may be all right for him, but it is anything but attractive to us." Elliot quickly determined that there wouldn't be enough water here for his caravan, so instead of making camp as originally intended, he gave orders to unload the camels for some rest. They'd start again in the afternoon for the next watering hole, which was said to be four hours farther. Elliot chose a large acacia and huddled leisurely with his hunters in its abundant shade. Dodson then shattered their idyll by blasting three beautiful turacos, *Schizorhis leucogaster*, from the highest branches of the very tree under which they were lounging. Conspicuous for its large size, its brilliant colors, and its loud, harsh pappap call, this species was singularly attractive.[24]

The caravan started again in midafternoon. Elliot followed thirty minutes later, his speedy mule quickly overtaking the lumbering camels. He traced the dry riverbed a long way, then scrambled over a desolate, hilly country until he reached the next water source, Haili, which consisted of two large pools of rainwater, one above the other. On arriving, the men rushed like lemmings toward the water. Elliot stopped them, giving strict orders that no one should go near the upper pool, but that all could wash to their heart's content in the lower one. Any man who disobeyed this rule would be fined five rupees, he warned. They pitched their camp nearby in a sheltered spot shaded by low trees. Their elevation was considerably higher. The air was still hot, but much more comfortable than it'd been in previous camps.

Near their new campsite was a somber assortment of graves. "It can hardly be called a cemetery," Elliot wrote cryptically, "although quite a number of persons have been buried there." A ring of piled stones encircled several graves. Elliot asked his men about this curious practice, but no one could tell him anything. They agreed that these burials had been done by a mysterious people who lived there long before. Elliot guessed that the deceased were probably "Galla . . . for that was the race which were driven out and are now found beyond the Shebeyle river." Although he mentioned nothing about it in his journal, it was probably here that Elliot pilfered a single human skull for the museum's anthropology collection.[25]

A villager visited camp and offered to show Elliot a place where he'd seen kudus and wild donkeys. He claimed to have guided other hunting parties from this same spot. Dodson planned to hunt early the next

FIGURE 7.3.

morning, while Elliot and Akeley intended to abstain. Dodson invited the would-be guide to accompany him the following day.

Akeley was feeling better and had handled the long journey in the saddle well. Elliot believed that he was now truly on the mend. During his illness, the curator noted, his appetite had never diminished, yet he'd grown noticeably thinner.

MORE DISCIPLINARY ISSUES

Another test of discipline arose. Dodson had gone off early with the guide. Elliot, meanwhile, was enjoying a late breakfast. A shikari and a gun bearer approached and asked for a shotgun to kill dik-diks, as there was again no meat in camp. Elliot consented, giving them the gun and a few shells. Others in camp took advantage of this permission too, including a tent boy, the cook, and two assistants who worked for Akeley, who all tagged along. A short while later, Elliot heard a shot, and the tent boy, Abdi, came rushing back to camp clutching the bloody hindquarter of a dik-dik that had been crudely divided in the bush. Elliot called to him and asked him to bring all the men back to camp, and soon enough all were standing before the disgruntled curator. Elliot asked the truants where they'd been. They explained. He then asked who'd given them leave to go, "to which they had no reply." He then gave them all a lecture about discipline and fined them four rupees each. As for the shikari and gun bearer, he told them that because they'd cut up the dik-dik themselves and not brought it to camp, they'd never be permitted to have a gun again. "They were a crestfallen lot," Elliot later reflected. He was careful to treat them all decently in the aftermath, and he wondered how long they'd sulk. "At all events they will find they cannot trifle with me," he noted vainly. He instructed the cook to roast the dik-dik meat and to give it to Dodson for his breakfast, as he'd eaten very little before he went out hunting. In this way, "the shooting spree," as Elliot called it, brought no benefit to any of the participants.[26]

Just then the guide returned to camp for a camel, gladly proclaiming that Dodson had shot another donkey. Soon, they returned with the carcass, which was an old male, almost red in color. Elliot wanted to save both skeleton and skin. They now had sufficient males, the curator thought. They still lacked females and especially a foal to complete a group display. Akeley and Dodson spent the afternoon roughing out the bones. Elliot was delighted to have so rare and valuable a specimen for his collection. "Not one is in any Museum in the World," he boasted.[27] Soon, he predicted, there'd be none remaining in the wild, either. The specimens he salvaged, and the habitat group Akeley would make with them, might one day be the only earthly trace of the Somali wild donkey, Elliot predicted.

Elliot watched several crows soaring in a thermal like vultures, something he'd never witnessed. Nor could he recall having read about this behavior. The birds rose in tight spirals with their wings fixed, only occasionally flapping once or twice, until they were high and out of sight. Elliot wondered whether this was a habit peculiar to this species.

The following morning, all the hunters went out at four o'clock, each taking a different route from camp. Shortly after setting out, Elliot spotted several distant antelopes. He tried to approach, traversing a great length of plain covered with sand, stones, and shrubs. He saw no other game. He found abundant donkey and kudu tracks, but none appeared to be of recent vintage. He also spotted many different birds, some with "rich, metallic plumage of startling colors." Numerous hornbills, for example, were flitting from tree to tree, perching, and uttering their distinctive croak. In flight they were "very handsome . . . the black & white of the plumage showing to great advantage, when the wings and long tail are spread." Elliot had written a monograph on hornbills many years earlier. He found it strange to see so many living animals—full of energy and vitality—that he'd only seen previously as dried skins in museum drawers. Although his feelings about this experience were contrary to his salvage mission, he was strangely moved by it, writing: "It was as though a museum of specimens had suddenly been resurrected and [imbued] with another joyous existence."[28] The reverie was brief, however, and old professional habits ultimately prevailed. He called for a shotgun, but his gun bearer merely shrugged and muttered an apology. The disappointed curator was unable to secure any specimens.

The heat finally chased Elliot back to camp, where he found his companions. Dodson had wounded a gazelle but lost it. Akeley, on the other hand, had succeeded in shooting a beautiful eagle, *Helotarsus ecaudatus*, on the wing. He'd also brought in a curious shrew, *Macroscelides rivolii*, which one of his men had noticed under a bush and killed with a well-aimed stone. It was the size of a small rat, with a pale gray coat and a long proboscis-like snout. Akeley described a running stream of clear water flanked by plentiful grass some distance away, near Sheik Pass, where he collected a hissing sand snake, *Psammophis sibilans*, and a small, spotted bush snake, *Philothamnus semivariegatus*.[29] Elliot considered moving camp to that place. He was tempted to venture farther into the mountains to begin the hunt for kudu, but this would require considerable time and a stock of new provisions for the men. With the long journey into the interior ahead of them, maybe it was time to beat a retreat.

They didn't go hunting that afternoon—the heat was too intense. Akeley, however, shot a dik-dik that dared too close to his tent. After removing the skin, he made a plaster cast of the entire body, showing the musculature. Casts like these would be useful when it came time to mount the animals for exhibit. Elliot visited the mess and took stock of his dwindling supplies. With provisions running perilously low, and game relatively scarce, he gave orders to pull up stakes and return to Berbera the next day by the shortest possible route.

This side trip, Elliot reflected, had been a great success. Among other things, they'd bagged four wild asses, three males and a female. This was an invaluable acquisition of an exceedingly rare animal flirting with extinction.

They started unusually late, moving out at five o'clock the following morning. Selon anticipated four marches to reach Berbera, but Elliot wanted to make fewer. The curator rode at the head of the caravan, and when they reached the place where Selon intended to stop, he inquired about the route. With only two or three hours to the next water source, he ordered the caravan to push on. They threaded their way through "a most dismal barren country washed & torn into huge crevasses like our own Bad Lands covered with sand & stones, and dotted over with thorn bushes which seem to live anywhere no matter whether there is any soil or not." No animals were seen.[30]

After a five-hour march, they halted for lunch at a spring-fed pool that had flooded an extensive area where "a fine crop of grass was growing" on the damp and boggy ground. Near the water, Elliot and his big-game hunters found numberless small frogs taking great, flying leaps. Making "considerable effort," they collected several specimens each of the three kinds present. These later proved to be two new species, *Phrynobatrachus hailiensis* and *Bufo garmani*, and one new variety, *Bufo viridis somalacus*.[31]

The midday sun burned fiercely. They pitched a tent for shade, drawing up the heavy canvas sides for better circulation, and took refuge beneath it. The heat was soon "something inconceivable, and at times it really seemed as if our heads would burst, sheltered as we were." The entire country "seemed slowly burning up."[32] When a slight breeze sprang up it was a great relief to the men. The horses and mules, meanwhile, took refuge in the shade of a lone tree.

Late in the afternoon they again pulled up stakes and pushed on, remaining in the saddle until after dark, when they stopped on a wide, sandy plain a dozen sweltering miles from Berbera. A heavy mist began to fall, so Elliot had his tent pitched. Akeley and Dodson slept on their beds in the open air. Elliot thought this imprudent and told them as much, but the heat was so terrible—even late at night—that they refused to submit to the torment of their tents.

Thus passed Akeley's thirty-second birthday.

Elliot wrestled with insomnia in the wearying heat. At two in the morning, shortly after he'd finally slipped into a dream, loading commenced. An hour later, they were back in the saddle. It was relatively pleasant riding until the sun rose and the heat returned with renewed force. Elliot noted the scarcity of game animals, but he "did not blame them for deserting." The expedition had descended to the coastal plain, where the heat was always extreme. "From the time the sun rises until it sets," Elliot wrote, "the maritime plain is no fit place for man or beast & it is pretty generally shunned by all." They passed a small party of Somalis heading for Berbera, carrying a month-old Pelzeln's gazelle. Elliot still lacked a juvenile to complete his planned group, so he bought it for a mere two rupees. It was "a beautiful little creature," he noted, "with the loveliest eyes imaginable."[33]

They arrived at Berbera at half past eight in the morning, on Wednesday, May 20. Elliot and his hunters were "pretty well done up with the heat."[34]

The young gazelle made itself at home on the veranda of the expedition's headquarters for what remained of its short life.

8

ALL DRIVE AND RUSH AND SENSATIONAL EFFECTS

Before embarking for Africa, Elliot had arranged for his assistant curator, Oliver Perry Hay, to take temporary charge of the Department
of Zoology during his long absence. Hay enjoyed a sizable raise to compensate for the increase in responsibilities. If he expected to have some
autonomy in running the department, however, he was disappointed. He
received explicit instructions from Elliot "to report to [Director Skiff],
and seek [his] counsel and advice in all important matters."[1] Yet, the point
where Hay's authority ended and Skiff's began was ambiguous, and this
caused inevitable friction.

Hay had excellent credentials. Born in Indiana in 1846 and raised in
rural Illinois, he'd forsaken a career in the ministry to pursue science. He'd
earned a bachelor's degree from Eureka College in Illinois in 1870 and
a PhD in zoology at the University of Indiana in 1887. He did a year of
graduate study at Yale in 1876–77. He taught science at several midwestern
schools before settling at Butler College, in Indianapolis, where he served
as professor of biology and geology from 1879 to 1892. At Butler, he was a
leading light among a talented group of naturalists at the Indiana Academy
of Science. His knowledge of comparative anatomy was extensive. Like
Akeley, he was a tireless worker, often laboring long hours every day of the
week. As a colleague, he was friendly and unfailingly kind, always ready
to lend a hand or offer an opinion when his help or advice were solicited.
Yet he could also be headstrong—even to his own detriment. According
to a biographer, once Hay formed an idea, "he held to it tenaciously and
was difficult to move from his final decision." Biologist William Perry Hay
agreed that his father could be mulishly stubborn.[2]

Seeking a wider field of opportunity for himself and his family, Hay left
Butler in 1892 and moved to Chicago. The following year, he accepted a
fellowship to continue his postgraduate studies in zoology at the University
of Chicago. There, he concentrated on the origins of vertebrate life and the

growth and development of living and fossil turtles, working closely with the university's embattled vertebrate paleontologist, George Baur. In December 1894, he wrote a letter to University of Chicago zoologist Charles O. Whitman, inquiring about an opportunity to write a second PhD thesis under Whitman's direction. Ultimately, he abandoned this somewhat quixotic goal to focus on a salaried position at the Field Columbian Museum.[3]

Hay began jockeying for a job at the museum within days or (at most) weeks of its founding, in the fall of 1893. The museum then existed on paper only, so patrons brushed Hay off, promising merely to "bear your matter in mind." Months later, the Palace of Fine Arts was a beehive of activity as a swarm of teamsters and temporary curators scrambled to put the museum's mishmash of new collections into some semblance of order. Hay wrote to Skiff on March 2, 1894, again expressing interest in employment. Unfortunately, Frank C. Baker had just been hired as the museum's first curator of zoology, so Skiff replied with regret that there were no openings. He noted, however, that Hay's application would be filed "for future consideration if circumstances justify." On July 25, probably because he'd just learned of Baker's June departure, Hay wrote again to Skiff to offer himself for the vacated position. But Skiff was then in New England negotiating with Elliot to fill the vacancy. Finally, in October, after a year of campaigning, Hay accepted a position as Elliot's assistant curator. His specialization in fishes, amphibians, and reptiles dovetailed nicely with Elliot's preoccupation with birds and mammals.[4]

THE ELEPHANT IN THE ROOM

Once they arrived at the museum late in the fall of 1894, Elliot and Hay could both see that there was a great deal of work to be done to raise the standard of its zoology exhibits. The curator preferred to have a direct hand in the rearranging. "I would ask that the rooms be left as they are until my return," Elliot wrote to Skiff from New York, "so that I could have some personal supervision in the matter & decide which rooms certain of my Departments should be located, in order to display the collection to the best advantage." He was unhappy with most of zoology's exhibits. For instance, he griped that the handful of fishes and reptiles on display were "of the usual type generally witnessed [in museums], shrunken, ill shaped distorted objects that misrepresent the species."[5] He had equally unflattering opinions about other displays.

To address this problem, Elliot planned to segregate the finest exhibit specimens from what he began to call the study collection. "It is not my purpose . . . to develop the Zoological Department on the lines adopted . . . by the long established Museums," he explained to Skiff. Specifically, he had no intention of continuing to exhibit "every specimen that comes to the Museum." There were very good reasons to segregate. After all, not every specimen was ideally suited for exhibition. Some were too common, others too dull, too scrawny, or otherwise unremarkable. Instead, Elliot would display a selection of the best, the largest, and the showiest specimens. He would prioritize animals with huge antlers or horns, long tusks or claws, sharp teeth, striking colors or patterns, and flashy displays of fur, feathers, and scales. Extinct or rare animals would take precedence. The total number of specimens displayed would decline. With respect to fishes and reptiles, for example, a few attractive specimens would suffice for exhibition. The rest would become the study collection, which would be housed separately in jars and in drawers in a dedicated laboratory and collection space inaccessible to the public. A separate study collection was essential for a research museum, Elliot recommended, as "a mounted specimen as a rule is of little use for scientific study."[6]

Another of Elliot's gravest concerns was the low quality of exhibit cases in his department. Like the Palace of Fine Arts itself, zoology's cases—which were obtained from Ward's Natural Science Establishment at the close of the World's Columbian Exposition—were never intended for permanent exhibition. Many were cheaply and shoddily made. The worst of them had broken doors that wouldn't close, or ill-fitted hasps that wouldn't lock. They were mismatched in style, size, color, and quality of materials. It would be foolish to go to the trouble and expense of expanding the collection and building elaborate zoological exhibits without providing appropriate housing for the specimens. "There is no use accumulating valuable material," Elliot argued, "if it is to be placed in a receptacle that does not preserve it." Cheap, carelessly made cases, were worse than no cases at all, and "should never be placed in any Museum possessing valuable materials," he insisted.[7]

Elliot crafted ambitious plans to renovate zoology's exhibits. Changes that could be made by the museum's own labor force at negligible expense were implemented right away. Indeed, one renovation had already been completed the previous June, when a small, synoptic exhibit of Galápagos specimens was unceremoniously removed from Hall 19 and returned to its owner, Hay's University of Chicago colleague Baur. The space vacated was used temporarily as an office for zoology.

Another major renovation began shortly after Elliot's arrival in Chicago. This was the complete deinstallation of two halls—22 and 23—that treated animals explicitly as commodities and were of dubious zoological value. Hall 22 housed the Section of Animal Industries, which comprised a number of animal-derived oddities obtained from the World's Columbian Exposition. Chief among these was a "valuable collection of tanned skins and leathers," an "extensive collection of footwear," an "interesting collection of leather articles from Jerusalem," and "skins from [the] Argentine Republic." In a nod to Chicago's booming meatpacking industry, there were also two detailed models of Chicago slaughterhouses, showing the modern, mechanized method of killing and butchering livestock and packaging the meat.[8]

Hall 23 harbored the Section of Fishery Industries, including models of whaling ships, examples of scrimshaw, sperm whale teeth, walrus tusks, and narwhal horns. Hanging from the ceiling in the center of the room was an original whale boat, outfitted for service, complete with six life-size models of sailors and their whale-hunting accoutrements. A vivid description from the *Guide to the Field Columbian Museum* noted that a determined-looking sailor at the bow stood poised "to drive [his] harpoon into a whale according to the olden practice."[9]

Over the course of 1895, these halls were emptied of their contents. Hall 23 was reinstalled with mounted vertebrate skeletons removed from an overcrowded Hall 27. (The vacated Hall 27 was then turned over to ornithology.) Hay played a lead role in transforming Hall 22 into an exhibit on ichthyology and herpetology—his areas of specialization—as the museum's zoological collections gradually grew in size, and as the fishes and reptiles were crowded out of Hall 20 (the Hall of Vertebrate Zoology) by the rapid growth in mounted mammals.[10]

Recognizing that his department would need considerable space for expansion in the coming years, Elliot fixed his roving eyes on the museum's roomy West Court. By rearranging the hodgepodge exhibits there into tighter configurations, much additional space was created. Some objects were removed altogether, including a large terra-cotta pavilion, which—now that the Field was evolving more and more into a natural history museum—was deemed no longer appropriate. Elliot gained more significant ground when he urged the removal of the museum's shaggy, oversize model of a Siberian mammoth. The model had become, by the summer of 1895, a "breeding place for countless moths, which feed upon the paste used in the construction of this historically incorrect specimen," Elliot complained. He wanted it gone. "In view of the danger incurred from its presence, & its small value as [a] true representation of the Mammoth, I would advise its removal from the building." In February 1896, shortly before Elliot left for Africa, the mammoth was deinstalled, and the space gained was used to exhibit three of Akeley's beautiful new mammal habitat groups.[11]

To accomplish large-scale renovations required a larger labor force, especially while Elliot and Akeley were away in Africa. To fill in for Akeley, the museum engaged Charles Brandler as assistant taxidermist. A mustachioed Austrian with a thick accent, Brandler had worked with Akeley at the Milwaukee Public Museum and was familiar with the latter's methods and style. Nevertheless, he was to serve in a backup capacity only, maintaining or repairing already mounted specimens as required. He was

not to attempt to do anything too technical. "Do not let any of the skins in the [holding] tanks in the storehouse be touched until I return," Elliot lectured Skiff. "I do not want anyone but an expert to attempt to mount them." Yet, with all the work to be done, more help would be needed. Skiff added Moritz Fischer to the staff as a preparator. The director also hired a young man, William O'Brien—a skilled tanner and furrier—to put the museum's skins into first-class shape. For a general laborer, he added the industrious E. Russell Cooper.[12]

By the close of 1895, the museum had committed to completely overhauling the Department of Zoology. According to Skiff, the "policy of the Museum for the next two or three years, will be to . . . devote itself almost entirely to the development of the natural science divisions."[13] And while botany and geology also received some attention, the bulk of museum resources flowed to zoology. For example, despite the precarious state of the museum's finances, the executive committee agreed to spend $6,000 on new mahogany casework throughout Elliot's department. Renovating zoology's exhibits began early in the new year under Elliot's direction and with Akeley's exacting standards guiding the work. After Elliot and Akeley left for Africa late in February, the baton passed to acting curator Hay.

HAY'S TRIAL AND ERRORS

The beginning of Hay's trial as acting curator was relatively rosy. He assumed command of zoology renovations in March. Then, in early April, museum trustee Owen F. Aldis invited him to join a fishing expedition of several weeks' duration to southern Florida and the Gulf of Mexico. Aldis would bear the expenses, affording the acting curator an opportunity to make marine zoological collections at almost no cost to the museum. Hay's first major expedition for the museum was an unqualified success. He returned from Punta Gorda, Florida, with more than a thousand specimens, including 686 fishes, 180 mollusk shells, 77 crabs, 16 sea urchins, 15 starfish, 10 shrimp, 4 shark skins, 2 dried sponges, 2 shark egg cases, 2 juvenile alligators, 2 sea squirts, 2 cow-nosed ray skins, 2 mollusk egg case masses, 1 mollusk egg case string, 1 sawfish skin, 1 shark jaw, 1 rough fish skeleton, and 1 mass of oyster shells cemented to the root of a mangrove tree. Several showy specimens, including "a splendid tarpon," were immediately mounted and placed on display in Hall 22, the new Hall of Ichthyology and Herpetology. The new material scaled up the hall's somewhat drab appearance.[14]

While Hay was chumming with Aldis in Florida, Skiff directed the remodeling efforts of the zoology staff in Chicago. Skiff was no scientist, but he'd gained considerable experience as an exhibit manager in the Department of Mines, Mining, and Metallurgy at the World's Columbian Exposition. Orchestrating exhibits was a role in which he felt supremely confident. To supervise the work, Skiff depended on John B. Goodman, building superintendent. For his labor force, he commanded a squad of new, temporary employees—all keen to land better-paying, permanent positions—who were doubtless eager to please the museum director and the superintendent as they shifted cabinets, stained woodwork, and spruced up the mounted mammal skins. A great deal of cosmetic work was accomplished under Skiff's watchful eye. "Very fair progress is being made in [the] Mammal Hall and in the West Court," Skiff wrote in a self-congratulatory letter to ornithology curator Charles B. Cory.[15]

Not all the work undertaken was merely superficial. For example, when ten flat-topped table cases were installed in the West Court to accommodate the rapidly growing malacological collection, someone had to rearrange all the specimens in systematic order. This was a task for an experienced naturalist. Yet, Skiff felt competent to oversee this work as well. "I have a man . . . uniting the Ward material that we had, with the Carpenter shells that Prof. Elliot bought in Canada," Skiff reported.[16]

The largest single space belonging exclusively to zoology was Hall 20, a cavernous gallery of more than 4,300 square feet. This space had been earmarked originally for vertebrate zoology. While Elliot and Akeley were in Africa, it was repurposed as a hall of mammals. First, all specimens were removed. Next, all the old, decrepit, Ward's casework was dismantled. The walls were then touched up with fresh plaster and calcimining. Finally, the hall was outfitted with twenty-one new, high-quality, dustproof floor cases. The new cases were large, with a footprint greater than fifty square feet each—they were designed to accommodate all but the very largest mammals. Their bases and framing were made of dark mahogany, with sides and top enclosed by plate glass. Every edge was tongue and grooved, with poisoned felt sandwiched in the joints to keep out dust and vermin. To repel sticky-fingered visitors, they were secured with Yale and Townsend bar locks. The cases were arranged uniformly in three rows of seven. Large carnivores and ungulates, including Akeley's recently mounted African antelopes, inhabited most of the cases in Hall 20. A smaller, adjacent room, Hall 19, was used to exhibit so-called lower mammals, including monotremes, marsupials, rodents, and bats.[17]

Once the new cases were installed, Brandler, O'Brien, Cooper, and the other assistants reworked a remarkable amount of museum material. In fact, the entire collection purchased from Ward's was "carefully restored, embellished, repaired, renovated, re-mounted, etc." The positions of some of Ward's mounts were altered slightly to look more natural. Their noses and toes were touched up with fresh, glossy paint. New, realistic-looking glass eyes were added. Once placed back on exhibit in their new cases in Hall 20, these specimens presented "an entirely fresh appearance." Some specimens were in such a sorry state, however, that they had to be set aside to await Akeley's return. The entire collection was also carefully repoisoned, probably with arsenic. This work was begun in March and completed in April. "I shall be able to throw the mammal . . . rooms open to the public about the 15th," Skiff reported proudly to board of trustees president Edward E. Ayer. "It will be the handsomest suite of rooms in the building." Skiff wasn't shy about hogging most of the credit for the appearance of the new mammal hall: "You have no idea how my scheme to repair, remodel and fix up the old Ward specimens has improved the collections," he crowed.[18]

Returning to the museum in May, Hay resumed immediate charge

FIGURE 8.2.

Hall 20, circa 1894. The largest hall in the Department of Zoology, this space was used first for vertebrate zoology and later, in 1896, for mammals. Skiff and Elliot agreed that the mounted mammals acquired from Ward's Natural Science Establishment were some of the poorest materials in the museum. The casework shown here was demolished in 1896, while the specimens were rehabilitated. Photograph courtesy of the Field Museum, CSZ8218.

of the department. At that time, the consolidation and rearrangement of the malacology collection was still underway. Skiff had placed Fischer in charge of this work. Fischer's task was to shift the museum's collection of some 12,000 mollusk shells into new cases, while simultaneously consolidating it with a collection of 4,039 new specimens—an important set of duplicates from the Philip P. Carpenter collection at the Peter Redpath Museum in Montreal. Hay took it for granted that Fischer was working under his supervision as acting curator, and presumed to provide instructions regarding the final arrangement of the shells. But Fischer "refused to receive directions," and an exasperated Hay wrote to Skiff late in May demanding satisfaction. "I have no further use for him," Hay wrote, "and desire that he be allowed to do [no] more work in my department."[19] Skiff took no action.

Meanwhile, Cooper had resigned his museum position in April. With labor in the department at a premium, Hay enlisted Brandler, the museum's assistant taxidermist, to clean some three thousand small mammal skulls in the study collection—a task Elliot had been especially keen to have done during his absence. Preparation of the skulls was necessary for identifying the skins to which they belonged. This type of work was important for science, but it had no impact on the most visible part of the collection, the exhibits. Skiff thought that using Brandler for such work was a waste of museum resources, and he wrote to Hay about a less costly alternative. Hay replied immediately, arguing that the work required skill. But he agreed to find a cheaper replacement as soon as possible. "I shall be very glad indeed to have another man at work in the department," he added hopefully. In mid-June, Hay wrote to Frank C. Baker, at the Chicago Academy of Sciences, to ask whether he knew of a man who would be willing and able to do the work well for very modest wages.[20]

IMPORTANT CONTRIBUTIONS TO KNOWLEDGE

While exhibit renovations were underway, the acting curator was also expected to do original zoological research. Although employed as an assistant curator in zoology, Hay aspired to become the museum's first curator of paleontology. His interest in fossils began at Butler, where he started working on a monumental bibliography of paleontological literature. He later did a single season of fossil vertebrate fieldwork in western Kansas in 1889 or 1890. At the Field Columbian Museum, there was no curator to take the reins of vertebrate paleontology. Hay, therefore, was free to dabble with the modest collection of fossils housed in the museum's Department of Geology. (The curator of geology, Oliver C. Farrington, was a hard-rock geologist who lacked training in vertebrate paleontology. Elliot and Hay, therefore, believed that these specimens belonged more properly in zoology.) This he did, publishing in 1895 and 1896 four papers dealing to some degree with vertebrate fossils, including three contributions to the museum's new zoological publication series. The first paper, which described the development of the vertebral column of the fish genus *Amia*, made no reference to any specific fossil material. The second briefly described a partial skeleton of the fossil turtle *Protostega gigas*. The third was a description of a new species of fossil fish, *Petalodus securiger*. The fourth described a partial skeleton of a fossil turtle, *Toxochelys latiremis*.[21]

Vertebrate paleontologist Edward D. Cope took notice of the museum's burgeoning research output. In a May 1896 editorial in the *American Naturalist*, he reversed his earlier opinion about the sorry state of science at the museum. "We are pleased to notice the excellent scientific work being done by the Field Museum of Chicago," he wrote agreeably. "The management has called to its aid a number of scientific men, and is publishing the result of their work in suitable style." He singled out two papers by Hay (on *Amia* and *Protostega*) as "important contributions to knowledge." To Cope, quality publications signified a sea change at the museum. "It seems that the Museum is not to be merely a show place," he argued, "but is to be a center of original research, worthy of the great city in which it is situated."[22]

Cope and other critics in the scientific community had expected the Field Columbian Museum would privilege extravagant exhibition over the less showy and more specialized work of scientific research. Anthropologist Frederic W. Putnam, for example, warned: "In Chicago all would be drive and rush and largely sensational effects. That is what they are now after, and it is natural in a place which has started out with great hopes and plenty of money and a feeling that money will do anything." Botanist John Coulter likewise cautioned in 1894 that "the danger has been the common danger of Chicago, namely, to make the Columbian Museum a big show instead of a center for scientific collections for study and work."[23]

Cope's new editorial laid some of these anxieties to rest. Skiff, naturally, was delighted by the favorable press—although he was not especially surprised. The director had been aware of the criticism of his leadership from the scientific community. Cope's previous editorial, which called Skiff out explicitly, had gotten the director's attention. Rather than suffer in silence, though, Skiff plotted to forestall any further criticism from that particular quarter. The director was familiar with Cope's embarrassing personal circumstances and turned his adversary's dire financial straits and colossal ambitions against him. He enlisted the help of a well-connected New York colleague, George Kunz, to encourage Cope to solicit a position as paleontologist at the Field Columbian Museum. "Can you in a very round-about way entirely concealing my interest in the matter get Cope the Phila[delphia] naturalist to make an application to me for position of paleontologist? It must be done very artfully," he explained, "and you may not find a way for several weeks. But there is no particular hurry."[24] Skiff wanted Cope to call at his office in person, bringing a letter of introduction from Kunz. Such a letter, the scheming director knew, could be kept as evidence of Cope's hypocrisy.

Skiff, of course, had no intention of hiring Cope. Yet a hopeful job applicant, he took for granted, could be expected to take a more favorable view of his prospective employer. Perhaps the applicant might even write and publish another, more positive, editorial about the museum? This was Kunz's thinking, also. In his reply to Skiff he wrote: "Prof. Cope . . . assured me he would give the facts about your museum."[25] Evidently, Kunz deduced that Skiff's interest in Cope was connected to the latter's editorials in the *American Naturalist*.

Skiff, who was extraordinarily cunning—"the outstanding feature of his character," according to William Henry Holmes, the museum's anthropologist—could be as vindictive as a wounded leopard.[26]

9

LIFE SEEMS MORE WORTH LIVING

May 21–June 2

Once back in Berbera, Elliot learned that Dualla had acquired only two more camels. Thus, the curator was still well short of the requisite number to carry his baggage into the far interior. But—unwilling to delay any longer—he decided to convey his outfit "by detachments." This meant ferrying half his belongings to some waypoint, either Hargeisa or Milmil, unloading, and then sending the camels back to Berbera to retrieve the rest. "It is very annoying & will cause me to lose much valuable time," he fumed. In the meantime, the expedition's livestock was taken elsewhere, as there was meager grazing around the city, and buying fodder "would be altogether too expensive."[1]

Elliot sat sweating at his desk at headquarters that evening and wrote a long letter to Skiff describing his recent excursion in excruciating detail. "Of all the God forsaken countries it has been my lot to see this was the worst," he wailed, "nothing short of Dante's description of the Inferno could do it justice." He bemoaned the bare, lava rock hills and the "dreary waste of sand." The only green relief was a sparse scrub of thorn bushes that tore their clothes to ribbons. The heat and the blazing sun were unrelenting. There was "not a drop of water save at long intervals & then either deposited in holes in the rocks by rain, or scooped out of muddy holes dug in the ground." They'd contended with dermestid beetles, ants, centipedes, scorpions "built like lobsters," venomous snakes, and other vermin. "It would seem as if the good Lord had made Africa the waste basket for all the foul and hideous things no other land would have." This, Elliot grumbled, was "a faint description of the Paradise we have been living in for the past two weeks."[2]

Yet, during this time of tribulation, they'd made a valuable collection of mammals. "We obtained over a thousand dollars worth of material . . . regarding it from a commercial point of view," Elliot boasted. By this, the curator meant that it would cost over $1,000 to purchase the same skins from a specimen dealer. But the collection's true value far exceeded that figure, he explained. Especially noteworthy was the Somali wild ass. "The museum will be able . . . to make a better exhibition of these creatures than all the other Institutions in the World put together." Somaliland harbored some of the handsomest animals on earth, among them the antelopes that Elliot intended to salvage prolifically over the coming months. "One sorrows to think that they must all vanish soon," he lamented unselfconsciously.[3]

UNAVOIDABLE DELAY

To beat the withering heat of Berbera, Elliot cultivated the habit of wearing his pajamas around headquarters. "We go about in as few clothes as decency will permit," he wrote, "but even when [we are] sitting quietly in a chair," the sweat "simply pours down our faces." During the day, there was no respite from the suffocating heat, other than sitting motionless in the shade or seeking a rare breeze. Sunset brought some relief. Yet, his room remained so stuffy at night that he often found it impossible to sleep. Some mornings he arose "as wet from perspiration as if I had been dipped in the sea."[4] One miserable evening, he had his bed placed outside on the veranda, where the nighttime air was a shade more comfortable.

Elliot was eager to leave town again as soon as possible, but there would be some unavoidable delay. Dualla, mumbling various flimsy excuses, maintained that the caravan wouldn't be ready to leave until after the weekend. Elliot suspected the real reason for delay was that a Somali holiday was approaching, and Dualla and the men wanted to remain in Berbera to celebrate. Though Elliot urgently wanted to get away sooner, he couldn't. His money was locked in the government safe, which—because of the holiday—was inaccessible until Monday morning. He was stuck.

The holiday was celebrated on Saturday afternoon. It began with a procession of singing and dancing Bantus from the south, who came marching through Berbera with harp and drum bearing red banners emblazoned with the star and crescent emblems of the Ottomans.[5] The enormous, four-stringed harp was tasseled with colorful ribbons, its sounding board a bulb-shaped box draped with decorated antelope skins. "The tone of this barbaric instrument," Elliot noted, "was soft and deep, not at all unpleasant." He offered to buy it, but its player demanded a princely 600 rupees ($289). "The dancing was as uncouth & barbaric as the instrument," according to Elliot, "& consisted chiefly in jumping up & down & circling about the head of the crowd. The singing was a chant that never varied, ending in [a] shrill chorus." The performers visited the most prominent houses in the government quarter, singing and dancing joyfully before each one, hoping to earn a present. When one was received, the procession moved on. Elliot's headquarters was visited in turn. "The men were

FIGURE 9.1.

A Somali holiday
celebrated in the
streets of Berbera.
Photograph courtesy
of the Field Museum,
CSZ6032.

black as ink, with fine teeth and a powerful well developed physique, and
looked well in their white tobes," he wrote approvingly.[6] The festivities
lasted all day and well into the night.

Meanwhile, when unpacking and inspecting the specimens collected
thus far, Elliot made a terrible discovery. Ants had infiltrated their collection of bird skins, doing considerable damage. The legs and nostrils of
many of the birds had been "badly eaten." About forty specimens had to
be discarded—an ironic outcome, given their salvage mission. Somalil-
and was a particularly difficult country in which to preserve museum
specimens. "If the sun and heat do not ruin a specimen," Elliot lamented,
"there are thousands of insects of all kinds waiting & willing to do it." The
mammal skins, on the other hand, survived in good condition. These had
been stored in brine-filled barrels impervious to insects. Elliot soon added
the month-old gazelle to the mix. "The baby gazelle has gone to join its
ancestors," he wrote pitilessly, "the little body is now awaiting the services
of the Taxidermist."[7]

Elliot feared the expedition's return to Berbera had "thoroughly de-
moralized the men." Their days in town were punctuated by what the
curator characterized as many "disagreeable occurrences." He was writing in his room one morning, for example, when the shikari who worked
for Akeley, who'd been relatively uncooperative over the previous three
weeks, arrived and informed Elliot that, as he'd done his work very well,

he now required a raise. The curator "told him his opinion of himself was of no consequence, and as I did not consider him worth what I was already paying, I would dispense with his services altogether." The shikari left, "very much crestfallen." He was immediately followed by an attendant, "who bluntly informed me that if he did not receive more wages, he would go no further. I told him he would get no more wages & bundled him out." Then a gun bearer and the assistant cook arrived "with tales of woe at home which required their immediate presence to soothe. I told them I would take it into consideration." His tent boy, Abdi, who'd been scheming with Dualla to be made head tent boy with a raise, arrived and announced that he had "a yearning desire for . . . home." Elliot told him that if he found a replacement he could quit. So, he did. "They dare not leave me without permission," Elliot explained, "for their names would be posted at the Government House, & they would never be permitted to accompany a caravan again. A severe punishment to them, for all the money they get comes from the caravans & expeditions."[8]

Dualla later attempted to intercede for Abdi. The former tent boy was thoroughly sorry, the headman explained. He'd made a fool of himself, and he wanted only to rejoin the expedition. Unfortunately, Elliot now had no position to give him. Nor did he particularly want Abdi back. It would be a mistake to take him, the curator suspected, "as he probably would always be trying to supplant my new boy, Ahmet Hirsi, who is a great improvement over the magnificent Abdi." Ahmet had been gun bearer for Elliot. In Ahmet's place, he hired Elmi Farras, a Midgan. Elmi had served on several occasions as Captain Merewether's shikari. The captain spoke highly of him, "especially as a most excellent tracker."[9]

Musa, their head cook, who'd made them many wonderful meals, also demanded a higher wage, or else he would quit the expedition. Elliot threatened to haul him before Captain Merewether, charged with desertion. He warned that the punishment would be a month in prison with hard labor and a fine. The cook was "very defiant for a time, but afterward came to me humble enough & said he would go." In fact, he'd already been to see Captain Merewether himself, where he'd received "such a reception as convinced him further resistance on his part was worse than useless."[10] Though he'd agreed to remain with the expedition, Musa was likely angry about his employer's coercive tactics. Nevertheless, Elliot believed he would have no further difficulty with the head cook. The second cook, meanwhile, reported sick, so Elliot hired a substitute on Dualla's recommendation.

At last, on Sunday evening, plans to return to the field had been put "in pretty good shape," but only after many vexations. Many of the hired men had used the expedition's return to Berbera to try to improve their working arrangements with Elliot, but the curator denied their requests as a matter of pride. "Every device & stratagem have been employed by many of the men to get the better of me & gain their various points," he explained, "but I am happy to think they have been worsted in every case, & now they know the kind of man they have to deal with. . . . I will permit no nonsense, & they must attend to their duties & obey my orders." Elliot tended to view the relationships between himself and the Somalis as more akin to master and servant than employer and employee.[11]

HEADING FOR THE INTERIOR

The caravan was ready to start again on Monday, May 25. Dualla, meanwhile, had acquired ten more camels, bringing Elliot's total to fifty-six. The new men, thirty-two in total, who hadn't yet signed their contracts, were brought that morning before Captain Merewether, who delivered a stern lecture about the importance of discipline. Elliot himself added that anyone present who was unwilling or unable to abide strictly by the terms of his contract should step out of rank and remain behind. But "not a man stirred." Marching back to Elliot's headquarters, the men each received one month's pay in advance. Under Dualla's skillful management, there was no confusion, and business proceeded quickly and efficiently. Once they'd been paid, the men dispersed. Elliot assumed that they were all headed straight to the bazaar to spend their money. He was "heartily glad to get rid of them."[12]

The men began loading the camels in midafternoon. They draped large, woven reed mats over the camels' backs, fixing them in place with great lengths of grass rope wrapped several times under and around the bellies of the beasts. Casks, bundles, and boxes were then tied tightly to the mats using more grass rope. The loads looked clumsy but—when skillfully arranged—always rode securely, seldom shifting or falling off. Dualla had purchased provisions for fifty men for two months, which was "all or possibly more than we can carry," in Elliot's estimation. The commotion of loading was tremendous. There were protesting growls from the heavily encumbered animals. Crowds of spectators surrounded the proceedings on all sides, shouting and waving to their comrades and loved ones. Women came to bid an emotional goodbye to husbands and sons, who were leaving on a mysterious odyssey filled with unknown hardships and dangers. Dualla had a reputation for only going on lengthy expeditions, which added to their anxiety. Elliot had been asked many times and in many crafty ways how long he planned to be away. Always he refused to give a definite answer. His intention was to stay out as long as possible with the resources at his command. If he fixed a time to return and then exceeded it, he feared, then there could be discontent among the men, maybe even mutiny. "It might be difficult to hold the caravan together," he pondered with a shudder.[13] As soon as they were loaded, the camels were sent off in small groups. All had started off by a quarter past four in the afternoon.

FIGURE 9.2.

Loading the
camels with
equipment and
supplies near
expedition
headquarters
in Berbera.
Photograph
courtesy of the
Field Museum,
CSZ5998.

Despite his eagerness to get away, Elliot lingered in town. He penned a hasty letter to Skiff explaining that he now had almost nothing remaining on his original letter of credit. He was holding the second letter of credit—for £800 ($4,000)—in reserve to meet his expenses when he returned to Berbera. He needed more money, he hinted. The success of the expedition depended on it. He scratched an expectant signature, then mailed his letter, satisfied that he'd done and said as much as he could for the financial well-being of the expedition. By now the afternoon was winding down. He mounted his mule and followed the tracks of his caravan, heading almost due south across the darkening plain.

The sun had nearly set; an ivory disk of moon was rising on his left. The weather was warm, but the ride was pleasant. As night came on, moonlight illuminated the desert landscape so that it was nearly as distinct as in the day. Soon, he began to overtake the caravan, which was strung out in a long, ragged line. At half past eight, he reached camp, which had been pitched in an open space encircled by a protective screen of low thorn bushes. Everything was done in an orderly fashion, "Dualla's skill in management being clearly shown." Every man knew his duties and did them expeditiously. There was no confusion, no one getting in anyone else's way. As they hoped to start again very early, the men went to bed soon after dinner. The scene from the open flap of Elliot's tent was "an attractive one, the camels lying in groups their loads near them, the men gathered about various small fires, & the bright light of the moon, causing every object to appear sharply defined, even the sand glittering like silver."[14]

At three in the morning, the bellowing of camels announced that loading had begun. Elliot and his hunters set out early, in order to see game in advance of the caravan. The elevation had risen steadily since they'd left the coast, and their path tilted still upward from the maritime plain. The shimmering sea was visible in the distance behind them. The cooler weather was already "a decided improvement" over the extreme heat of Berbera. Soon after daybreak they began to see antelopes. Elliot encouraged Akeley and Dodson to shoot them. They acquired two Pelzeln's gazelles and one dik-dik as a result. These animals were small and wary, Elliot wrote, but the men's rifles were more than equal to the long shots. "I wouldn't be without my Lee Metford for anything," Dodson confided to his parents. "It's a daisy."[15]

After marching for three hours, they entered a rough and broken country in advance of distant mountains. By late morning, they'd arrived at Daragodleh, where they found wells sunken into the sand of a dry riverbed. The water "was of the usual pea-soup color, but sweet." They pitched their tents and sent the camels away to feed. The plan was to remain in this spot "until early, very early, tomorrow morning." The sun was hot, but the air was considerably cooler at higher elevation, "and life seems more worth living." Although there were few trees in sight, birds were seen in great numbers, including species different from those of the coastal plain. During the morning march, for example, they passed a hawk perched on a small tree just off their trail, singing a delightful series of notes. Unfortunately, the man carrying Elliot's shotgun had stopped somewhere to pray, and so the curator lost the opportunity to secure "this warbling bird of prey."[16]

In the afternoon, the hunters ventured out to see whether there was any desirable game nearby. Akeley hunted gazelles. Armed with shotguns, Dodson and Elliot went after birds and other small game. Akeley returned with nothing. Elliot managed to bag an eagle, *Aquila rapax*, a dik-dik, *Madoqua swaynei*, and a hare.[17] Dodson brought in a pair of rare sand larks.

The late evening was "glorious." The moon was nearly full, and the "entire landscape bathed in a soft light."[18] The temperature was cool and pleasant. Although they planned to make an early start, the hunters were slow to go to bed. Later in the night, a cold wind commenced that drove Elliot crawling beneath a blanket for comfort.

The next morning, they were off again early, ahead of the caravan, riding pleasantly for several hours under a bright moon. Once the sun rose, however, a strong headwind blew in from the south, which made riding "anything but agreeable." They were approaching the Golis Range, and their road wound circuitously through the foothills, ascending gradually all the time. Game was almost entirely absent. They encountered a small stream with a murmur of running water, "a most unusual sight . . . for this part of Africa."[19] They stopped to let the animals feed on the grass growing along the water's edge.

While they were loitering, a villager approached and offered to sell Elliot a juvenile kudu. As the young of all game animals had proved difficult to get, the curator made an offer. So, the man returned to his village to retrieve the captive animal. When the speediest of the camels began to arrive, Elliot and his party set out again for Laferug, their next oasis. An hour later, they'd arrived. "This is the most attractive spot I have yet seen in this barren land," Elliot wrote. "Of course sand is everywhere, but there are trees and numerous birds some of great beauty, & the water is good."[20] They pitched their tents on a low bluff overlooking a dry riverbed. There, they dug small holes in the sand and found abundant water. A village was close behind them. A wall of mountains that they soon had to scale loomed ahead.

Wanting to make the most of what remained of the morning, Dodson ventured out with a shotgun and returned with nine birds and two bushy-tailed, stripeless ground squirrels, *Xerus rutilus*. Elliot learned that at the next camping ground, a lion had been making a nightly nuisance of himself. Unfortunately, they were still on the reservation, and "as I promised not to shoot any lions within its limits, I regret very much that we cannot pay our respects to the troublesome gentleman."[21]

FIGURE 9.3.

A comfortable campsite at Laferug. Elliot (*center*) is writing in his tent, while Dodson (*right*) is skinning birds. Photograph courtesy of the Field Museum, CSZ5941.

In the afternoon, Akeley and Dodson went back out. Akeley returned with a dik-dik of a different species, larger and darker than the other specimens they'd taken. Elliot suspected that this was the true Phillip's dik-dik, and that the dik-diks shot previously were likely Swayne's. Dodson brought in a long-necked female gerenuk, *Lithocranius walleri*, a female Speke's gazelle, *Gazella spekei*, and a small dik-dik. Elliot noted that all the dik-diks collected thus far were males. They'd seen females, together with some young, but hadn't gotten near enough to these timid animals to take a shot. Elliot speculated that this was where the lowland and upland gazelles overlapped in distribution, the former becoming scarcer as they gained in elevation.[22]

While they were enjoying their supper, the enterprising villager returned, bringing his young kudu. It proved to be one of the lesser kudus, *Strepsiceros imberbis*. It was "a most beautiful creature, its reddish body being covered with numerous narrow white stripes. It was four or five months old & tame as possible, and drank its milk from a basin with great satisfaction."[23] This was a very rare and difficult specimen to obtain, and the thrifty curator considered it a bargain at fifteen rupees ($7). Elliot originally planned to move camp in the early morning hours, but they now had so many specimens on hand that required care that he decided to remain another day, at least.

The next morning, they were all up early, as usual. Elliot designated

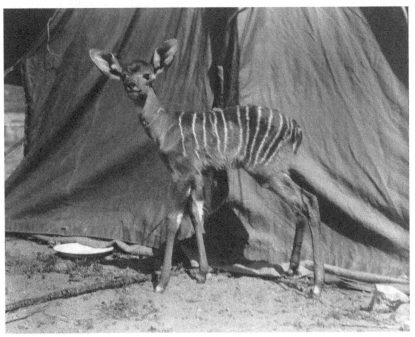

FIGURE 9.4.

A juvenile lesser kudu. All the juvenile captives were remarkably tame. Photograph courtesy of the Field Museum, CSZ6086.

Akeley to remain in camp. Dodson and Elliot, meanwhile, climbed a steep mountain path opposite the campsite, guided by a villager who'd seen wild donkeys the previous day. At the same time, another villager took a shortcut to try to find the animals and let the curator know their position. Elliot and Dodson arrived at an open plateau covered with broken, red, iron-rich rocks, and rode for a considerable distance. The going was rough and their progress correspondingly slow. They came across a herd of up-land gazelles and had dismounted to stalk them when the villager came running, shouting that he'd seen donkeys only a short distance ahead. No more attention was paid to the gazelles, and instead they went after the rarer, more valuable animals. After stalking with great care and diffi-culty over the rocky landscape, they finally sighted the donkeys feeding in a meadow some three hundred to four hundred yards away. A closer approach was difficult. Elliot noticed a gully with a clump of shrubs at its head, which would bring the hunters nearer to their quarry if they could only reach it undetected. The companions conferred and decided to try this approach. They crawled forward stealthily on hands and knees. Reaching the gully, they proceeded slowly toward the cover of the shrubs. When they arrived, however, they found that their prey had moved and were now something over two hundred yards distant. Worse, the donkeys stood gazing at them uneasily, having spotted the hunters the moment they arrived at the shrubbery.

Elliot had already decided that he'd shoot first and Dodson was to fire immediately after. There was one large donkey directly in front of the hunters. At that distance, it appeared no wider than a pencil. Near the large donkey was a smaller one, the rump of which could barely be seen through the bushes. As Elliot particularly wanted a juvenile for his museum group, he aimed at the smaller animal and fired. It dropped in its tracks. The remaining donkeys scattered when Dodson fired and missed. Elliot jumped to his feet and fired at another animal, wounding it, so that it soon reduced its pace to a slow walk. Approaching the spot where the donkeys had been grazing, he passed the juvenile lying dead on the ground. It was a male, maybe a year old. He followed the trail of the wounded donkey, which led over the crest of a hill. He and the others reached the spot where the donkey disappeared, but it proved impossi-ble to track the wounded beast over the stony ground, and they lost it. Returning at once to the dead donkey, Elliot sent a runner to camp for a camel to bring the specimen in, while Dodson, with the help of the shi-karis, proceeded to skin it.

Elliot rode back to camp elated. As he descended the mountain, he met a pair of camels going up to retrieve the donkey. In camp, he found to his surprise that Akeley had completed all his tasks but for the young kudu, whose lifeless head was even then embedded in plaster. Careful, exacting plaster casts had been taken of the heads of all four animals, and their skins had been processed. Akeley mentioned that soon after the others had left that morning, a man appeared on a nearby ridge and shouted that there were donkeys nearby. Unfortunately, he was then too busy to chase after them.

Dodson returned in midafternoon, leading a camel with a fine Speke's gazelle buck draped over its back. This animal he'd met and dispatched on his way back to camp. A second camel trailing behind carried the juvenile donkey. Processing the donkey and the gazelle occupied the rest of the afternoon. Elliot had the skeleton of the latter preserved as well as its skin. Akeley took some photographs of their picturesque campsite, then capped off his afternoon by relaxing in his tent with a drink.

That night the men held a festive dance, with raucous singing and shouting, in order to keep predators away, as three villagers had warned of lions nearby. In the evening, while the expedition was at dinner, some of the men said they heard a lion roaring in the distance. How, Akeley wondered, would he fare in a confrontation with the most dangerous game? Would he keep his cool? Would he prevail? Or, would he be hurt or killed?

Elliot gave orders to move camp again in the early hours of the morning, then went to sleep. Later that night the temperature plunged, so he took refuge again under his blanket.

FIGURE 9.5.

Akeley relaxes with a drink in his tent at Laferug. Photograph courtesy of the Field Museum, CSZ6097.

WONDERFUL VITALITY

They were away before four in the morning, following a path that wound among the foothills. The moon, now on the wane, was nevertheless shining brightly. The predawn air was cool and pleasant. They continued to rise with every step and were now more than a thousand feet above sea level. Game was conspicuously absent. They reached a forest that extended into the mountains. It was thinly populated with acacia and mimosa trees and had broad paths running through it. Beautiful vines of vivid green with large, fleshy, heart-shaped leaves clung copiously about many of the trees, forming leafy bowers, while many birds with beautiful plumage and unfamiliar songs "flitted among the branches." There was a small village nearby, which appeared to be rich in livestock. They didn't stop but proceeded through the woods until they came to a broad, dry riverbed. There they met some Midgans, who informed Elliot that they'd just seen a lion in his lair. They encouraged Elliot to hunt it, but the curator regretfully declined. One Midgan mentioned that there were plenty of lesser kudu in the woods, so Elliot directed Dodson to accompany him. The curator, meanwhile, continued marching up the river until he reached a well called Mandera. He ordered camp to be pitched on what would have been an island had there been any water in the riverbed. A clump of beautiful shade trees rendered this "the prettiest place we have met with."[24]

The birds near camp were numerous and diverse, some, Elliot noted, "having most brilliant plumage." It was odd, the ornithologist reflected, to hear so many unfamiliar notes, including "the toc-toc of the Hornbills [and] the scoldings of the Turacos." Having done most of his previous hunting in museum drawers, Elliot had never heard many of these bird-songs. This signaled to the naturalist that he was in unfamiliar territory. Perhaps the local avifauna was unknown? Thanks to his museum's unusual division of labor, the pursuit of mammals occupied nearly all of Elliot's time and attention. He regretted that he had no mandate and thus little incentive to devote much time to birds. And he especially regretted that he wasn't in charge of his museum's bird division. "This would be a great field for a collector, & a splendid series of birds would reward an energetic man," he expected. Reptiles were likewise abundant. Someone captured in camp a small, bluish, stump-tailed chameleon, which was later designated a new species, *Rhampholeon mandera*.[25]

Late in the morning, a runner came charging into camp asking excitedly for two camels and announcing that Dodson had killed a pair of lesser kudus, a bull and a cow. Hours later, Dodson returned to camp triumphant, bringing his specimens. "Without exception," Elliot wrote with obvious delight, "the Lesser Koodoo is the handsomest antelope known." The bull sported a pair of long, spiral horns and a bluish coat crossed with abundant white stripes. The cow, on the other hand, was hornless, wearing a reddish coat with white stripes. There were myriad other small marks and patches of black or white scattered over the skins of both animals. The bull was enormous—not much smaller than a wild donkey. These were wonderful acquisitions, very valuable, especially as this species was so rare in museum collections and dwindling in the wild. With the young animal procured at Laferug, Elliot was pleased to record that they now had "the chief materials for a group." A proud Dodson described the lesser kudu as "a great prize and extremely difficult to get. There was a jubilee in camp when I brought them in."[26] Processing the specimens occupied Akeley and Dodson until dark.

They spent several more days at this productive locality. One morning, Akeley killed a lesser kudu bull, the head and horns of which were found to be longer and wider than Dodson's specimen. The animal was quickly skinned and the meat divided among the men. Dodson shot four more females, including two older individuals and one juvenile. They now had more specimens than Elliot needed to build a habitat group, so some of these would doubtless be used as study or exchange specimens.

The lesser kudu, Elliot remarked, inhabited the slopes of the foothills, preferring ground covered with thorny shrubs and trees and five-foot-tall aloes crowned with bright crimson and golden flowers. It was very speedy, bounding away quickly when startled and invariably heading for a thick cover of shrubs. He complained of having almost no time to aim and fire when hunting this animal, which was "not always satisfactory from a sportsman point of view." Still, it was a beautiful thing to behold. He was moved by the sight of "a fine buck in all the pride of his strength and beauty flying over the bushes," taking surprisingly athletic leaps.[27]

The curator had better success hunting warthogs at this locality. Shortly after the men had set out one morning, a shikari spotted one dashing into a thicket. Elliot went with one man to the other side of the bushes, while the others attempted "to drive the brute out." A gun bearer handed Elliot the Winchester. Soon they heard a rustling in the bushes, and "a magnificent boar trotted out" less than twenty feet from Elliot's position. With head, mane, and tail erect, it moved slowly and deliberately, its monstrous tusks glinting in the sun. Taking quick aim, Elliot squeezed the trigger. Alas, only the metallic clack of the hammer was heard. The magnificent boar, meanwhile, disappeared into the brush. "The careless Shikarri," Elliot griped, "had forgotten to throw a shell into the barrel before giving the gun to me." But the chase was still on. Elliot and his gun bearer immediately took up the warthog's track and followed it a great distance. The curator, at long last, caught sight of the hog scampering through the bushes maybe 150 yards ahead. His first shot missed. But his second caught the fleeing hog in the hip, and it fell, grunting rapidly. The hog jumped up again and scurried some twenty yards before its legs suddenly gave out. When Elliot finally reached it, the hog was supporting itself upright on its forelimbs. It was the very "impersonation of rage, and tried to get at me champing his tusks together." But the fight wasn't finished. "As he seemed to have no idea of dying I shot him with the Lee Metford behind the left shoulder & knocked him over, but he rose again to his knees & dragged himself along striving to get at me." Elliot then took his Winchester and put another bullet behind the boar's right shoulder, which killed it instantly.[28]

The warthog, *Phacochoerus africanus*, was an "enormous brute possessed of wonderful vitality," its great head covered with a constellation of fantastic "warts & excrescences."[29] Its hairy mane was fully two feet long, and the great tusks each measured twelve inches in length and spanned eleven and a half inches from tip to tip. As no Muslim would handle a pig, Elliot sent for Dodson to come skin it. The curator then cooled his heels a long while, waiting and watching the sun sink. Fearing that he'd have to leave the carcass overnight, he'd begun to build a crude zareba of thorn bushes over it, when Dodson arrived at last. Immediately, they set to work, Elliot providing what assistance he could. Just before sundown, they got the skin free. The more complicated head they didn't attempt to skin. Tying their prize to a long pole with rope, the two men each heaved an end and walked with it to camp, arriving just as darkness descended.

Lions roaring near camp that evening unsettled the expedition's livestock. At one point, another stampede seemed imminent. At least three lions answered one another from different sides of camp. The menacing roars, and a rustling in the bushes, caused unease among the men, as they'd built no zareba and their camp was spread over a wide area. Someone fired two shots in the air to scare the lions off. They heard no more that night or the next. "The roar of a lion at night is a most impressive sound," Elliot thrilled, "as it reverberates through the forest or over the plain & it is wonderful." He lamented that time was running out for this magnificent predator. "Like most wild creatures," he wrote, "the lion is rapidly passing away."[30] He was keen to add one or more to his specimen barrel before this should happen. Leopards were apparently common near camp, also, though no one had seen one yet. A villager reported that a young goat was carried away the previous night by one of these sneaky beasts. Elliot wanted some of these for his specimen barrels too.

Elliot reprimanded Dualla one morning. Akeley had come to the curator's tent to say that the men were killing sheep in alarming numbers. They'd had meat every day since leaving Berbera. Yet, when game was scarce, serving meat once a week was considered sufficient. Elliot sent for Dualla, asking whether any sheep had been killed and, if so, on whose authority. Dualla admitted that he'd ordered sheep killed to feed the men, as this would conserve their rations of rice and ghee, which the expedition had in limited supply. Elliot informed him that he didn't intend to serve meat to the men every day, and that thereafter no sheep were to be killed without his orders. Dualla "expressed regret at what was done . . . but was mad at getting a lecture before the men." The curator, for his part, was especially sensitive to any threats to his authority. Yet, such a conflict was probably inevitable, with Elliot in nominal charge over Dualla, who wielded real authority over most of the men in camp. "I see I shall have to keep as firm a hand on him as on any of the others," Elliot bristled. "In many ways [Dualla] is a most excellent and competent man, very intelligent."[31] But, the curator noted warily, he would require watching.

LIKE GIANTS REFRESHED

They passed a tranquil Sunday skinning kudus and making a plaster cast of the shoulder muscles of one of the smaller females. While Akeley was waiting for plaster to set, a bold pair of warthogs made an appearance opposite camp. He bolted after them, catching sight of them scampering over a hill. He fired one shot, which caused the hogs to separate, each running in a different direction and escaping. Meanwhile, a jungle cock strutted into a thicket near Elliot's tent. The curator grabbed his shotgun and flushed it out, killing it on the wing with his second shot. The men were delighted, expressing their admiration with boisterous shouts and laughter.

Near camp, on the right bank of another dry riverbed, rose a towering pinnacle called Dagaha Todoballa, or Rock of the Seven Robbers. Rising more than a hundred feet above the surrounding landscape, it made a striking landmark. The highest peaks of the Golis Range were still to the south of camp. One of these, Gan Libah, topped out at more than six thousand feet. This mass of rock appeared to be rent by fissures and deep ravines and was capped with thorn trees and a thick, green mane of aloe and cedar trees. Many villages were then migrating to the high plain beyond the mountains, passing their camp in long, overburdened processions over the previous two days. Reports suggested that there'd been showers of rain and cooler temperatures on the upper plain. A stubble of new grass was said to be sprouting among the stones.

Elliot commemorated the first of June by tabulating their results to date. In roughly three weeks of collecting, much of this time in what he described as "almost game less country," the expedition had obtained forty mammal skins, worth an estimated £2,000 ($10,000). He considered this "very good" productivity to start, but he wanted to raise his rate of capture. The rapidly disappearing donkeys and kudus were especially valuable acquisitions. "I hope the present month will turn out better," he added.[32]

The hunters went out that morning in different directions. Elliot had sent the day before for three Midgans to lead them, but only two answered the summons. So, one each went with Akeley and Dodson, while

Elliot went out with a shikari. Elliot walked for five tiresome hours, seeing nothing, before returning to camp. Soon, a Midgan appeared, asking for a camel. Akeley, he reported, had killed three gerenuks, a male and two females. The camel returned with the gerenuks two hours later. Akeley then arrived, carrying the carcass of a young piglet slung over his shoulder. He'd spotted a boar sow with four or five young and had killed one of the juveniles, as these were much wanted for a group display. Dodson came in later and reported that he'd wounded a greater kudu bull, but that it escaped. Even so, there was sufficient work in camp for the remainder of the day, and Elliot doubted that they would be able to move camp the following morning, as planned.

Elliot wanted to hunt greater kudu but was unsure where to go. There was apparently no water at the place he'd planned to camp next. Nor was there water between this place and the plateau at the top of the pass, which was too far to travel from their present camp to hunt. He decided to move camp to the foot of the mountains on the afternoon of the following day. He would let Dodson go out in the morning to hunt greater kudu and come to the new camp in the evening to report on its relative abundance. If kudu were plentiful, Elliot would remain at that place, sending back each day for water from the wells at his present camp. The water supply, of course, dictated all movements, and the lack of water could and often did upset "many well-laid plans."[33] At this season, which was supposed to be wet, there should've been plenty. But rain had been uncharacteristically light, and Elliot couldn't find water in the places or the quantities he'd expected.

Dodson set out with four Somalis the next morning, shortly after four o'clock, riding his mule to the base of the mountains and then climbing from there on foot. They hunted from daybreak until early afternoon without seeing a single wild animal. Frustrated and hungry, they came across a villager leading a sheep, which Dodson bought for three rupees ($1.50). They dressed, cooked, and ate it on the spot, picking the bones so clean that "the Vultures had a sorry time afterwards." They started off again "like giants refreshed."[34]

Using his binoculars, Dodson spied a greater kudu buck a mile away, on the far side of a deep gorge. To reach the animal, he and his Somali assistants climbed hastily down and across the gorge, running some three miles in total, Dodson's pulse racing "like wildfire" with "the heat and excitement" of the chase. They closed within a hundred yards, screening themselves behind a bush and waiting for the buck to clear a boulder, when one man accidentally dislodged a stone that went crashing down the slope. The huge buck looked up, its "beautiful head and horns" turned in Dodson's direction for an instant, then "off it went at a grand gallop wonderfully fast on such ground." Dodson raised his rifle quickly and fired.

The bullet "struck with a loud plunk in the flank." Away Dodson raced, heedless of his own safety, to catch the fleeing animal. He spotted it trying to make its escape and fired again. The buck fell dead. Elated, Dodson sprinted toward it again, his tired legs eventually giving out on the steep slope, and down he tumbled, crashing hard right into the carcass, his head planted squarely in the dead animal's gory flank. This is how the Somalis found him when they caught up a moment later. Together they began to skin the "magnificent beast as big as a horse."[35]

Akeley, meanwhile, was occupied with overhauling the skins and skeletons in anticipation of moving camp. Elliot was pleased to find that they'd obtained two camel-loads of specimens since leaving Berbera a week earlier. While Akeley was packing, the curator shot "a splendid vulture with a white head, belly & thighs . . . plumage blackish brown & his neck was bare but the head covered with white downy feathers." Its wingspan exceeded six feet. Unfortunately, with all the work in camp, he feared they might have to leave this specimen behind. Birds were difficult to carry, as they had no boxes to store them. Moreover, bird skins dried very slowly in that climate—something Elliot found impossible to understand. Large specimens, when packed for transit, would get "terribly mussed." However, Akeley, after packing all the mammals, set to work and skinned the vulture, *Lophogyps occipitalis*, expeditiously.[36] Elliot would carry the great bird flat in his personal baggage, and any damage would be repaired in Chicago.

In midafternoon, after their invariable siesta, the men began loading the camels. Elliot mounted a mule and Akeley a horse, and they rode off with their Midgan guides. They went together for a great distance, but, seeing no game, they separated. Soon, Elliot saw a gerenuk feeding at four hundred yards. Too far, his men thought. The curator fired anyway and saw sand kick up beyond his target. The gerenuk sprang away safely. They went on and spotted another fine buck of the same species standing about 200 yards away, staring at them. Elliot fired. The gerenuk sprinted a short way before falling dead. At the same moment, Elliot heard Akeley shoot and saw him stalking through the bushes nearby. The Midgan slung Elliot's buck on his shoulders and carried it to where Akeley was now tarrying. There, they found the taxidermist taking a photograph of a male, eight-month-old lesser kudu he'd just killed. The bereft mother stood vigil nearby, but Akeley ignored it. Elliot was pleased, as he wanted another juvenile to complete a habitat group of this species. The Midgans carried the carcasses to camp.

They found their new camp pitched at the foot of Jerato Pass. As they rode in, a grinning Dualla came out to greet them, announcing that Dodson had shot a greater kudu male and needed two mules to be sent to bring it in the next morning. Elliot, however, worried that it might spoil in

the heat and sent word that the skin must be brought in at once. He sent food and water, as Dodson had been out without provisions since early that morning. The gerenuk and lesser kudu had just been skinned by the time Dodson finally arrived with his prize, at half past nine that night, his men belting out a victory song. It was a splendid specimen, well worth a hard day's work. The horns rose in a double spiral curve of forty-eight and a half inches. The skin was handsome, though its white stripes were few and faint.

Dodson was exhausted. Elliot, on the other hand, was overjoyed to finally have a rare specimen of greater kudu, *Strepsiceros kudu*, in his possession. He hoped this would be the first of many. He popped open a bottle of warm champagne to celebrate, which revived Dodson in body and spirit. With the skinning completed and the materials preserved, they retired to their tents in the small hours of the morning "in a contented frame of mind."[37]

10

A BUILDING AS NEARLY PERFECT
AS POSSIBLE

In Chicago, a movement was afoot to redress the deplorable condition of the Field Columbian Museum's building. Over the winter of 1896, landscape architect Frederick Law Olmsted (celebrated designer of New York's Central Park) drew a plan for the permanent improvement of Chicago's Jackson Park, former site of the World's Columbian Exposition. Olmsted's plan called for a new network of walking paths and drives, with the museum—one of only a few buildings remaining from the exposition—as the park's central landmark and "point of departure." In the spring, Chicago architect and planner Daniel Burnham (lead designer and director of works at the exposition) sketched a great lakeside greenway linking Jackson Park on the south side with Grant Park downtown. Both ambitious plans hinged, in part, on the status of the museum and the state of its crumbling building.[1]

At a mid-March meeting of the museum's executive committee, Director Skiff offered a report on the condition of the building, which had been ravaged by winter storms. The museum edifice, he emphasized, had "pretty nearly fallen to pieces, and looks very badly on the outside." To repair the damage, and to spruce up the museum's appearance, the director requested a special appropriation of $15,000. He also asked permission to invite an architect to inspect the building with an eye toward making a more professional recommendation on upkeep. The committee members debated Skiff's report at length. They approved the expert consultation, but they deferred a decision on expenditures until their next meeting.[2]

Skiff enlisted the aid of Ernest R. Graham, an architect in Burnham's firm. Graham, together with Edward C. Shankland, the firm's engineer, inspected the building carefully later in March. Their report was even more dire than Skiff had feared. "The experts are almost unmeasured in their alarm at the condition of the roofs and the skylights [and] the plaster ornamentation," Skiff noted grimly. There was concern that the collapsing plaster veneer was endangering visitors who ventured too close to the building. Graham and Shankland concluded that the falling plaster couldn't be prevented: "They say that the exterior of the Building cannot be repaired, and recommend fencing it in to keep the public away." On the other hand, the figure they quoted for repairing the skylights—nearly $50,000—was through the roof! Some trustees were so alarmed by the tone of the report that they agreed to accompany Graham on a show-and-tell inspection, scrambling up ladders and scaffolding in their top-coats and bowler hats, while the architect pointed out conspicuous cracks, loose plaster, dry rot, and corroding structural elements. The dereliction of duty that the building's woeful condition implied was embarrassing for all concerned. Yet still no extraordinary action was taken to shore up the ailing structure.[3]

Meanwhile, the institution's financial health was also suffering. Several regular subscribers to the so-called McCormick fund, which covered the museum's operating expenses, had neglected to honor their annual pledges. By late March, the museum was overdrawn at the bank to the tune of $16,000 in outstanding contracts and other obligations. Under the circumstances, the executive committee was reluctant to spend. Its members rejected the costly interventions recommended by Graham and Shankland as unwise. Why spend so extravagantly on a temporary structure, they wondered? Why throw good money after bad? Not fully committed to the current structure, or even its location, the executive committee wanted to know what improvements would be absolutely necessary to ensure the building's safety and stability for the next five years. It then appointed a select subcommittee—Director Skiff and a single trustee, Owen F. Aldis—to consider the short-term viability of the building.[4]

Aldis was born the son of a prominent judge on June 6, 1853, in Saint Albans, Vermont. He received a bachelor's degree from Yale in 1874, then studied law in Washington, DC. He moved to Chicago in 1875 to develop and manage new office buildings for wealthy East Coast investors. In this role, he was instrumental in developing many of Chicago's earliest and most innovative skyscrapers, including the Montauk, the Monadnock, and others, often working closely with architects and engineers on the design and with builders on the bricks and mortar. It was probably inevitable that Aldis, a member of the grounds and buildings committee for the World's Columbian Exposition, would become intimately involved with the Field Columbian Museum. He was a close friend, neighbor, or business associate of many of the museum's patrons, including Edward E. Ayer and Martin A. Ryerson. His older sister Helen was married to Bryan Lathrop, whose sister Florence was married to Henry Field, younger brother of Marshall Field, namesake of the museum. A wealthy and scholarly man with wide-ranging interests in history, literature, and languages, he was devoted to cultural causes in Chicago. For his entire professional life, however, he was immersed in the business of real estate and construction.[5]

In May 1896, Aldis inspected every nook and cranny of the museum's dilapidated building, from its dirt-floor basement to the decorative tip of its central dome. Not satisfied with his own examination, he hired two expert craftsmen—a plasterer and a builder—to go over the building a second time. These inspections turned up numerous problems. In the basement, for example, there was evidence of dry rot on at least half the wooden posts that supported the weight of the main floor. Beams for reinforcing the floor that'd been put in place willy-nilly by the museum's building superintendent, John B. Goodman, were insufficiently supported. Aldis recommended hiring an engineer to reexamine every beam and post and reinforce them as necessary. There was also considerable rust on the iron columns that supported the roof and domes. To his surprise and horror, Aldis found a workshop in the basement with lumber, paint, and other flammable materials haphazardly stored. He wanted this removed immediately. Steam pipes, wherever they passed through the floorboards, weren't carefully insulated from the wood. Water from the building's downspouts, Aldis feared, was undermining the brick walls of the foundation. He wanted the entire roof to be retinned and painted. He wanted the three domes replastered. He wanted all the skylights leaded and carefully painted.

The biggest problem, though, was the decorative plaster veneer on the building's exterior, which, as it disintegrated, was falling off the building in boulder-sized chunks. Budget considerations and a tight deadline had forced world's fair officials to utilize the cheapest materials and to choose the most cost-effective, expedient methods in constructing the exposition buildings, including the Palace of Fine Arts. One expedient was a material called staff, an inexpensive plaster mixture made from gypsum and water and reinforced with hemp fibers. Staff could be easily molded and cast, bent, sawed, or bored to make decorative columns, cornices, balustrades, friezes, and more. Its natural color was a murky white that mimicked marble. Staff would harden in a mold in about thirty minutes. It could then be nailed directly to wooden framing or the wooden furring strips that lined the brick walls of exposition buildings. A decorative veneer made from staff gave the exposition buildings the look of classical architecture, but without the expense and time of stonework. While staff was well suited for ornamenting temporary buildings, it deteriorated over time in Chicago's winter weather.[6]

Aldis wanted every square inch of plaster to be carefully inspected. Wherever it was found to be cracking or loosening, it needed to be refastened to the building with long screws or nails and heavy washers. Every decorative projection, finial, or statue found loose needed to be fixed in place with nails or heavy wire or removed altogether. Wherever plaster had fallen off, the exposed brick needed to be scored and covered with Portland cement, then whitewashed. The plasterwork inside the building was also crumbling, especially the decorative cornices in the exhibit halls. Aldis recommended removing them.[7]

In the eyes of the executive committee, the outlook for the museum in the spring of 1896 was fairly bleak. Their most important asset, the building, was gradually falling down around them, which jeopardized the entire museum enterprise. Meanwhile, their bank balance was so depleted that they were forced to make difficult choices and to limit spending on essential repairs to their ramshackle building.

This was the institutional context in which Elliot's pleas for additional expedition funding were received in April and May. That the museum ultimately provided the funding to the curator shows how much the executive committee believed in Elliot's salvage mission.

PERMANENT IMPROVEMENTS

As a consequence of Aldis's no-nonsense report, repairs and permanent improvements made to the building, beginning in late May and continuing through the fall of 1896, were unusually extensive. A temporary crew of craftsmen retinned and painted the roof, installing an entirely new roof drainage system. They replaced cracked and broken glass in the skylights. They strengthened the main floor with stout wooden beams and posts rooted to new concrete pads. They scraped and repainted all the structural ironwork on the building, inside and out. They closed off several unused entrances with iron railing and installed iron grillwork on all windows.[8]

Repairs to the plaster on the building's exterior were particularly painstaking. A team of eight workmen removed much of the building's ornamentation and fixed in place any plaster accessories that seemed in imminent danger of collapse. They patched up the building's many bald spots. They pulled down flimsy plaster statues of Minerva and Augustus Caesar at the main north and south entrances. Skiff anticipated that the crumbling exterior would continue to cause the museum much annoyance and expense. Yet he hoped future damage could be mitigated by the museum's regular force of workmen. Much depended on the weather. An especially severe winter might cause extensive damage, which would require additional, short-term laborers to repair. Barring some catastrophe, the director estimated that the building's exterior could be maintained in its present state for at least five years using only the regular force of the museum and at an estimated expense of about $3,000 per annum.[9]

When major problems with the plaster exterior persisted, however, whole sections of it—particularly from the troublesome cornices—were removed and replaced with a patchwork of sheet iron. Thin iron sheets were also substituted for broken glass to repair wind damage to the main dome. The dome was so high above the main floor that this exchange made no noticeable difference in the illumination of the museum's exhibits. Skiff resolved to carry out this plan of substituting iron for glass in any parts of the roof where glass wasn't necessary for lighting the exhibits below. Repairing the crumbling plaster on the building's interior was, likewise, a never-ending drain on museum resources. The columns, the coves, the decorative cornices, and all the fretwork, especially, required constant attention and upkeep.[10]

Improvements were also made to the museum's steam heating plant. A new, independent line of pipe was installed to carry steam to radiators in the building's east annex. The idea was to make it possible to heat the far eastern end of the museum without raising the temperature of the main building, where zoology specimens were displayed. Unfortunately, extremely cold nighttime temperatures continued to hamper nearly every section of the museum. One especially cold winter night, some of zoology's formalin-filled specimen jars clouded over and spoiled when the temperature plunged too low in an exhibit hall. Substituting alcohol for formalin solved this problem for the museum's fishes, but its staff and visitors continued to suffer grievously during the coldest winter months.[11]

Vermin was an everlasting problem. Recurring plagues of moths, woodworms, and dermestid beetles in all areas of the museum required constant vigilance and remediation. Rats were also an issue. In 1894, as other buildings at the world's fairgrounds were being demolished for scrap, rats fled to the Field Columbian Museum in alarming numbers. "We have been in danger of being overrun by rats," Skiff remarked to a newspaper reporter. A permanent poisoning force was established, under the direction of the building superintendent, to target any section of the museum showing signs of infestation. By the fall of 1896, Skiff reported that pests were under control, yet the work of the poisoners continued throughout the building and at all times. Skiff, himself, made a careful study of the problem, soliciting advice from other institutions about the most effective methods to pursue and the deadliest poisons to deploy.[12]

A morbid fear of conflagration was pervasive at the museum. The infamous Great Chicago Fire of October 1871 had consumed much of downtown, and from the still-warm ashes a new and modern city was raised. The fire hastened the rise to prominence of many of the museum's patrons by exhausting and demoralizing the older generation of civic leaders. Multiple fires in January, February, and July 1894 destroyed nearly every building remaining from the World's Columbian Exposition. Therefore, museum patrons and staff alike were expected to maintain a constant vigilance against fire. Woe to the staff member—such as Akeley—who appeared to take this mandate too lightly. Aldis's inspection of the building in 1896, for example, was done with an eye toward fire prevention. "As you well know," he reminded Skiff, "the building . . . is extremely dangerous from fire."[13]

Indeed, fire danger was never far from Skiff's mind. Early in 1894, he'd hired Captain Michael O'Brien to organize a firefighting brigade at the museum. O'Brien recruited seven professional firefighters to guard the building around the clock, cooking, eating, and sleeping like mummies in the museum. Three were on duty during the day, and all seven at night. The firefighters were equipped with three carts, each carrying 300–500 feet of hose, which connected to one of nine fireplugs, eight inside and one outside the building. They also had a 55-gallon Babcock chemical fire engine, a 24-gallon Miller chemical fire engine, and 28 Babcock fire extinguishers. For good measure, 150 buckets of water were distributed throughout the building. (Fire buckets had to be kept near the radiators to prevent the water from freezing.) Finally, in the event of catastrophe, a fire alarm box outside Skiff's office connected directly to the city's fire department.[14]

Skiff enforced aggressive fire prevention policies. The use of coal and oils was banned inside the building, for example. Electrical wires—except those that powered the building's arc lights—were also forbidden. The director "positively prohibited" smoking.[15]

In September 1896, Chicago's fire marshal, Denis J. Swenie, made a personal inspection of the building. To Skiff's surprise and relief, he judged the museum to be "in first-class shape as regards cleanliness and freedom from rubbish, or anything that would invite a fire." His written report emphasized that the museum's continued safety depended on maintaining the present policy of "doing everything possible in the way of prevention." Swenie recommended the procurement of additional fire protection equipment. Skiff, therefore, purchased hose, couplings, and pipes for ten new firefighting stations, as well as ladders, fire axes, and pike poles. He ordered several additional trapdoors cut in the museum's main floor for speedier access to the basement in the event of a fire there. Swenie was likewise satisfied with the operation of the museum's fire brigade. The bimonthly drill of the guards and the firefighters, and the hourly, round-the-clock inspection of the building, including the entire basement, the roof, and the main dome, continued as before.[16]

In the Departments of Zoology and Ornithology, it had become clear that maintaining a taxidermy laboratory apart from the museum—at Fifty-Sixth Street and Jefferson Avenue—was exceedingly inconvenient. To solve this problem, a new, fireproof brick building for taxidermy was constructed in an inconspicuous corner to the north of the east main entrance of the museum. It was "almost concealed from view," according to the museum's annual report, and provided "ample room for work. . . . Its proximity to the Museum itself simplifies and systematizes the work." The new lab was piped for water and steam and included a specially built vault to accommodate giant metal vats for storing mammal skins. After the taxidermists moved into their new space, several improvements were made, including a cement floor with a drain in the cleaning room. The taxidermists had to share this new building with the painters and carpenters, who'd been booted from the museum's basement. The Jefferson Avenue space was eventually taken over by the museum's new osteologist, George V. Bailey, who began in April 1897. It was used for macerating, cleaning, and mounting vertebrate skeletons.[17]

Skiff also approved extensive renovations to the Department of Zoology office and laboratory. Zoology's "suite of apartments," as the director called it, was perched on the second gallery, two floors above the south door. During Elliot's absence in Africa, this working space was furnished with storage shelves, poisoning boxes, and containers of all shapes and sizes to house the rapidly expanding zoology study collection. The following year, the space was enlarged and entirely refitted. A large stock of improved specimen storage cans was provided for the laboratory. Likewise, a new storage room was outfitted with immense racks for specimens preserved in alcohol. Additional space was found for Elliot's office, as well, by enclosing the outside gallery within it. Although Elliot found the enlarged room still too small, he could nevertheless pursue his work "with much additional comfort and satisfaction."[18]

Lastly, Skiff had a local telephone system installed in the museum to provide "prompt and easy . . . communication" among the various departments and offices throughout the building. The director, whose duty it was to oversee the entire operation of this sprawling complex from his central office, found the telephone to be "of the greatest utility and economy."[19]

The museum had hired a temporary team of carpenters, plasterers, painters, and contractors to fix the most serious problems noted in Aldis's report, spending $21,751 on labor and materials—considerably more than the director's original budget of $15,000. Yet Skiff and Aldis, both, were satisfied with the results. "The Building is as nearly perfect in condition now, as it appears possible to make it," Skiff confidently reported in the fall of 1896.[20]

Renovations to the museum building—especially the new lab and office spaces—were an immediate boon for work in zoology. Yet, the long-term consequences of the building's inexorable decay would later fall squarely on Elliot's shoulders.

11

A GREATER MISFORTUNE THAN WE DESERVED

June 3–11

The hunt for greater kudu continued for days, although Elliot held back. Hunting these elusive animals was "very hard work," the curator explained, "and requires that a man should be in the prime of life, and able to withstand great fatigue." Greater kudu lived high on the steep slopes of mountains, "amid defiles and ravines" cluttered with loose stones and thorny brambles. To hunt there was to risk life and limb. "It is an incessant tramp over the most difficult ground," he wrote, where a tumble could be fatal. He therefore left the heroic pursuit of "these fine antelope" to his younger comrades.[1]

They followed a pattern: Akeley and Dodson ascended the mountain at daybreak, while Elliot remained below. Each day, the curator ambled around in the punishing heat, often stopping at a village to buy milk. He spied an extraordinary number and diversity of birds on these walks. These were of all sizes, shapes, and colors, from dull finches and starlings to tiny sunbirds with their "glorious metallic hues"; from hoopoes, gaudy parrots, and pied hornbills to birds of prey. The beautiful hornbills, he later wrote, formed "one of the characteristic features of the landscape." He was fascinated to find cattle hosting dozens of red-billed oxpeckers, *Buphaga erythrorhynchus*, which obligingly picked bloody ticks from their hides. "The birds alighted wherever they took a fancy," Elliot wrote, "on any part of the body, head, neck back or sides, & I saw two alight on a heifer's nose & investigate the interior of its nostrils, without the animal making the least effort to dislodge them." The nosy birds were pale brown with red bills and bright yellow eye patches.[2]

Elliot observed a bird that had a peculiar form and color and that struck him as most unusual. This bird looked "very beautiful on the wing," its chestnut color "showing so conspicuously in flight." He tried but failed to obtain one. Dualla later brought him a specimen of this same bird that he shot near camp. Examining the specimen closely, Elliot recognized a starling, with a gray head and neck, the body a blue metallic color, and a long tail of the same hue. The primary feathers were pale chestnut, except for the tips, which were steel blue like the rest of the plumage. Alive, the bird was wary and didn't tolerate a close approach. When shot at, it wisely escaped and remained hidden among the shrubs. Its flight was rapid and performed in a straight line by a quick flapping of wings. It was an entirely new species to Elliot, but, having no bird books on his person, he was unable to identify it in the field. He later concluded that it was the rare Somali starling, *Amydrus blythii*.[3]

Elliot forged on doggedly one morning, footsore and hot, but saw nothing save the rump of a greater kudu bull disappearing in the distance. Suddenly, a female lesser kudu leaped from the bushes at a hundred yards and cantered away casually. Taking quick aim, he fired, and the bullet struck the unlucky animal in the back of the head. It dropped in its tracks, and the shikari raced up to cut its windpipe. According to Islamic tradition, a blessing was supposed to be said at the same time, but Elliot noticed that this was often omitted unless someone was watching. His Midgan gun bearer carried the carcass to the shade of a tree, and Elliot sent a man for a camel to bring it to camp. He left a guard, "lest some prowling lion or leopard should find it in our absence."[4] He took the remaining men and continued, hoping to find another male kudu. Although he walked a great distance, he saw only females and juveniles.

Returning to the carcass, Elliot found no sign of a camel. The gun bearer then offered to carry the heavy burden to camp, although it was far. He shouldered it with the help of two other men, and then the party started walking, but not, the curator suspected, in the direction of camp. The head shikari, "a cunning old fox assured me we were going right so I followed & the rascal led me to the village where we got milk in the morning." Elliot insisted on going directly to camp, but the shikari—who'd evidently made up his mind to spend some leisure time in the village—protested that it was very far. The curator announced that he was going to camp, and that the shikari should come along as quickly as he could. Elliot hadn't gone far before the shikari came trotting after him "in no good humor."[5] After a long, hot ride, they reached camp, meeting the tardy camel on its way for the kudu only a short distance from the tents. Akeley and Dodson were already there—they'd brought back nothing.

One of Dodson's morning hunts gave him more than he'd bargained for. He noticed a solitary man lurking mischievously in the hills near camp. A shikari explained that the man was a thief who was watching for an opportunity to nab one of the expedition's animals. Dodson decided to give chase, calling out to the fugitive to halt. But this fleet-footed man "fairly flew over the rocks."[6] Dodson and his men were gaining on the would-be thief, and his capture seemed almost certain, when the rubber sole suddenly split from Dodson's boot. This brought the entire pursuit to a standstill, as the men refused to go on without their leader. Thus, the fugitive escaped. Elliot was just as pleased that the man had been chased off. Since he hadn't been caught stealing anything, the curator could hardly have punished him, anyway.

In camp one hot afternoon, Dodson came to Elliot's tent for a rifle. He'd noticed some subtle movement in the dirt. When he fired a bullet into the ground just in front of it, a small animal was fatally flung from its burrow, pinwheeling, its body perfectly intact. This proved to be a specimen of the naked mole rat, *Heterocephalus glaber*, a most unusual rodent. Only one specimen had as yet reached Europe, Elliot thought, and that had been brought from Ogaden. It was a little larger than a common mouse, with hardly any body hairs, save a few stragglers on its head and back. The hind feet, on the other hand, were armed with stiff bristles that enabled the animal to plow through the loose, sandy soil like a swimmer through water. The incisors—two in each jaw—projected forward nearly horizontally, well beyond its lips. Altogether, it was "very peculiar & not very attractive." Dodson plopped the specimen bodily into formalin for future examination. Though the naked mole rat wasn't thought to be nearing extinction, Elliot considered this "a most happy prize as the creature is rare & its underground life affords but few opportunities for its capture."[7]

On another long trek through the foothills, Elliot saw a black and yellow fox with a heavy coat of fur. He leaped from his mule to shoot it, but the animal sprang nimbly into the bushes and disappeared. Later, he spied numerous dik-diks. One of these inquisitive animals stopped to look him over from fifty yards. Too close, as it turned out, for the curator killed it with his Lee-Metford. Unfortunately, the powerful rifle destroyed the specimen. The bullet struck the animal, which had been standing broadside to Elliot, in the center of its small body. It blasted away the pelvis and everything behind it. The curator saved the head for a personal trophy, apparently taking the same pleasure in macabre animal souvenirs as his sport-hunting rivals. In a later bid for permission to keep this specimen for himself, Elliot explained to Skiff that such specimens had little or no value to the museum. "To him who shot them," however, "the value consists chiefly . . . as trophies."[8] He returned to camp without securing any additional game.

While Elliot was busy admiring his trophy, a Midgan came into camp for a camel and a mule, saying that Akeley had killed a male greater kudu with tremendous horns. Elliot sent the animals, along with a supply of drinking water, as theirs was exhausted. About two in the afternoon, Akeley arrived, followed shortly by the camel, bringing a fine young bull of about four years bearing an excellent coat with distinctive stripes. Skinning the specimen commenced immediately. While this was ongoing, a man arrived saying that Dodson had killed a young bull of the lesser kudu. This had been a mistake, Dodson having identified the animal as a female of the greater kudu, which he'd wounded and was tracking. So, Elliot sent another camel to retrieve Dodson's animal. The kudu that Akeley killed had only one broken femur, so Elliot decided to preserve its skeleton as well as the skin. It was late in the afternoon, with half a dozen men at work on the specimen, before the work was complete. At dusk, Dodson came in with a camel carrying two kudus, both the lesser kudu and the female greater kudu that he'd wounded, which he had succeeded in killing after a three-hour chase. Having obtained three exhibit-quality specimens of the greater kudu, including two males and a female, Elliot decided that the next day they'd move camp to the plain above. Dodson's two kudus were skinned that evening, and it was late in the night before the work was finally done.

Elliot was gratified with the uptick in the expedition's productivity. "The month has certainly opened auspiciously," he wrote. "Of the Greater & Lesser Koodoo I never had much hopes of getting sufficient specimens for groups, as they are very difficult animals to shoot."[9] Now the outlook seemed promising for building habitat groups of both rapidly diminishing species.

In gratitude for their invaluable service, Elliot gave welcome certificates to the Midgan guides, which testified to their keen abilities as hunters and

FIGURE 11.1.

These Midgan hunters were the best Elliot had ever seen in any land. Photograph courtesy of the Field Museum, CSZ6134.

trackers. The Midgans subsisted almost entirely by the chase, never traveling without their long bow and poisoned arrows. A Midgan, he wrote, could follow the trail of a wounded beast with the "unerring sagacity of a hound & is always in at the death." Their method of tracking an animal was truly "wonderful." More often than not, they tracked on the run, sometimes over difficult ground, immediately distinguishing the correct footprint from numberless similar ones. This, Elliot felt, perfectly exhibited "the marvelous ability possessed by these men, equaled only by those beasts that pursue their prey by scent, & which they have inherited through a long line of ancestors who hunted in order to live." Two of these men, Jama and Hait, "were excellent specimens of their class, & I have never seen in my long experience better hunters in any land."[10]

That evening, the men kept up such a commotion singing and dancing that sleep was "out of the question." The Somalis, Elliot opined, were "the noisiest race in the world. They talk incessantly and at the top of their voices, generally all who may be present taking part in the conversation at the same time."[11] To Elliot's ears, the racket they made was an unintelligible cacophony of words and sounds that really got under his skin. The curator regularly stationed five guards around camp at night, requiring them to call out every ten minutes that all was well. They took pleasure in shouting this out as loudly as possible. Hot, insomniac nights would echo incessantly with their irksome calls.

ASCENDING THE PASS

There was so much work to do packing specimens that Elliot decided to stay put until the following afternoon. The bulk of the caravan could depart in the morning, though, for the ascent from camp was steep and difficult for the overloaded camels. For efficiency, Elliot had divided the camel men into five companies. Every man had to know his company's number. Each company had certain equipment and supplies allotted to it, so that there'd be no confusion in loading, the companies knowing exactly where to find their items. At night, when the animals were brought into camp, each company had a designated spot to take its camels so that everything could be done in an orderly fashion.

They made an unusually late start. It was half past six in the morning before the first camels began to ascend Jerato Pass. The tortuous path wound in "a succession of curves & zig-zags" to the top of the mountain forming the face of the ravine where camp was pitched. Other companies followed in short order, and the path was soon crawling with slow-moving animals struggling up the steep grade. The flock of sheep—somewhat reduced by Dualla's liberality—was last to depart. This left only their personal baggage and riding animals at a campsite that "so lately [was] a scene of bustle & activity." Elliot was relieved to be "rid of the noisy crowd."[12]

Nine of the sturdiest camels returned from the top of the pass to retrieve the remaining baggage. Soon after lunch, these were loaded up and started on their second ascent. The sun, by then, was very hot, and—as Elliot knew they could easily overtake the slow-moving caravan—the hunters took shelter in the shade of a large tree and remained for a leisurely hour or more. Meanwhile, clouds gathered steadily to the west. With the sun now mercifully obscured, Elliot gave the order to start. The ascent, which Akeley and Dodson had made many times over the last several days, was very steep for several hundred feet, until it led to a good road switching back and forth at a comparatively low grade, ascending gradually all the while. A thunderstorm swept harmlessly by as they rode, shedding only a few heavy dollops of rain.

They reached the top of the pass, where the view of the mountains behind them presented a glorious sight. Numerous sharp peaks reposed below. Only Gan Libah, with its long, flat summit, towered above. Looking off to the south, they spied "a vast sandy plain stretching as far as eye could reach, covered with thorn trees & bushes[,] a few low hills rising here & there among the sea of green." A terrific thunderstorm was booming to their right, "covering the sky with its black pall, the strong wind that preceded it raising whirlwinds of sand that seemed to travel onward with great force & rapidity like miniature cyclones."[13] Rain was falling in immense sheets over a vast territory to the west, which portended potable water and fresh grass. They followed a path from the summit, which descended gently. It was broad and well worn, having been trod by a procession of camel caravans traveling between the interior and the coast for countless generations. Soon they reached a stingy well. The curator, however, had already decided to push on to Adadleh, where there was rumored to be better water.

They soon began to overtake the lumbering caravan and rode along its entire length, examining the camels to see how they were withstanding the difficult journey. Most seemed to be faring well, save one scrawny animal, which struggled to carry only a few blankets. It wasn't hard work that had used any of them up, Elliot knew, for they'd enjoyed several days of rest in nearly every camp they made.

A low hill two or three miles ahead indicated the well of Adadleh, and they weren't long in reaching it. Just as they arrived, a single shot rang out. Elliot thought Europeans must be present, for only the men attached to their caravans were permitted to carry guns. Instead, he found only a few men working for the English hunter Henry F. G. Barclay, who was sending several camels back to Berbera with the horns and skins of the animals he and his partner, Lionel C. G. Sartoris, had killed. Barclay and Sartoris had been hunting near Hargeisa, where they'd already taken eight lions. "They certainly had been very fortunate to find so many," Elliot noted with envy, no doubt believing that his salvage zoology mission was more deserving than their sport-hunting lark.[14]

Elliot's expedition pitched camp on "a clean sandy knoll in the midst of thorn trees scattered about which gave the place the aspect of an old orchard." From their tents they looked out over a vast extent of plain. The air there was "decidedly cool," and as night approached, they unpacked their neglected sweaters and jackets. Akeley, who was seldom idle, developed a dozen photographic plates in camp. These turned out beautifully, showing "the scenery, animals, & people of this part of Africa" to advantage. His animal death portraits would be of tremendous value to the taxidermist when mounting skins back at the museum, "as they show the outlines & anatomy very clearly."[15]

The curator's barometer measured 3,300 feet of elevation. Elliot thought it felt higher, judging from the coolness of the air. They crafted a small zareba to protect their flock of sheep from prowlers. Camp, on the other hand, was guarded only by the sentries, whose shrill calls were heard through the long night.

OMINOUS CREEPINGS

The following morning, a Sunday, Elliot remained in camp. Akeley went out early to hunt. The curator predicted that game would be scarce, as the animals had likely followed the rain out to the open plains ahead, seeking the new grass that must be springing up. Dodson, who felt unwell, slept late, then went on a short ramble to shoot birds near camp. The weather was like a day in September back home, "the air fresh, cool & bracing."[16]

Akeley returned to camp later, bringing a very fine female gerenuk. He'd seen several, all very wary, but no other mammals. Dodson found no birds, but he did bag a rabbitlike hyrax, *Procavia brucei somalica*, and a little, chinchilla-like animal, *Pectinator spekei*. These lived in narrow clefts among the rocks and were very timid. One was rarely seen without the other species being close at hand. Unfortunately, both specimens proved troublesome. A considerable portion of the left side of the skull of the former had been shot away, while the thin, delicate skin of the latter proved impossible to remove without tearing.[17]

Early that morning, Dualla sent four men to scout ahead for water. He hoped to locate some hospitable pools farther on, as this place—Adadleh— was the nearest known water source to Toyo Plain, which was still some thirty miles away, a two-day march. It was necessary for Elliot's party to visit Toyo, as that was the best place to find hartebeests. It would be a major undertaking to carry all the water the party required from Adadleh. In fact, it'd be necessary to keep the camels constantly on the move, shuttling casks back and forth. Elliot wanted to give them rest. He was hopeful that rain had pooled in many places and that his men would find several dependable prospects on the plain ahead. But they didn't. Given the parched conditions, there seemed no option but to leave the greater part of the caravan at this spot and take all the water they could carry—possibly as much as ten days' worth—for a smaller group. After much deliberation, Elliot decided to leave the next afternoon.

Meanwhile, Dualla reported that two tent boys had been fighting. Elliot considered this "a very serious breach of . . . discipline." He ordered the culprits to be brought before him, and "opened court" in front of his tent. "There were several witness to the combat," Elliot related, "all of whom were present, & formed a dusky ring in front of my chair, with an outer ring of interested spectators . . . looking over their shoulders." The tempest arose from a squabble over tea. It began with some rough play, then escalated when one tent boy slapped the other's face. They both then "clinched & rolled on the ground," until they were pulled apart with difficulty by Dodson and others. Elliot fined the tent boy who struck the first blow three rupees, while the other he fined only one. "It was difficult to get at what really brought forth the blow," Elliot wrote. He gave them both an unsparing lecture on the enormity of their offense and caused it to be known throughout camp that any man who struck another for the duration of the expedition would be fined ten rupees. This was the only kind of punishment they really felt, Elliot suspected, "for the great fault of a Somali is avarice, and to take his money . . . hurts him more than anything else." Fighting among the men was something that must be firmly suppressed. So many rival tribes were represented in Elliot's command that any fight could quickly escalate. This could lead to "a small war . . . , with . . . disagreeable results." Elliot was adamant that he wouldn't tolerate any fighting. "Judgement given, court adjourned & the men scattered . . . to talk the affair over, & for once were pretty quiet."[18]

The following morning dawned with a chill, "causing the men to hug their tobes closely."[19] They were soon busily engaged sorting the packages that were to go forward from those to be left behind. They cooked food for the entire traveling party for two days, so as to save the water they carried for drinking. They buried six barrels, containing their pickled skins, in the damp sand of the riverbed in order to keep them tightly sealed. Finally, the men built a sturdy zareba around the camp for better protection while the parties were split. Loading commenced shortly after lunch. By three o'clock, the caravan started out. Their trail began with a tedious trudge through deep sand, as they wound their way among thorn trees and scrub. Soon, they came across a wide and uninteresting plain spread sparsely with short, wiry grass. On the way, Dodson stalked and fired uselessly at a Speke's gazelle, the only game they saw during the long march. Large flocks of sheep and goats were here and there, feeding on the grass, guarded by watchful shepherds. At intervals they passed several

FIGURE 11.2.

A cool morning
in camp with
men and camels.
Photograph
courtesy of the
Field Museum,
CSZ6147.

FIGURE 11.2. A cool morning in camp with men and camels. Photograph courtesy of the Field Museum, CSZ6147.

small caravans of people moving to the plain, leading heavily loaded camels and donkeys. A thunderstorm rumbled to the west, its dark clouds presenting a stirring sight, charcoal gray with bright silver edges, and carrying at its crest a glittering rainbow, something Elliot had never before seen in Somaliland.

An hour before sunset, Elliot felt strangely tired. An ominous sensation creeping up and down his spine warned that something awful was in store. As they were then considerably ahead of the caravan, the curator ordered a sudden halt. The camels arrived later in scattered groups. Tents were quickly pitched, and Elliot went at once to bed. Soon, he was racked with a high fever. He took quinine and hot tea and swaddled himself in a cocoon of blankets, after which perspiration came and broke the fever. Although not of long duration, his illness was severe. He tried to take his own temperature but found that the mercury in his thermometer had risen to the very top of the tube, indicating the intense heat trapped inside his steamer trunk. He gave up the attempt, then slipped into a deep, convalescent sleep.

The next morning, Elliot felt "quite shaky after my rough treatment last night." In no shape for marching, the curator decided the caravan should remain in place for the day. He stayed prone on his cot until noon. Rising, he discovered that a shikari and a tent boy had also suffered from fever in the night and were "pretty miserable" that morning. Elliot rejected the possibility of malaria, "as it is very difficult to find water at all in this land."

He diagnosed a malady he called "sun fever."[20] With the cooler temperatures prevailing at higher elevation, he expected no recurrence.

Elliot remained at his tent, although he didn't get much rest. All day the camp was visited "by crowds of natives who squat on their haunches holding their two spears upright in front of them & their shields on their left arm. So they sit & talk energetically for hours seeming never to weary. I expect a Somali can talk more & say less than any other man," he wrote.[21]

A strong wind blew all day, sweeping drifts of sand over their campsite. Despite his illness, Elliot was eager to get away from this inhospitable spot. Meanwhile, Dualla had heard reliable rumors of a pool of rainwater on their line of march. With the curator's consent, he emptied the water barrels into every available pail, pot, and container, and sent them on camels back to Adadleh to be refilled. Elliot planned to move out again the next day.

Akeley and Dodson, meanwhile, set off at first light to hunt. They both returned at midday, the former bearing a fine oryx bull, *Oryx beisa*, and the latter bringing a female Speke's gazelle. They reported a dearth of game in this part of the country and an abundance of villages. The two taxidermists spent the afternoon processing the skins of the animals brought in that morning. They also saved the oryx skeleton. Akeley, Elliot noted, had killed his oryx at five hundred yards with his Mannlicher rifle. "Very long shots are going to be the rule I expect on these plains," Elliot wrote, "as there is no cover and stalking [is] in many cases impossible."[22]

Sometime in the night, a hyena attempted to steal into camp. The expedition's sheep, of course, would be easy pickings for a hungry hyena. Even horses, Elliot noted, had been "known to be killed by these sneaking brutes who tear half the belly away with one bite."[23] The guards cried out and drove the hyena off, but for a long while it remained within earshot of Elliot's tent, howling and making its awful cackle. It was too dark to see much of anything, so they could do the hyena no harm, despite the worst intentions.

ANYTHING BUT HAPPY

Early the next morning, Dualla visited Elliot's tent to see whether he was well enough to travel. Feeling somewhat stronger, the curator decided to march. "I thought we might as well try it & I would go as far as I could which would be a gain anyway," he reasoned. Elliot started out at four in the morning. The caravan followed shortly thereafter. They soon entered an area that resembled a well-manicured park, with trees and thorn bushes standing about in attractive clumps, except that the ground was covered with sand or red earth "baked hard as stone." How anything could grow there was a mystery. "Yet the foliage appeared fresh & bright as if watered every day," Elliot observed.[24]

After five hours of tiresome marching, Elliot decided to make camp, feeling he "had done quite enough for the day." He estimated that they'd covered "two-thirds the distance between Adadeleh & Toyo," and that they'd reach the latter the following morning. Storm clouds gathered in the afternoon. A vast wall of sand, "hurled to a great height by the wind," loomed on the horizon. By dark, rain had begun to fall in sheets, accompanied by "vivid red lightning" and deafening claps of thunder. The storm "was most welcome," as it promised a supply of potable water. The men, as was their custom, were in "a great state of excitement rushing about to get something to catch water."[25]

The rain soon dwindled to a steady, prolonged drizzle, which drummed a soothing lullaby on Elliot's tent. The curator had purchased two spendy milk goats the day before, one of which was accompanied by a nursing kid. With the onset of the storm, the shepherd brought all three animals into Elliot's tent and "proceeded to make himself comfortable" for the night. "This," Elliot explained, "was rather more than I cared for, so I arranged with D[odson] to take one goat & the kid & the boy . . . into his tent & I would keep the other goat, which stood contentedly chewing the cud all night at the foot of my bed."[26] Meanwhile, a tent boy spread a blanket on the floor beside his cot, and the second shikari curled up in a corner, and thus they passed the stormy night. Akeley, too, had eight men sheltering in his tent.

Rain fell through the night. By morning, the camel mats were soaked and heavy. They couldn't march until the mats dried, so it was after ten in the morning when loading finally commenced. Although it was only five or six hours to Toyo, Elliot—who was still feeling unwell—wasn't confident he could go the whole distance. Nevertheless, he started the march at the front of the caravan, feeling "weak & shaky." The road led through the same kind of parklike country they traveled the day before. This became "very monotonous," though, as they could see only a short distance, the trees blocking the vista and camouflaging any game.[27] Fortunately, the rain of the previous night had cooled the air, and a brisk breeze blew all day. They plodded steadily and silently forward, searching for game animals. They were out of meat.

After four hours in the saddle, they spotted game. Elliot, still feeling weak, sent Akeley and Dodson after it. "I would not have got off any horse for all the antelope in Africa," he proclaimed, "for my main object in life was to stick in the saddle until Toyo plain was reached & the camp site selected." The hunters returned empty handed, and the march resumed. An hour later, they finally broke through a line of trees, and the plain stretched before them to the horizon. "A mirage of a beautiful blue lake was playing over it, tantalizing us with most coveted thing in this land," Elliot wrote. He dismounted under the shade of a tree. The caravan came up shortly, and they made camp. Elliot was relieved to be out of the saddle: "Right glad I was to get there," he wrote.[28]

Soon, some villagers approached. Salutations were exchanged, hands shaken. A member of Elliot's party asked about water, and the curator was delighted to learn that in one place, apparently, there was an abundance. "Awfully most sea," his interpreter said. They inquired after hartebeests, which they were especially keen to collect, and were told that these animals were everywhere on the plain. "A bright young fellow wanted to go as guide so A[keley] & D[odson] started at once to see what they could do before night," Elliot wrote. Storm clouds gathered again in the afternoon, and Elliot expected rain imminently. Dodson eventually returned to camp alone. As the sun sank, there was no sign of Akeley. They felt certain he must've seen a hartebeest and given chase. When Akeley returned soon after dark, Elliot learned that his guess was correct. The taxidermist had seen five hartebeests and managed to wound three. The distance at which he was compelled to fire—from four to five hundred yards—precluded accurate shooting in the twilight. One hartebeest was "very badly hurt," and Akeley hoped to find and retrieve it in the morning. Though the promised storm bypassed camp, they did enjoy "a fine display of lightning towards the eastern horizon." A fierce wind howled over the plain, dropping a gritty, red sand. Everything in camp was "simply buried under it. It makes life here anything but agreeable."[29]

The next day, Thursday, June 11, Akeley set out very early, taking a guide. Finding and following the trail of the wounded hartebeest, they soon discovered it lying peacefully in the grass. As soon as the animal saw its tormentor, it snorted and staggered partly to its feet, but it couldn't stand up to flee. It was a female that evidently "had a bad night of it" fending off predators. In one spot nearby there had apparently been a struggle with a hyena. Akeley assumed that the would-be scavenger had gotten the worst of it, possibly even a nasty gouge from the hartebeest's horns, for the hyena "had walked straight away" and "taken no further interest in the proceedings." He sent immediately for his camera, hoping to take a photograph of the hartebeest, *Bubalis swaynei*, in a seated position. But it rolled over and died before the camera could be brought up. "It was a fine beast," Elliot crowed, "but light colored, much lighter than I expected." They spent the morning preserving the skin and skeleton.[30]

A guide had claimed that hartebeests were more plentiful at Gelalo, in the center of Toyo Plain, so Elliot decided to relocate. After lunch, the camels were brought in for loading. The curator was still feeling somewhat ill, and "rather dreaded the four hours ride, but it had to be done, & there was no use losing time as water was scarce." They rode, the vast plain stretching away in every direction, "flat, bare and featureless." Elliot called it "even worse than our prairies at home." A cluster of trees rose distantly from a watery mirage. This was their objective. "It was a dreary ride," Elliot wrote, "with no living thing in sight save ourselves & a few birds, and we plodded wearily along under the hot sun." Fortunately, a cool breeze was blowing, which made the ride more bearable. Another late afternoon thunderstorm was brewing in the distance. "Hour after hour passed & the trees seemed to be no nearer," he griped, "but every road has an end & at length we began to draw near ours."[31] The moment the tents went up, a steady drizzle descended.

Elliot was sitting miserably in his tent, listening to the rain thumping on the canvas and watching lightning flash through the open flap, when Dualla approached and announced that he had some very bad news. "Let me hear it at once," the curator resolutely spoke. Dualla then explained that a man had just arrived from Adadleh to report that their six barrels of preserved mammal skins, including every specimen collected since last leaving Berbera, had been carried away in a flash flood that swept suddenly through the dry riverbed where they'd been buried. Three of these barrels had been recovered, intact. One was found "in a smashed condition."[32] The remaining two were still at large.

Elliot was aghast. What could've happened? The barrels had been filled to the brim with specimens soaking in a heavy, briny solution. After the men had replaced the heads of the barrels by hand, they were no longer watertight, and the brine leaked. With the idea of getting the wood barrels to swell shut, Elliot had his men bury them in the wet sand of the riverbed. He never imagined that a flood could excavate and carry away such heavy objects. It was a calamity.

The first priority, Elliot decided, was to determine the extent of their losses, which could only be accomplished by a personal inspection. Someone needed to ride back to Adadleh to evaluate the damage and to salvage everything possible. The messenger had added that gangs of men had gone down the river in search of the two missing barrels. Elliot knew, of course, that if the barrels were broken, the brine would be washed out of the skins and all their specimens would soon spoil in the heat. After conferring and commiserating with the others, he decided to send Dodson back, starting him away as soon as possible. Dodson was to make the journey as quickly as he could, and Elliot would await his report at their present location.

The bad news disheartened them all. The material they'd collected—with much toil and sacrifice—was "magnificent," and "to lose any of it, especially when every precaution that we could think of had been taken to preserve it, was a greater misfortune than we deserved." The irony of losing these rare skins when they'd come to Africa to salvage them for science must have been a bitter pill for Elliot to swallow. Still feeling weak and sickly, the curator turned in that night "in anything but a happy frame of mind."[33]

12

HIS LIFE OR MINE

June 12–20

Dodson sped away at three o'clock, hoping to reach Adadleh late that same morning. Elliot remained in camp to rest "after the fatigue & trouble of the day before." Akeley went hunting and met with a small herd of wary hartebeests that kept their distance. After firing two misses at the herd, he spied a solitary bull. The bull, which stood calmly facing him, allowed Akeley to come within three hundred yards. Taking careful aim with Elliot's Lee-Metford, he placed a bullet in the very center of its chest. The bull ran fifty frantic feet, then fell dead. It proved to be "in every way the finest specimen we have yet obtained, very dark in color, & marked with very black patches on shoulders & hind quarters, & a fine knotted pair of horns."[1] A camel brought the specimen back to camp, where the skin and skeleton were painstakingly preserved in the shade of a tree. While Akeley was thus engaged, and Elliot rested in his chair, camp was visited by Sultan Nur of the Habr Yunis and his royal retinue.

The sultan's visit reminded Elliot of the necessity of having gifts on hand to present to distinguished guests. He dispatched a man to Adadleh to tell Dodson to bring back cloth, and also to send five camels to Berbera for salt, extra barrels, and sacks, for treating and storing specimens in the field. These critical articles were getting scarce. Elliot stayed up late writing instructions for Berbera and letters home. He also wrote to Lieutenant Colonel William B. Ferris, a British official at Aden, asking that export duties on the material he intended to ship from Berbera be waived for the benefit of science. He expected a favorable reply.

Some days earlier, a young boy had appeared in camp. Only two feet tall and stark naked, he carried a tiny tobe in his arms. His head was shaved, "only a small circle of wool being left to go around the skull and a tuft on top." His tearful eyes were running with matter and swollen almost shut, so that he could hardly see. Elliot had examined the boy, who "objected strongly." He gave the boy's father crystals of boric acid with directions on how to use them. He was glad to see the boy back in camp again that morning, his eyes healed and "shining like black beads."[2]

With Elliot recuperating in camp, Akeley spent several days hunting hartebeests with his Somali crew. One day he killed two females, "an old & young cow." The former had a young calf that Akeley wounded, but it managed to escape with the rest of the herd. It was late, and the taxidermist was too tired from the hunt to skin his captures, so they were left lying on camel mats until morning. Another day he shot a bull, a fine, richly colored animal, "but with rather smooth horns not deeply annulated as are those of most bulls." He preserved the skin and bones, so they now had three valuable skeletons of this rare antelope, two males and a female. An old Midgan sat curiously by and watched as Akeley skinned the bull hartebeest. He then turned to Elliot and asked, "What would he live on if the expedition killed all the game in his country?" Elliot, who took it for granted that Africa's game animals were hurtling headlong toward extinction, deftly dodged responsibility for contributing to the problem. He pointed to the man's scanty, white beard and said that "there would be game on these plains long after he had gone to Allah, at which he smiled & grunted an assent."[3] The same Midgan promised to show Elliot a place where he would find lions.

Clouds gathered one afternoon in the west, while Akeley was out hunting. A rising wind, lightning flashes, and the crackle and boom of thunder told of another heavy thunderstorm approaching. "The heavens were black as ink, and clouds of red sand whirling in every direction," Elliot wrote. He was anxious about Akeley, as it was so easy to lose one's way in the darkness that was quickly falling, and a night spent exposed on the plain meant a thorough wetting and possible fever. Elliot ordered a lantern to be hung in a tree, but the wind immediately snuffed out the flame. He had the flimsy lantern placed on a box under the shelter of a tent, but he knew there was little chance Akeley would see the dim flicker of its candle. The storm suddenly "burst over us & the rain poured in torrents, & everything outside the tent was wrapped in a black pall, the only time anything was visible was during the flashes of lightning." After waiting a long time for Akeley's return, Elliot concluded that he must've taken refuge in a village or was content to remain on the plain all night. He ordered signal shots fired, the sound of which "was swallowed up in the howling . . . wind."[4] Then he settled down to a solitary dinner. Much to his relief, Akeley returned soon after the curator finished eating. He'd missed camp altogether and walked right past it—scanning eagerly in all directions—when an opportune flash illuminated it. He was soaked, of course, but he soon settled down between his blankets, with no ill effect.

The following morning, Elliot could see hartebeests grazing through the open flap of his tent, and he decided to join Akeley for the day's hunt.

This was the first time he'd picked up a rifle in a week, and his hands were unsteady. Two bulls were feeding some distance apart, and Elliot aimed at one while Akeley went after the other. Elliot closed within four hundred yards, when his bull ran a short distance, then stopped to look back. He fired and missed. Immediately, the bull bolted toward its companion. Akeley sprinted after them. Elliot's bull stopped to look again, giving Akeley a clear shot from three hundred yards. Akeley fired, and the bullet struck the hartebeest in the face. It dropped "like a clod." The other bull quickly ran off. Sending for a camel, they sat down by the carcass to wait. This was a young animal, judging by its teeth. Elliot estimated that it was barely two years old. Its horns, though "soft and rather loose on the core," were well developed.[5]

A shikari found and collected a hartebeest skull. This animal must have been killed by an African or by some wild beast, Elliot assumed, "for a European would certainly have taken the head of so scarce a species as a trophy." Another shikari, meanwhile, captured a young hare, *Lepus sp.* "It was a mite of a thing, and lay quietly in my hand too frightened probably to move," Elliot wrote, before giving it to Akeley, "who added its skin to our growing collection."[6]

While Elliot was resting in camp, two "very curious" wattled starlings, *Dilophus carunculatus*, alighted near his chair, the first such birds he'd seen in Africa. "Fortunately," he crowed, "both were secured." The back of this bird's head and its throat were blue-black; its front was yellow. The plumage was a slaty gray, with white rump, black wings and tail, and a splash of white on the wings. The bill was white, the feet flesh colored, the eyes black. Elliot observed the two birds feeding noiselessly on the ground before blasting them with shot. "The wattles were very conspicuous at all times," Elliot wrote, "the two pendant from the throat flapping about as the birds moved their heads, while the two on top of the crown stood upright."[7]

Grass grew plentifully around camp, and the livestock was left free to graze. The mules and horses, after getting their fill of grass and water, wandered off one night. Elliot's own mule was especially restless, having done no work since arriving at their present camp. Some men left at daybreak to recapture the missing animals. Elliot slept late that morning. Emerging, at last, from the shelter of his tent, he was surprised and delighted to find three large pools of rainwater in the midst of camp. This unexpected bonanza would be sufficient for several days. The wind was blowing furiously from the west, and everything in camp had to be lashed down tight. Late in the afternoon the truant animals were brought back to camp. Henceforth, they would "have to suffer the indignity of being hobbled at night."[8]

GREETED WITH GOOD NEWS

Dodson returned at last from Adadleh with good news. Following a heroic salvage effort by all hands, the expedition's only losses were four gerenuks and four lesser kudus, two species of which they had an abundance of specimens. Elliot was much relieved. "All the really fine material which I had gathered for groups for the museum was all right, that which was lost being specimens for exchange," he explained. This was still a bad blow, of course, and represented "quite a little sum of money." Yet he was relieved that the losses hadn't been much worse. Dodson made clear and Elliot recorded that Selon, the hardworking and resourceful second headman, deserved most of the credit, "for his energetic action in saving the majority of the material."[9]

Dodson brought with him two Soemmerring's gazelles, one of which had already been skinned and preserved. He also brought an Englishman named Mr. Derby to camp. Derby, with a partner, Norman B. Smith, had been in Somaliland for several months on a sporting tour, killing three lions, four rhinos, and three elephants. Elliot hoped to match or exceed their results with the largest and most dangerous game. Their camp was three hours away. Now headed home to Britain by way of Berbera, they were eager to sell camels and surplus equipment. Elliot agreed to send Dualla to shop.

All three hunters ventured out for hartebeests the next morning. Elliot took a shot from a great distance and only managed to graze a cow's back, which sent a herd of eight animals sprinting toward Akeley, who fired a miss. Akeley's bullet ricocheted off something, then struck the ground harmlessly near Elliot. The curator's shikari picked up the spent bullet. "Thinking it was too warm a corner," Elliot left and went after a solitary bull. Riding away, he noticed Dodson following a cow accompanied by a calf, which at that distance looked minuscule, but which ran much faster, easily keeping up with its mother and often plunging ahead. Elliot stalked the bull for a long time. When the curator got too close, it would run a short distance and then slow to a walk. Occasionally, it stopped to look back. Elliot kept up a rapid pursuit. At length, the bull, "as black as a coal & very large," took a final look. Taking his rifle from the shikari, Elliot fired. The bullet struck it in the spine, and the bull dropped. It rose, then fell again immediately. When Elliot caught up, there was just time to cut the windpipe "before he breathed his last." This was a "splendid" specimen, "the largest and deepest in color of all we had killed."[10] The span of its horns measured nineteen inches from tip to tip.

When Elliot returned to camp, Akeley was already there, feeling tired and peaked. Dodson came in soon thereafter, bringing the speedy calf the curator had seen him chasing. "It was a curious looking little creature & was quite a contrast to the big bull as they lay side by side," Elliot wrote.[11] They estimated the calf's age at about ten days. They now lacked only a half-grown hartebeest to complete a habitat group of these animals.

Dualla had gone to Derby and Smith's camp to peruse their belongings.

While Akeley and Dodson were skinning the bull hartebeest, he returned, with Smith in tow. The Englishman, who joined Elliot's party for lunch, had decided to negotiate directly with Elliot, rather than leaving such gentlemanly business in the hands of a headman. This, Elliot wrote, "was a sensible thing to do." Smith had twelve camels, two donkeys, and eight water barrels to sell, all of which Elliot wanted. After some pleasant small talk, Elliot bought the camels—subject to Dualla's approval—at fifty-five rupees each ($26), which was "twenty rupees cheaper than the cheapest of those I had bought in Berbera & much cheaper than I can buy from the natives even here." He also paid thirty rupees ($14) for the donkeys and three rupees each for the water barrels. "I think I have made a very good bargain," Elliot boasted, "& am that much stronger in my transportation department." Dualla later picked out the twelve camels from Smith's caravan—every animal worth having, in the headman's judgment. Unfortunately, there were no good camel mats and no ropes to be had. He brought eleven animals back with him, leaving one to help Smith and Derby get to Adadleh, as the camels they had left "were hardly capable of carrying anything."[12]

A visit in camp from another European was a novelty, and always welcome. It was a chance to swap stories, trade goods, and gather useful intelligence about hunting conditions. The campfire camaraderie was also an opportunity to make new friends. Smith, however, got under Akeley's skin. He boasted of having shot seventy-five Soemmerring's gazelles before bagging one with a record-large rack.[13] To what end, Akeley wondered silently? If, as Elliot frequently claimed, Africa's wildlife was truly facing extinction, how could this wasteful, destructive mode of hunting be acceptable? The encounter with Smith undoubtedly gave Akeley much food for thought, though his own hunting, and that of his scientific companions, seldom seemed to trouble his conscience.

Just as they were finishing lunch, a body of horsemen appeared riding over the plain. This proved to be another band of Habr Yunis warriors, coming from a distant village. These riders "made a very brave appearance on their . . . ponies gaily decked in . . . bright colors." They gave a performance similar to that given by Sultan Nur's group. "A fine body of irregular cavalry these fellows would make," Elliot thought, "as they are bold riders & valiant fighters & would be most useful as scouts."[14] Akeley took photographs as they were performing on their horses. One rider had a young, captive hartebeest, which Elliot bought for five rupees.

Elliot and Dodson went hunting the next morning, while Akeley stayed in camp to process the young hartebeest. The curator rode a great circuit. He captured a large tortoise, *Testudo pardalis*, tying it with rope such that it was imprisoned in its own shell. He sent it back to camp on the shoulders of an attendant. Shortly thereafter, he saw a lone hartebeest and was just moving in its direction when he noticed that Dodson was stalking the same animal. So, he gave it up. After riding for another half hour, he saw a herd of some thirty hartebeests, including individuals of all ages. He especially wanted a half-grown juvenile. He began to stalk, but the herd wouldn't allow him to close the distance before the animals began to withdraw in a tight, protective body. A few brave individuals stopped to look back. Among these was one that appeared to be the ideal size, so he fired. It ran, fell, then got up and walked a short distance before falling again. When Elliot arrived, he witnessed a few convulsions before the animal died in a paroxysm of blood. He'd shot it through the neck, severing an artery. He was amazed that the mortally wounded beast had been able to run at all.

They all went out again the next morning, each hunter going his own way. Elliot pursued a solitary hartebeest. As he was drawing near, he saw a wounded bull come over a rise to his front. The bull limped, making slow and painful progress, and Elliot considered going after it, when he spotted Akeley come over the same rise, following the wounded bull. Elliot stopped to watch and admire. Akeley fired at the bull, apparently missing, just as four other hartebeests came over the same rise of ground. These were an easy rifle shot from Akeley, who took advantage of the opportunity by dropping a young cow in its tracks. An older cow, meanwhile, had run toward the bull, and Akeley—not pausing to relish his last successful shot—kept at it. In a few more moments he knelt and took a long shot at the cow and hit it squarely in the center of the forehead. The hartebeest dropped like a log. There were now two animals down, and a third was obviously wounded. Knowing now that Akeley was likely to get the wounded bull, also, Elliot went after his own animal, which had wandered a long way off in the meantime. But, after following it for an hour, he gave up the chase. The curator returned to camp with a full magazine.

Dodson, meanwhile, had wounded a young female hartebeest with a long shot. Because his horse was "exceptionally fast," and the hartebeest appeared to have a broken leg, he determined to run it down. Slinging his rifle over his shoulder, he started off at a quick canter, circling far around the wounded animal in order to chase it in the direction of camp. The hartebeest sprinted away, gaining ground quickly. After about a mile, the wounded animal appeared to flag somewhat. "Striking spurs home," Dodson gradually closed the distance on his horse, "both the Pony and quarry straining every nerve." The ground was pockmarked with burrows, so Dodson had to steer carefully. At last, he caught up, and "now unslinging my rifle I . . . hit it a little back [and] stopped its speed." He fired twice more, missing. But his next bullet brought the hartebeest down "head over heels." Dodson's black horse was white with sweat, and the skin had been rubbed off the insides of the Englishman's knobby knees. Yet it was a "glorious run of about 3 and a half miles and worth it all."[15] The rest of the day was spent preparing the skins. Since they now had fifteen hartebeest specimens in pickle, Elliot decided to move camp the following day to another locality farther south, one that was said to be teeming with Clark's gazelles.

The hartebeest frequented the grassy plain and was never found among the bushes. There was typically little cover. This made stalking "practically impossible as the animal sees you even more quickly than you him." Whenever one or more were seen, it was best to walk directly toward them if they were grazing or walking away. If they stopped to look, however, it was best to walk away from them in a slanting direction so as not to lose too much ground. Often, when the hunter was almost near enough to take a shot, a hartebeest would begin to run. Normally, this was kept up for only a short distance, but the work of stalking would have to begin again. By "the exercise of much patience, & the possession of considerable staying powers one at length gets near enough to shoot, anywhere from 250 to 500 yards." A lightly wounded animal, even one with a broken leg, would often escape. According to Elliot, "it is practically useless to follow such a one, as it is rare it permits a near approach or any favorable opportunity for a shot." It was ungraceful looking, its rump slung lower than its withers, so that it appeared to be standing foremost on a footstool. It ran with a kind of lumbering canter, yet it had the most speed and the best endurance of any antelope, "always going just fast enough to keep a certain distance from its pursuer and graduating its pace by his." Solitary bulls were the easiest to approach. The larger the herd, however, the more difficult it was to get near them, "as the courage of the entire number is only equal to that of the most timid, and as soon as one begins to run, all are off at once." It shouldn't be considered a game animal, Elliot concluded, as there was "little or no sport in shooting them." Having killed one, "if it had not been for scientific purposes, I should not have tried to shoot another."[16]

Hard rain commenced overnight, promising a stormy day. But it stopped at dawn, and the day was instead pleasantly overcast and cool. Elliot decided to send all the skins and bones taken there—more than four barrels of the former, and a great quantity of the latter—back to Adadleh for safekeeping. The entire morning was devoted to packing bones in various sacks, and all the remaining skins—including four killed the previous day—in a barrel. The barrel leaked somewhat, and they had neither salt nor sufficient water to refill it. Around one in the afternoon, the camels purchased from Smith were brought in, and the specimens accumulated since they left Adadleh were loaded on their tired backs. These left to join the main caravan. One of the new camels had an abscess on its back as big as a fist, despite Smith's assurance that there wasn't a sore back in the entire lot. Elliot sent word to Selon to lance it. Elliot wanted this animal to be caravan ready by the time he moved on from Adadleh.

As soon as the new camels were off, Elliot gave orders for the rest of their reduced caravan to prepare to march. They moved out in midafternoon, crossing the plain in a southeasterly direction. The wind blew fiercely, as it had done nearly every day, and the blowing sand was a nuisance. They saw great numbers of hartebeests, but they let them be. It seemed strange to Elliot to ignore the hartebeests—so rare in museums—when only a few days before they were anxious to get them. "But it makes all the difference," he reflected, "whether the barrels are empty, or whether as in our case, four of them contain fifteen skins."[17]

They kept moving until the limit of the plain was reached, at which point they entered the bush country, much like that on the other side of Toyo Plain near Adadleh. They had scant information as to the whereabouts of Clark's gazelle, or the dibatag, as it was called by the Somalis. Since Elliot and his companions were, by now, a long way ahead of the caravan, they stopped to wait. Once the caravan caught up, they altered their line of march more to the south. They kept moving until nearly sundown, when they made camp. After talking things over with Dualla, Elliot decided to make another short march the next morning to get farther into the bush, so as to be nearer to the dibatag's natural habitat.

A RED-LETTER DAY

They made another early start. Dodson went off by himself, heading southeast. Elliot and Akeley, on the other hand, left together and stayed close to the line the caravan was to follow. They rode for a long while without seeing anything and then decided to separate, each tracing the arc of a great circle, but in opposite directions, ultimately meeting back at the place where they expected camp to be made early that afternoon.

Elliot soon spotted a dibatag sprinting through the bushes ahead of his path, its dark, brushy tail carried stiffly upright and pointed somewhat forward, its head bent slightly backward over the withers. He tried to stalk the animal, but it was very wild. He took an improbable shot, missing badly. Going on, he soon saw a Soemmerring's gazelle feeding, but it bolted just as he was about to fire. Elliot nimbly skirted around a shrub and caught sight of his quarry. He fired, hitting the animal and bringing it to a stop under a nearby tree. The curator approached cautiously, though he thought the animal incapable of putting up a fight. His shikari urged him to shoot again, which he did, knocking the small buck to the ground for the last time. The animal "had two holes in him," Elliot regretted, "either of which would have caused his death." They skinned and crudely butchered it, taking the skin and some meat and placing them behind the saddle on Elliot's mule. Then they went on. Elliot saw two more dibatags, but despite "all our caution & trying our best to get within shot, we could not."[18] The animals escaped without even having heard the whistle of a bullet. If all dibatags were as wary as these, Elliot ruminated, then he and his hunters would have a difficult time completing a group.

Tired from the hunt, Elliot and his attendants now sought their campsite, whose location they had only the vaguest idea of. They kept on, however, trying to follow their imaginary arc, "which had been pretty badly twisted with the many turns we had made following the Dibatag & Aoul." As they seemed to be making no progress toward camp—wherever that was—Elliot was about to propose that they head back to the previous night's camp and then follow the camel tracks, when they stumbled on their own tracks from earlier in the day. These they traced back until they came across the heavier tracks of the caravan, which led them to camp in short order. They were camped considerably higher than Toyo Plain, "which lies like a yellow sea below us stretching away to the distant hills we passed on our way from Adadleh."[19] Yet they'd descended considerably from the highest limit of Jerato Pass. Much to the curator's surprise, the new campsite was surrounded by a zareba. He soon learned the happy reason: Dodson had killed the expedition's first lion, *Felis leo*, and Akeley—who'd arrived in camp empty handed—had gone out to help his comrade skin it.

Earlier that morning, the men of a passing caravan had shown Dodson fresh lion tracks. Excited by the prospect of hunting truly dangerous game, Dodson took the precaution of checking that his guns—a .500 Express rifle, a .307 Lee-Metford rifle, and a shotgun—were all in working order. He followed the tracks, which soon led to a dense patch of tall grass laced with thorn bushes. There they found the abandoned and half-devoured remains of some small animal. They raced around to the far side, where "a rustle apprised us of something coming and a loud growl told us what. Instantly my boys in fine order dropped on one knee each side of me," one with the loaded .500 rifle and the other with the buckshot-charged shotgun. Dodson had his Lee-Metford rifle in his anxious hands. The "huge head" and torso of a lion now appeared from the brush twenty-five yards away. Its "lip curled" as it "emitted a most savage growl." The lion appeared ready to spring when "it came like a flash" through Dodson's racing mind: it was "his life or mine."[20] Taking quick aim, he fired once, and down the lion went, flailing and raising an angry cloud of dust. Thinking it might only be stunned, he rushed in and fired four shots into the lion's writhing body, ending its struggle. Dodson soon determined that his first shot had struck the lion in the eye, and would've been fatal, had he only been a bit more patient. He was photographed standing in triumph over his prey.

Akeley and Dodson returned with the lion's skin and skeleton by lunchtime. It proved to be a well-grown male, probably not quite three years old, but having a fine skin topped with a modest coiffure of yellow mane. Measuring eight feet ten inches from the tip of its nose to the tuft of its tail, it was "quite big enough if it comes to a tussle, fists against claws,"

FIGURE 12.1.

Dodson, his
shikari, and a gun
bearer pose with
the expedition's
only adult
lion specimen.
Photograph
courtesy of the
Field Museum,
CSZ5964.

Elliot wrote. The curator thought it looked overweight and concluded that it'd been living large on sheep. Some local villagers reported that a lion would sometimes leap over their zareba, seize a sheep in its jaws, and then quickly jump back out with its bleating prey. The lion had become such a scourge that the villagers were clearing out. When word spread of Dodson's deed, however, the village elders came and offered him their "Chieftanship," along with ponies, camels, sheep, and wives. Dodson thought this was all "humbug," but Dualla assured him that the gesture was sincere. To Dodson, the "whole business" was "very amusing," but he declined the honors "as gracefully as possible."[21]

The lion and gazelle skins were soon salted and stowed in barrels. And although Elliot bagged none of the animals he was especially keen to get, he regarded the day as one not "entirely lost." In the evening, as he was sitting outside his tent to enjoy the cool air, Dualla approached, bringing all the men, most of whom squatted expectantly behind the headman. Dualla, as their leader and spokesman, explained that it was customary when a lion was killed to give the men a gift. "In a savage country," Elliot considered, one "generally finds it best to conform to the customs prevailing." He ordered five sheep slaughtered for the men and promised the shikari who led Dodson to the lion some bonus money. At this, the men "salaamed" and dispersed. "They have had more meat than they could possibly eat since we left Adadleh," Elliot griped, "but they cannot get over their hankering for the miserable tough mutton this land produces."[22]

13

A DISGRACE TO SCIENCE

Meanwhile, trouble was mounting in the Department of Zoology during Elliot's long absence. In early June, acting curator Oliver Perry Hay sent Skiff the draft of a paper he'd written for publication in the museum's journal. Skiff, who took issue with the polemical tone of Hay's manuscript, wrote a diplomatically worded letter expressing his concerns. "I have hastily glanced at your copy," he wrote, "and beg to ask your opinion as to the appropriateness and wisdom of conducting a controversy through our publication series." The director's opinion was that "controversial contributions" should be submitted to other journals, reserving the museum's publication for research based on museum collections or expeditions. "Will you be kind enough to give your views about this[?]," he politely asked.[1]

Hay was irked. The privilege to publish research free from administrative interference was something he took for granted. Together with many of his curatorial colleagues, Hay rejected the very idea that Skiff was qualified to pass judgment on his science. And he certainly didn't want to be dragged into a debate with Skiff over the propriety of publishing controversial ideas. Impetuously, he replied: "I would have preferred not to discuss either the wisdom of conducting a controversy through our publications or the controversial character of my paper." Hay then asked to withdraw the paper. He added: "I do not believe that any publication . . . can be long conducted without sometimes accepting controversial papers. A scientific paper can hardly be written without opposing somebody's views. . . . Few men can write papers without making mistakes, and these mistakes must be corrected."[2]

Hay seemed to think that Skiff would profit from his guidance regarding the management of the museum's new journal. The director felt otherwise, and he fired Hay later in June. To Harlow N. Higinbotham, chairman of the executive committee and Skiff's most steadfast ally in the museum's administration, the director explained: "I have the honor to inform you that O. P. Hay was dismissed from this Institution because his services were unsatisfactory and could be dispensed with." Despite all the renovations that were underway in zoology, Skiff insisted that it wouldn't be necessary to hire a replacement, "as there is nothing to do in the Department except to care for the material already installed." Nevertheless, Moritz Fischer, who'd been added to the staff as a preparator in March and whose work had been so unsatisfactory to Hay, was elevated to acting curator.[3]

Now, Hay was furious. "I have been forced out of the Museum by that man Skiff," he fumed to his colleague George Baur at the University of Chicago. "I have endured more insults from him and his tools," he continued, "than I ever received from any other set of men. And I might say that every decent man in the Museum suffers in the same way." Hay vowed to try to drive the director out. "It is a disgrace to science," he concluded. "Skiff is wholly unfitted in every way for such a position. . . . I may, if the Trustees do not do me justice, write up the whole affair in the scientific papers."[4]

There was no love lost between Baur and Skiff either. In 1891, when he was still a young faculty member at Clark University, in Worcester, Massachusetts, Baur had dreamed up the idea of a salvage zoology expedition to the Galápagos Islands. He believed that the archipelago constituted a vast natural laboratory offering an unparalleled opportunity to study the mechanisms of evolution. A representative collection of plants and animals from each of the islands would be necessary to conduct such an investigation. Unfortunately, the islands' native biota had already been depleted by pirates, whalers, and settlers from Ecuador, who were introducing invasive animals and plants and driving many endemic species to extinction. And the problem was only getting worse. Thus, to be really useful, a collection would have to be made right away. "Such work ought to be done," Baur advocated, "*before it is too late*. I repeat, before it is too late!" Delay too long and "the natural history of the Galapagos will be lost . . . forever." Baur visited the islands later that same year with Charles F. Adams, a young but well-traveled Illinois taxidermist. Together they amassed a large and diverse collection of vertebrate animals.[5]

By 1894, Baur was looking to sell a portion of that collection for $3,500. In March, he offered it to the Field Columbian Museum as a loan—with an option to purchase—to be exhibited in toto for one year in one of the museum's smaller galleries, the cost of transportation and installation to be met by the museum. Skiff gladly agreed to these terms. The collection, received at the museum on April 1 by Frank C. Baker, was placed on exhibit in Hall 19. It consisted of a giant tortoise, two iguanas, a bat, a rodent, and a preponderance of birds, including a rare lava gull and an equatorial penguin, among other specimens. Baur provided copy for the *Guide to the Field Columbian Museum* that argued his controversial theory that the islands had once been a single, large landmass connected to

Central America, and that the present islands had formed by subsidence. This theory was at odds with virtually all previous scholarship—since Darwin—which held that the islands were volcanic in origin.[6] And Skiff, so it would seem, was strongly averse to scientific controversy of any kind.

In June 1894, before Elliot or Hay had even been hired, Skiff suddenly reneged on his arrangement with Baur. Apparently, with the onset of warmer summer temperatures, Hall 19 began to stink. Skiff, assuming that the Galápagos collection was putrefying, ordered Baker to demand that Baur remove it. "I have to state," Baur wrote testily to Skiff after retrieving his specimens, "that the smell is not caused by the birds, which are in best condition and absolutely dry, but by the room itself." Baur suspected an ulterior motive, but the director denied it. Skiff blamed Baker—who'd resigned his position at the museum only the day before—for the misunderstanding. "I stand ready to make any reparation in my power," Skiff offered to Baur by way of apology.[7] The specimens, however, never returned to the museum.

Hay recognized Baur as a potential ally in his squabble with Skiff. He let his colleague know that the foul odor lingered in Hall 19 long after the specimens had been removed. Hay encouraged him to explain his experience in writing and share it with board of trustees president Edward E. Ayer, who was beginning to lose confidence in Skiff's administrative abilities. "If you should see fit to add your testimony regarding the treatment of your collection it would have weight," he suggested.[8]

In any event, Hay was culled from the museum staff before the end of June 1896. Elliot had lost a capable assistant curator, someone he trusted and respected. Months later, when the bad news finally caught up to him in Africa, an aggrieved letter relating Hay's version of events reached the curator first.

14

SCRIMMAGE WITH A LEOPARD

June 21–July 1

The first full day of summer fell on a Sunday, the expedition's traditional day of rest. Elliot stayed in camp to audit his accounts. These needed careful scrutiny, as the curator was "constantly called upon for all kinds of sums . . . & if I did not keep a rigid oversight upon my cash book I would soon get into a fearful snarl."[1] Updating his ledger reminded him that he was quickly running out of money.

Akeley and Dodson both ventured out. Elliot preferred no hunting on Sundays unless the expedition needed meat. But his hunting companions found an entire day spent idly in camp intolerable. Dodson returned around noon with two gerenuks, an adult male, which he'd mistaken for a dibatag, and a juvenile, maybe ten days old. Because he shattered one of the juvenile's legs, it'd have to be mounted lying down. Akeley came back with nothing.

Later that evening, a villager brought a live gerenuk only a few weeks old, which Elliot bought for two rupees. "It is a most comical looking little thing," he wrote, "with its curiously shaped head mounted out the end of its long slender neck, and the legs disproportionately long for its slim body." It was in "an exhausted state," although it soon recovered. Elliot liked to imagine that the juveniles he acquired went into his collection willingly. "All the young animals brought me . . . are wonderfully tame," he wrote, "& exhibit not the slightest fear of us, and submit to be handled without struggling or making any effort to escape." Elliot gave the animal over to Akeley, "who will probably soon have its skin in the pickle barrel."[2]

On Monday, a Berbera courier arrived with a welcome batch of letters from home. Among other things, Elliot received an unexpected note from William W. Masterson, the American consul at Aden. The consul informed Elliot that he was holding a bank draft for £600 ($3,000), awaiting the curator's instructions. Elliot was elated! He wrote immediately to have the draft sent to the government treasury at Berbera, subject to his orders. With no letter from Skiff explaining this windfall, Elliot could only guess as to the cause. "It shows that you have not been idle but have looked out for the wanderer in African jungles," he wrote to the director in gratitude. This new surplus changed the entire tenor of the expedition. "I am at ease in my mind now as regards finances, & I think I see my way clear to go pretty much where I wish," he wrote. "I believe I will send to the Field Museum the finest mammalogical collection . . . ever received at one time by any institution. Our success so far has been phenomenal & astonishes us, & we can only hope it may continue." They'd been "very successful" in obtaining a complete series of hartebeests, he reported. "A selection from the lot will make a magnificent group." He also wrote with pride about Dodson's lion. Elliot suspected that his was now the most expensive private museum expedition ever undertaken, and he was humbled and grateful. "Please thank the Ex[ecutive] Com[mittee] for their last remittance," he wrote. "I shall do my best to see that they have made a profitable investment."[3]

Giraffes had long been one of Elliot's chief desiderata, yet getting them was proving to be a tall order. He'd about given up hope of obtaining any, he confessed to Skiff. But the new remittance now made it possible to stretch his expedition beyond the Shebelle River, where the largest African mammals could be found. Some zoologists suspected that the Somali giraffe was a different species than the one known from South Africa. Elliot wanted to shoot one, describe the new species himself, and retain the type specimen for the Field Columbian Museum. What's more, the director of the British Museum, which housed one of the finest zoological collections in the world, was willing to give Elliot nearly anything from his duplicate specimens in exchange for a giraffe skin from the Horn of Africa. "I will see what can be done for the Field Museum first & then think about exchanges," the curator wrote reassuringly.[4]

Other letters received were less momentous, but no less welcome. One from Elliot's wife related that she was now residing contentedly in Montreal. Another informed him that twenty specimen barrels he'd ordered had arrived. The men Elliot sent to Berbera from Adadleh brought eleven back with them. The curator now had thirty-one barrels in his baggage, which would "go a long way in holding the skins I shall reserve especially for the [Field] Museum."[5] Elliot spent a contented afternoon writing replics and business letters to Berbera and Aden.

They moved camp the next day, marching about an hour and a half to the east. All three hunters started out together ahead of the caravan, taking the same path the camels would follow. Only thirty minutes from camp, they spotted a dibatag. Elliot dismounted immediately to stalk it. He crept within three hundred yards and fired, wounding it. The shikari thought he'd missed his shot, but the others "heard plainly the smack of the bullet." Having been obliged to run "quite a distance in a stooping posture & then to shoot off hand, as the bushes were too thick, & high for me to see the animal," he aimed too low.[6] The wounded dibatag started on a full run and was joined in flight by two others. Eager to land a dibatag, at last, they sprinted after the fleeing animals. To get a better vantage, one shikari scaled an enormous termite mound. Suddenly, three dibatags bolted from behind some bushes to the front. Racing to that spot as quickly as possible, Elliot found a small, dark puddle of blood.

Dodson and Elliot took up the trail of the wounded animal with their shikaris, while Akeley went after the others. "Now began a splendid exhibition of tracking," Elliot wrote. "The men followed the wounded animal with the certainty & sagacity of blood-hounds, no matter how it twisted & turned & crossed the tracks of other antelopes & gazelles, picking out the right foot print from among all the others." The men soon found a trail of blood. The wound, apparently, was bleeding freely, and the animal's demise seemed all but certain. Dodson and his attendants now split from the group. Elliot's party followed the dibatag's whimsical track carefully for another hour, when suddenly the animal leaped up from behind some high grass and bounded away as if unhurt. They followed and soon found it hiding in some grass, only its head and neck visible. Elliot took careful aim and shot it again fatally through the neck. Examining the carcass, *Ammodorcas clarkei*, Elliot determined that the first bullet had broken its foreleg at the knee. It was a young doe, gracefully formed and very different from the gerenuk to which it was closely related. The curator was delighted to secure a dibatag after many fruitless days. They put the specimen on his mule and rode off, feeling jubilant.[7]

When Elliot returned to camp, the men saw what he had and there was "much clapping of hands & expressions of pleasure," the curator proudly noted. "One would have thought I had killed a lion at least."[8] Akeley and Dodson arrived earlier, having failed to collect anything. As they were on the extreme western limit of the dibatag's normal range, and there seemed to be so few in the vicinity, Elliot decided to head farther east in hope of finding a place where the animals were more plentiful.

They made a late start the next morning. Elliot rode ahead with Akeley and Dodson and all their shikaris, hoping they might find dibatags on the march. They went through another parklike country with acacias and other thorn trees growing in tidy clumps, with a pale green, wiry grass sprouting out of the sand. Occasionally, they would find a small flower resembling the morning glory, which added a splash of bright color to the drab landscape. They saw only one dibatag and, as usual, it was off running immediately, making an easy escape. During the march, Dodson shot a nightjar, *Caprimulgus donaldsoni*, recently described by Elliot's friend

Richard B. Sharpe from a specimen in Arthur Donaldson Smith's collection.[9] They marched until shortly after ten in the morning, then halted. When the camels arrived, camp was quickly established.

The expedition nearly lost a camel. On the march, camels were often tied head to tail in a long train, the connecting ropes kept taut. On this occasion, something arrested the lead camel momentarily. A connecting rope sagged to the ground, and a camel stepped over it. When the lead camel marched off again, the line was suddenly pulled tight. The unfortunate camel's leg was drawn up to its chin. The lead camel kept marching steadily ahead, and the stricken camel probably would've been maimed had not a quick-thinking camel man leaped in and cut the rope with a single slash of his knife. "It was a touch & go business," Elliot wrote, "and I was as near losing a valuable camel as I care to be."[10]

Elliot considered pressing on in the afternoon, but Dualla located a reliable source of water nearby. The curator therefore decided to remain, giving the headman a chance to replenish the expedition's dwindling supply. "Water is the great [desiderata] in this land," Elliot explained, "and some we had obtained lately was perfectly vile, the Somalis having permitted their sheep & other animals to do as they pleased in the pools, so that I found it impossible to wash my face in it, much less have any cooking done with it." Elliot was wise to some unexplained funny business going on with their water supply of late. That afternoon, he examined the contents of the casks. Finding that two contained very good water obtained at Jerato Pass, he ordered these sent to his tent. He put a private mark on them and prohibited anyone from touching them without his permission. A camel was to be set apart to carry these casks. "I shall attend in future to my own water supply," he resolved.[11]

Healthy drinking water was critically important to the expedition's success, yet much of it was obtained—by necessity—from muddy holes or open pools swimming with slime, algae, or worse. Elliot ordered his water to be skimmed of any obvious solids and then boiled before he would drink it or cook with it. Other expeditions to Somaliland treated their drinking water with alum, or aluminum potassium sulphate, which clumped with unwanted solids and then could be removed by simply fishing or sieving it out. Some used filters. The Count Potocki expedition of 1895–96 brought a three-month supply of Apollinaris mineral water in glass bottles from London. Several extra camels were needed to carry this burden into the interior. Some of Potocki's friends had a good laugh at this apparent extravagance. Yet, neither he nor his European companions came down with dysentery or other waterborne diseases.[12]

Dualla brought two villagers to Elliot who claimed they could guide the expedition to a place where dibatags were plenty. The curator asked them to return at first light. He planned to send one guide each with Akeley and

FIGURE 14.1.

A camel train crossing a dry riverbed. Photograph courtesy of the Field Museum, CSZ6059.

Dodson, while he remained in camp. "I would not go out as I had killed the only Dibatag we had," he wrote with smug satisfaction.[13]

SCARED ALMOST TO DEATH

Early the next morning, an agitated Dodson stole into Elliot's tent and shook him awake. The guides were frauds, he complained. Dualla then arrived, bringing a more sympathetic point of view. He explained that the men who'd offered to serve as guides belonged to a village that had moved away that morning, and that they'd been obliged to depart with their families. Elliot told Dodson to go out with Akeley and see whether they could find dibatags on their own. If not, they would march in the afternoon. After his hunters departed, Elliot went back to sleep. Rising later, he ate a solitary breakfast, then passed the morning pleasantly reading.

A long and wearisome morning ensued for Akeley and Dodson, who searched high and low but found nothing of interest. They'd just abandoned their fruitless hunt when one of their syces pointed excitedly. There, maybe a hundred yards away, three lions prowled silently across a sandy stretch of ground. Gun bearers leaped on the mules and galloped around the lions to box them in. When the lions took refuge in a thicket, beaters went in to drive them out. Akeley, scared "almost to death," tightened the grip on his rifle. His pulse quickened; his mouth went dry. He felt a sharp, achy hollowness in the pit of his stomach. "A roar came from immediately in front of me," he later wrote, "and I saw a lioness in mid-air as high as my head, springing, thank heaven, diagonally away from me." But the big cat spotted the hunter, and it landed facing him, preparing to pounce. Akeley raised his rifle. Just at that moment, a male lion leaped over the female's back, startling Akeley, who jerked his trigger reflexively, merely wounding the lioness. Both animals then retreated into the brush. Akeley fired twice more in their direction, silently cursing himself for missing a golden opportunity. To drive the lions out, the hunters set the dry grass alight. Two male lions then rushed out of the smoke and flames and made good their escape. They couldn't find the wounded lioness.[14]

Akeley had never seen a living lion outside a cage, and these made an indelible impression. "They looked as big as oxen to me," he wrote. How did he feel about the encounter? "I had failed to get a lion," he reflected, "but I felt satisfied none the less, because the lions had likewise failed to get me. That one moment in that day, when I saw the lioness in the air, I'll never forget, for I realized that death was but an instant away."[15]

Hungry and tired, both hunters returned to camp for a late lunch. They told Elliot excitedly about their memorable morning. News of lions on the loose—one of them wounded—stirred the curator from his lethargy. He wanted another cat for the collection. "Skins of these for a group would be a fine thing & very valuable," he wrote. He was also eager to prove his mettle against the most dangerous game. A letter to Skiff about hunting lions had been full of youthful bravado: "I know this young fellow has got to shoot very straight when he has to meet a charging lion, but when the beast calls me I hope to have, in the classic language of Poker, a *full hand*." Immediately after lunch, they all set out again with their shikaris and two men to serve as beaters. After an hour's search, they found paw prints and followed them fruitlessly for some five miles, until nearly sundown, when they finally gave up the ghost. Elliot was especially disappointed. The lion "was a cowardly beast," he later wrote sourly of the chase; "[it] never stopped in any grass no matter how thick & dense. Several times we were very close . . . & once a camel man saw [the lioness] as she bounded out of a thicket when we were not twenty feet from her."[16] They returned to camp tired and deflated. Elliot gave orders to march in the morning.

Again, they started ahead of the caravan, each hunter taking a different path to "see what game the country afforded." After an hour, Elliot spied gerenuks grazing and got down from his mule to stalk. He saw fresh and plentiful oryx tracks and heard sporadic gunfire in the distance. He assumed that Akeley "had fallen in with some." The gerenuks wouldn't let Elliot approach, so he took and missed a distant shot. As he rode on, oryx tracks became so thick that he again dismounted and, together with Elmi, his shikari, began to follow. Elmi soon spotted an oryx, and they stalked it carefully. As they both drew nearer, they saw many more animals—obviously there was a herd to their front. Creeping cautiously forward, taking advantage of every shrub, they came within two hundred yards of several oryx that were "bringing up the rear." The animals had already noticed them approaching, however, and were peering cautiously in Elliot's direction. Taking quick aim, he fired at the nearest animal, when the entire herd bounded away. A few bold oryx stopped for a moment, and Elliot took another shot. With the wind whistling and howling, he couldn't hear the bullet's thump. When the men were only five hundred yards away, the herd halted and glanced back at the hunters, presenting "a fine sight as they stood in a long line with heads up & ears cocked forward." Elliot noticed one oryx skulk away by itself—perhaps it was wounded. Forging on, Elmi suddenly pointed to a dead tree behind a termite mound, saying, "killed him." There, Elliot saw an oryx splayed on the ground, cowering close to the trunk. The animal was evidently badly hurt but was keeping its head up alertly. Fearing a charge, Elliot crept behind the wounded animal and shot it fatally in the shoulder. "I do not know that the shot was needed," he explained, "but it is always best to be on the safe side." The specimen was a "splendid" female with sharp, lancelike horns of thirty-four and a half inches.[17] They skinned it there, and the hide, head, and some of the meat they packed behind the saddle on Elliot's mule.

The day was now getting old, so they set off to find the caravan. In an hour they struck the camel tracks, which they followed. Soon the trees thinned out and the patches of grass began to widen. After marching steadily for another hour they came to a grassy plain called Gubato, about seven miles across. There were no men or animals to be seen on the plain, but the track led across it, so they followed. "It was a tiresome ride," Elliot wrote, "& I was glad when we reached the trees on the opposite side."[18] There were many villages scattered about the far side of the plain, and many camels grazing. It was now impossible to distinguish their caravan's trail in the bewildering assortment of prints. Elliot stopped at a village to get some camel's milk. Inquiring there, and at several other villages, he at last located his camp, the tents even then taking shape under a welcoming copse of shade trees.

Neither Akeley nor Dodson had as yet reached camp, but Dualla informed Elliot that they'd sent for camels. The strongest camels at this point were still burdened with gear, however, so they quickly unloaded one animal and sent it back out. Two hours later, Akeley and Dodson came in bearing two young male oryx. "Their horns were about a foot long & they were very pretty specimens," Elliot wrote.[19]

Local villagers congregated in large numbers under some shady trees just outside the expedition's zareba. They squatted there for many tedious hours, talking. "What the subjects can be that cause such a flow of language it is difficult to imagine," Elliot wrote contemptuously, "as there is little or no outside news, & their thoughts can only run on their little affairs which scarcely vary." In fact, every day was so alike that Elliot knew that he'd soon lose track of the date—even the day of the week—if not for keeping a journal. Yet, from these villagers, Elliot learned that the Habr Yunis encamped nearby were preparing for a violent foray against an Ogaden tribe. Consequently, the scene around camp became "very animated." Several warriors were riding about on their horses with much bravado. They would sometimes gather in boastful groups to discuss "the momentous affair." When a number of them, gaudily attired for combat, rode up to the expedition's zareba, Elliot asked Akeley to get a photograph. But as soon as the taxidermist raised his camera, the superstitious warriors beat a hasty retreat. "It would only be necessary to fight such fellows with cameras," Elliot wisecracked.[20]

Elmi, the shikari who worked for Elliot, had been born nearby. He claimed to know the area intimately. He confidently assured the curator that he could lead the hunters to dibatags the following day. Elliot wasn't so sure. Nevertheless, he elected to remain at least another day, instead of marching as he'd intended. He hired two additional guides for Akeley and Dodson, hoping one of them, at least, would find and collect another one of these rare and elusive animals.

The guides arrived as scheduled the next morning, and the hunters set out early. Elliot asked a Midgan tracker to join him. The Midgan led them south. They scoured the countryside fruitlessly all day, then returned to camp. Elliot was glad to be back "after such an uninteresting hunt." Unless Akeley or Dodson had found a place where dibatags were plentiful, he intended to march away as soon as possible. Food rations for the men were getting perilously low. Elliot had sent for a ten-day supply to be forwarded from Adadleh, but that was yet to arrive. He faced the choice of waiting in place for supplies or buying sheep to feed the men during the march—a "rather expensive" option.[21] At dusk, Dodson returned to camp bringing a young oryx he'd shot, but still no dibatag. If they couldn't get dibatags within the week, the curator reluctantly decided, he'd simply have to give them up.

Later in the evening, Elliot was lounging disconsolately in camp when he heard faint singing. Akeley must be returning, he thought. As the singing grew louder and clearer, the curator recognized a "song of rejoicing." He knew his taxidermist must have a dibatag. Soon, Akeley appeared, leading a donkey bearing a buck dibatag and a doe gerenuk. He'd wounded the dibatag early in the day and had tracked it laboriously for some six hours, at which point both animal and men were too exhausted to continue. The chase ended with the animal's live capture, but it died shortly thereafter. "It was a fine buck," Elliot wrote, "& I was very glad to obtain it, for we now have a buck & doe," which was nearly sufficient to represent the species in a habitat group.[22] Akeley saved the dibatag's skin and skeleton. He also brought in the tattered remnants of a lizard, which he found coiled inside a termite mound. Mistaking it for a snake, he shot it in the head, which ruined the specimen. So tired were Akeley and Dodson that they couldn't process the specimens brought in that day. The night, fortunately, was cool, so the carcasses were left lying on a camel mat overnight.

PERFECT IN EVERY RESPECT

The next day was Sunday, June 28. The hunters slept comparatively late, rising after the sun. Akeley and Dodson went to work skinning specimens. While they were working, a Midgan brought a lion cub to sell. The cub, which Elliot estimated at about four months old, was "fairly tame but rather frightened."[23] Young lions were bringing high prices in Berbera, where there was demand for them from foreign zoos and menageries. The Midgan wanted two hundred rupees ($96) for the cub, but Elliot finally bought it for forty ($19). Like all juveniles, lion cubs were very difficult to obtain, and Elliot needed one for a group.

Late that afternoon, Dodson took a short walk with the shotgun, bringing back four birds, including three satin breasts, *Cosmopsarus regius*, and one hoopoe, *Upupa somalensis*. The latter was considered a good omen by the Somalis.[24] Elliot decided against the long march to Ake—their next destination—as the expedition had only eight days' rations for the men, and it'd take so long to get there and return to Adadleh that there'd be only two days remaining to hunt. Instead, he planned to march the next day to a place where Akeley and Dodson had seen several dibatags the day before, to hunt them for the week, and then return to Adadleh. With luck, he hoped to get two or three more specimens.

That evening, Akeley packed another barrel with nine skins, including Dodson's lion, two juvenile gerenuks, two young oryx, and the female dibatag. This was "a very valuable barrel worth a good deal of money,"

Elliot boasted. The curator shot the lion cub with his .22 rifle, as Akeley wanted to skin it. "The little thing never knew what hit it & died without a struggle," he noted blandly.[25]

The next day they made an early start, the hunters leaving ahead of the caravan. After a short distance, Akeley stopped to help Dodson fix his saddle. All the attendants were ahead with Elliot, and for some distance no one noticed that the young hunters were no longer following. As soon as it was discovered, the men who worked with Akeley and Dodson walked back. Sometime later, as Elliot was riding, he noticed his own men looking back and talking to one another in furtive whispers. Turning around, Elliot saw only Akeley and Dodson following. He stopped and waited. When his hunters came up, they said they'd seen nothing of the attendants. All of them dismounted, waiting to see when the missing men would arrive. But none appeared. So, Elliot sent Ibrahim, his Somali gun bearer, back to find them. When Ibrahim also failed to return, Elliot decided to find the caravan's tracks and follow them to camp, as hunting was impossible under the circumstances. After tramping a long while, they heard signal shots, which they answered with two of their own. Three of the men soon appeared. Elliot was incensed, concluding that the men—instead of riding back for Akeley and Dodson, as expected—had ridden to the nearest village. "Of course," Elliot wrote, "they were full of excuses & said they had lost our tracks & other nonsense of the same sort. I gave them a sound lecture & fined the lot & told them if such a thing occurred again I would fine them three times as much as I did now."[26]

They found their camp among some trees. They'd seen a good many dibatag tracks on the march, but no animals. While Elliot was stalking a gerenuk near camp, a rifle discharged behind him and the animal immediately vanished. Dodson's Lee-Metford had gone off accidentally, as a gun bearer was loading it. These rifles, Elliot grumbled, had "splendid shooting qualities, but the mechanism for carrying a cartridge from the magazine to the barrel is . . . defective, & it is a very poor rifle to trust to, especially when after dangerous game."[27] Akeley and Dodson both went out again in the afternoon, but Elliot remained in camp. Dodson returned with a dik-dik, apparently different from any they'd yet obtained.

Each hunter went his own way the following day. Dibatag tracks were plentiful, and at last Elliot and his men startled one, which sprinted away at full speed, "with head & tail up," disappearing behind a thicket of bushes. They followed as rapidly as possible, when suddenly Elliot heard two shots to his front. Continuing on, the curator soon saw Dodson grinning broadly and waving his hat, and he knew the dibatag had been taken. Dodson related that he'd stopped to rest under a tree when he saw the animal racing straight for him. Spotting him, it froze, giving Dodson the perfect opportunity to shoot. They skinned and butchered the animal on the spot.

Much to their surprise, they found another animal in utero—"perfect in every respect"—that might've been born that night or the next day, Elliot judged. He was delighted to get a flawless baby dibatag. Dodson took the specimens triumphantly back to camp. Elliot, meanwhile, continued to hunt, but without success. He soon returned to camp. "These Dibatag," he wrote in frustration, "are the most difficult animals to get . . . wile, wary, and scarce even in their own country. The natives say that very few are killed & yet they do not increase."[28] Elliot concluded that the species was definitely doomed, making his mission to salvage specimens for the museum all the more urgent.

Later that evening, a man came from Adadleh bringing letters from New York from as late as June 1. He also reported another sick camel that was likely to die. This brought the number of camels lost to four—a serious matter. Elliot would have to buy more in Berbera, as he was unable to carry all the expedition's supplies, equipment, and specimens even now. He sent six camels away with two casks each to fetch water. These were expected to return in a few days. He would then start back for Adadleh, a three-day march. The afternoon was passed preparing the new skins and packing them away in a barrel.

The next day was the first of July. Elliot went out with Ibrahim, Elmi having asked permission to visit his people. Shortly after leaving camp, Elliot saw three saddleback jackals, *Canis mesomelas*, trotting in the distance. He shot one with the Lee-Metford rifle and took it back to camp. Starting once more, they went a long distance without seeing anything, when suddenly Ibrahim pointed out some dibatags. Elliot jumped immediately from his mule. They stalked the dibatags a short way, when they made out five animals altogether—a male and four females—trotting ahead amid some dense shrubs. They followed as quickly as possible, carefully stalking, sometimes crouching, and at one time crawling on their hands and knees. They crept within 250 yards, but their prey was so alert that Elliot couldn't assume a shooting position unseen. Even with this excess of caution, the buck was somehow aware of the curator's presence. It stood and looked at him steadily as he crouched uncomfortably behind a bush. For several excruciating minutes, it stared without so much as twitching. Suddenly, the buck trotted off, leading the does away. Elliot again took up the track, and after sighting them several times, he finally got within two hundred yards. He hid behind a giant termite mound, which Ibrahim climbed for a better vantage. The dibatags were grazing placidly, apparently unaware of their peril. Elliot noticed that the buck was advancing toward a narrow clearing in the bushes. So, aiming his rifle at the gap, he waited until the moment when the buck's shoulder appeared, then he fired. The buck bounded immediately away, "head & tail up."[29]

Elliot never heard the smack of lead, so he asked Ibrahim what he'd seen. The gun bearer, who'd been watching the buck continuously from his lofty perch, said, "Yes kill him, old man, big man, kill him," meaning, according to Elliot, that he'd killed a large buck. Ibrahim saw the buck fall in a thicket of bushes after running maybe twenty desperate yards. "He was quite dead when we reached him," Elliot wrote, "the ball having passed through the body a little behind the shoulder. He was a splendid animal by far the finest we have yet obtained."[30] They loaded the specimen on a mule and started for camp.

On the way back, they met Dodson, who'd gotten very close to a female dibatag but missed his shot. Akeley came in soon thereafter, likewise unsuccessful. He reported a curious scrimmage with a leopard, however, which he'd interrupted as it was stalking a gerenuk. The leopard took refuge in some thick grass, and as one of his men was trying to drive it out, it sprang, and the man had to fend it off with a club. Akeley, meanwhile, was watching for the leopard on the other side of the thicket. Hearing the commotion, he ran around, hoping for a shot. But the leopard had dashed away and was making for another patch of thick grass, "growling at a great rate."[31] The men were so intent on the chase that they kept getting in Akeley's line of sight, preventing him from shooting. It was only as the leopard gained the patch of grass that he was finally able to take a hasty shot. The leopard gave a tremendous growl and sprang awkwardly to its side, leading Akeley to believe that his bullet had found its mark. But, though they tracked it a long way, they never saw the animal again.

15

AN INEXPLICABLE FIX

At his Field Columbian Museum office in late June, Director Skiff received two letters from Elliot, which he read with great interest. The first was dated May 20, the day the expedition returned to Berbera from the Somali donkey hunt. It gratefully acknowledged receipt of a letter of credit for £800 ($4000). The museum's executive committee had approved this sum after receiving Elliot's letters from New York and London, which argued persuasively that his original appropriation was woefully inadequate for an extended trip into the interior, where Africa's largest and rarest animals could be found and salvaged. Elliot had received the good news by telegram just before he'd landed at Aden. By May 20, he had finally received the funds. To Skiff's dismay, the curator's letter then detailed the financial hardships that continued to hound his expedition, despite the generous supplement. "I am just now struggling with the transportation problem," Elliot explained, "figuring how to make one camel carry the load of two." He then had forty-two camels in his caravan, which was "at least twenty short" of the number needed to carry his equipment and supplies, not to mention specimens. He squeezed the director for more money: "I think I shall be obliged to buy more [camels], & shall trust that if I get financially stranded the Ex[ecutive] Com[mittee] will see me out. It would be a bad thing," he added ominously, "not to have enough money left to pay my men & have a vacation in jail at Aden in consequence!!"[1]

The second letter was dated May 25, the day the expedition left Berbera for the interior. This letter continued the refrain of financial difficulties. "The chief delay has been the camel question," Elliot explained. "I simply could not pay the ruling prices & have been obliged to wait & try to get the beasts at odd times when ever a chance offers." He had, by then, purchased fifty camels, "as many as I can, perhaps more than I ought from a money standpoint, but I was forced to do so, for the simple reason that I found with less I could not move at all." The money from his original appropriation was nearly spent, but he hadn't yet touched the £800 letter of credit.[2]

At last, Elliot arrived at the crux of his letter: he urgently needed more money. "I am hoping to receive a telegram from you . . . for I suppose you would show my letter to some member of the Ex[ecutive] Com[mittee]. If I am to work out my salvation on the [money] remaining I shall make a short trip [only], doing all I can with the time at my command," he warned. "The force of circumstance has been against me . . . but I would conquer in spite of it all & make the expedition a grand success," if only he could obtain sufficient funding. "I dare not stay in the jungle a moment after I reach the point when only money enough is left to pay off my indebtedness," he explained. Elliot was bringing only fifty hired men with him, and he noted that "it would be foolhardy and risk everything in the command if I went with less. I should have thirty more camels & the same number more men, but we must cut our cloth according to the price you know, & I have done the best I can with mine."[3] Without more money, the curator's letter strongly implied, the results of the expedition would be disappointing.

Strangely, neither of Elliot's letters made mention of the additional £600 ($3,000) that Skiff had wired him on May 13.

The director had done exactly as Elliot had hoped he would. Reading of his curator's lingering concern about the inadequacy of his budget, Skiff—on his own initiative—had approached Harlow N. Higinbotham, chair of the museum's executive committee, and pleaded Elliot's case. The full committee had only recently, on May 5, approved the additional $4,000 (£800) appropriation for the expedition. Nevertheless, under the circumstances as Skiff represented them, Higinbotham felt justified in personally approving another $3,000, which came as an agreeable surprise to Skiff (and Elliot). It was, on the other hand, very disagreeable to learn now that Elliot apparently hadn't received it. Skiff panicked.[4]

The director couldn't imagine what had become of the museum's money. He was certain the curator's careful instructions had been followed to the letter. Forgetting, for the moment, the distance between Aden and Berbera, and the difficulties and delays of communication, Skiff calculated that Elliot should've received the wire transfer no later than May 16. "This distresses me very much indeed. . . . I am so upset by my fears that the money has miscarried in some way that I hardly know what to do," he anguished in a letter to Elliot. He knew the wire had been sent—he'd paid the $14 cable fee more than a month before. "It is really too bad to be in such an inexplicable fix. I shall try to hunt it down today, [even] if I have to go to considerable expense in cabling," he wrote.[5]

Skiff visited the Northern Trust Bank of Chicago, where he learned that $3,000 had indeed been transferred from the museum's account to the Bank of Scotland at London. A cable to the Bank of Scotland confirmed that the money had then been sent to Elliot through its Aden branch. The bank was to notify the American consul at Aden, William W. Masterson, who was to notify Elliot "by native runner." Skiff then wrote a letter to the consul, asking whether these instructions had been followed. If the answer was yes, Masterson was asked to notify the director by mail. If it was no, Skiff asked Masterson to send a costly telegram advising him immediately of the facts of the case. Each day that passed with no cable only deepened the mystery.[6]

Adding to Skiff's misery, he was stricken in July with a painful affliction of the eyes—probably scleritis. He'd had the same debilitating ailment at the opening of the World's Columbian Exposition, in 1893. Now, suffering from swelling, extreme pain, and sensitivity to light, he spent most of the month of August at home, sequestered in a dark room. Regular conferences with his personal secretary, Marvin J. Welch, kept the director informed of the business of the museum. He returned to his office on August 29.[7]

What Skiff didn't know was that his wire transfer had gone through. Elliot had learned belatedly of the good news through a letter from Masterson on June 22. Elliot had warned the director emphatically that he needed to know about any supplementary funding before he left Berbera for the interior. Now Skiff feared that all his begging and scraping, all his haste and added expense, had been for naught.

16

OMELETTE À L'OSTRICHE

July 2–15

Before setting out to hunt, Elliot examined some traps that'd been set the night before. He found, to his chagrin, that a clever jackal had somehow pilfered the meat from one without springing it. Judging from tracks found close to the zareba, he concluded that a leopard had sniffed around another but was evidently too wily to trust it.

The curator went back to the place where he first saw a dibatag the day before. He was nearing that spot when Elmi indicated some gerenuks. As these animals appeared to differ from specimens obtained elsewhere, Elliot decided to go after them. He made a short stalk, spending some awkward time on hands and knees. The gerenuks—four females and one large male—were partly obscured by bushes. Taking careful aim from cover, Elliot fired at the buck, striking it in the hip. The does sprinted to safety, but the buck remained. He fired again, knocking the animal down. When he approached, the wounded buck "roared like a bull," before expiring. In life, this animal stood slightly taller than the ordinary gerenuk and was lighter in color, "having a good deal of the purple hue characteristic of the Dibatag, & with a white stripe above the eye from the horns to the nose."[1] They heaved the curious carcass onto the curator's mule.

Elliot returned to camp for lunch. Dodson was already there, having seen nothing. Akeley was still out. Elliot had heard the crack of Akeley's Winchester earlier that morning. About an hour later, the tardy taxidermist arrived, bringing a half-grown dibatag and three gerenuks, including an old and a young female and a newborn, each with the distinctive white stripe along the face and the light-colored coat. Now that he had a superabundance of gerenuk specimens, an ecstatic Elliot contemplated mounting two separate groups to illustrate their range of variation. He had also salvaged six rare dibatags of various ages, which would make up "a perfectly unique group," one not to be seen anywhere but the Field Columbian Museum.[2]

Akeley reported that he'd seen "a good deal" of game of various species that morning, including a bull of the lesser kudu. This piqued Elliot's interest in distribution and diversity. This kudu, it seemed to Elliot, was "entirely out of its range here, & it is difficult to understand just where it came from." This region had seldom been hunted by Europeans, "& it would not surprise me if some new species & subspecies existed here & farther to the south."[3] Would they take any new mammals in this area? Had they already? To describe a new mammal species from Africa would be a feather in Elliot's cap. Should that species then go extinct, its type specimen would be an invaluable addition to the museum's zoology collection.

Five carcasses provided ample work for Akeley and Dodson for the rest of the day. The camels, meanwhile, had returned with their water casks filled. With plenty of water for the journey and an abundance of rare antelopes in the specimen barrels, Elliot decided to return the following day to Adadleh. Because dibatags had proved so difficult to obtain, they'd been on this side trip longer than intended. Nevertheless, he judged the trip a great success, and a victory for salvage zoology.

That night, Elliot tied up a sacrificial sheep near the traps, with the idea of catching the wary leopard that'd been prowling around the campsite with impunity. He went out before bed to see whether anything had molested the bait, "but all was quiet." In the small hours of the morning, however, a sentry came to the tents and reported that he'd heard the sheep bleating in terror. Dodson went to investigate and found that a striped hyena had been caught by two paws in as many traps. It had killed the sheep and "nearly eaten it up." Dodson "killed the brute with buckshot & the men brought it into camp." Akeley and Dodson skinned the hyena the following morning; it proved to be a male with a fine coat. The striped hyena was relatively rare, and Elliot "was glad to get a second specimen."[4]

Though they faced a series of long marches, the hunters set out late, long after the caravan had already departed. They soon reentered Gubato, marching its entire length. They sighted a herd of oryx, and Elliot stalked, shot, and badly wounded one of their number. The wounded animal made off in a direction opposite their destination, however, so he didn't pursue. The day was overcast, which made for a comfortable ride. The morning was relatively cool, and a brisk breeze blew. Near midday, they reached the far side of the plain and entered parklike terrain again. They marched another two or three miles, finding their camp pitched at the edge of Toyo Plain.

Dualla reported to Elliot that their *Tuna* chums Barclay and Sartoris were camped an hour away. A social call was in order, the hunters decided. However, shortly after lunch, the British adventurers paid them a visit. They'd been very successful, Barclay and Sartoris boasted, taking a king's ransom of ten lions and two rhinos. They were now headed back to Berbera with their booty. They decided to march together the following day.

As Elliot sat writing in front of his tent that afternoon, "a great commotion occurred among the camelmen & a number seizing their guns started away on the run." At the same time, "great flocks of sheep & goats were driven by on the full run towards the villages amid great shouting & uproar." The cause of all the excitement, Elliot learned, was war. Some tribe, possibly from Ogaden, had raided an Eidegalla village near camp. The raiders had allegedly stolen many camels. It was thought that some of the camels taken belonged to the expedition, and for this reason some of the expedition's men had launched an armed pursuit. Quickly mounting his pony, Dualla rode away in the direction the looters were supposed to have gone. Elliot regretted that he had at his disposal only a mule, which was not swift enough to take part in the chase. "If I had a horse I should go out to see the fight," he wrote. If any of his camels were missing, the cocky curator vowed to "march on the tribe & make them pay for troubling my camp."[5]

Dualla returned around sundown and reported that he and his riders had nearly overtaken the looters, who'd been frantically driving some 1,500 captured sheep. Dualla's party fired several warning shots over their heads. Turning back and seeing their determined pursuers, the raiders "whipped up their ponies & fled." The headman then turned the flock of sheep around and returned to camp. "It was quite an exciting episode while it lasted," Elliot wrote, "& proved how easy it is in this land to change in a moment the peaceful watching of the flocks into the turmoil & bloodshed of war." The men had been anxious to join the pursuit. But, excepting those responsible for the camels, the curator wouldn't permit anyone to leave. The rest of the night passed quietly, Elliot noted, "the looters probably never stopping . . . until they had placed a wide distance between themselves & the owners of guns."[6]

INDEPENDENCE DAY

The next day was a Saturday, July 4. They set out across Toyo Plain before five in the morning. Barclay's camel train was soon seen traveling along a more-or-less parallel path, which gradually converged with Elliot's line of march. The wind was mercifully cool, but there were no clouds to shroud the sun. Elliot fired a single round at a hartebeest that had been ogling him at five hundred yards. The animal then "cut such queer antics" that he thought it was wounded.[7] But it scampered away unharmed.

They camped on the far side of the plain in a shady copse of trees. To honor America's independence, Elliot ordered the expedition's flags unfurled: an outdated American flag at Akeley's tent, a British Commonwealth flag at Dodson's, and a Field Columbian Museum African expedition flag for Elliot. Barclay invited Elliot and his hunters to a holiday

FIGURE 16.1.

Camp on July 4. Dodson (*left*) and Elliot (*center*) stand by their tents. Photograph courtesy of the Field Museum, CSZ5937.

dinner. "I should have been the one to have given the invitation," a chagrined Elliot later wrote, "& would have done, save for the fact that our supplies have practically all given out, & I had nothing to offer." Elliot determined to repay the favor with a dinner invitation at Adadleh, where the main body of his caravan was encamped, and where Barclay's party was also headed. The Englishman asked Elliot to bring a table, chairs, and dinner service, as his outfit was only meant for two. They shared "a most bountiful & jolly dinner," doing "great justice to the different courses." Sartoris wrote the menu on a scrap of paper, with the Union Jack and the Stars and Stripes sketched in ink across the header. The American flag, "by error"—or so Elliot wrote—had been drawn with its blue field of stars in the lower right-hand corner, "like a signal of distress."[8] The fare included hors d'oeuvres of pickles and cocktails, a fish consommé, fried sardines, mutton fillet, a meaty stew with rice, and, for the entree, an ostrich egg omelet.

The omelet, Elliot noted, was a great success. Although it was made from a single egg and shared by all five men, there was some left over. A villager had found the egg that afternoon on the plain and brought it to Elliot. It had evidently just been laid. Elliot gave it to Sartoris with the suggestion that it would make a good omelet.

After dinner, they drank coffee and shared a bottle of scotch. They sat talking together around a campfire for several hours, consulting books and maps and enjoying "very fine cigars." Late in the evening, Dualla sent a message to Elliot saying that he'd prefer to start the caravan very early the next morning. He planned to make a long march, leaving an easy distance to Adadleh. Elliot, anxious to get to Adadleh quickly, acquiesced, "though . . . a three o'clock start was not a pleasant prospect."[9] With that, the party reluctantly broke up.

The two caravans started out almost simultaneously. The predawn air was pleasantly cool, and the moon was shining brightly. They all rode together, threading their way through a maze of burrows. Soon, they entered the parklike bush. They rode until ten o'clock, when they stopped, at last, under the shade of some large thorn trees for lunch. At half past one, they set out again and marched to the edge of a plain, where they made camp.

On the march, Sartoris killed two small warthogs—too runty to serve as hunting trophies. He gladly gave them over to Elliot, who was eager to have them to complete a group. These were the last days of shooting for the Englishmen, as they lacked permission to hunt on the reservation. Thus, they planned to make the rest of their journey to Berbera as rapidly as possible. As promised, Elliot invited them both to dinner the following day.

The next morning, both camps started out at five o'clock. A relatively short march was in store, so there was no need to hurry. Barclay, however, left early to hunt a pair of Speke's gazelles for the British Museum. The rest of the party rode in company. During the march, Sartoris spied a jackal and killed it with a fine shot from more than two hundred yards. The skin was so badly torn, however, that he left it behind. They saw no other game while crossing the plain, nor anything in the woods beyond. Elliot, though, shot two satin breasts, *Cosmopsarus regius*. This bird "was always a most conspicuous object among the trees, its brilliant plumage flashing in the rays of the sun," the curator later wrote. "Altogether it is probably the most beautiful bird as yet known to inhabit Somali-land."[10]

AU REVOIR AT ADADLEH

Their old camp at Adadleh was deserted, Selon having moved after the flood to a superior site farther up the riverbed. There, Elliot "found everything in good shape except the camels[,] which were much thinner."[11] Worse, five had died and one was ailing. The desiccated corpses were heaped in a grisly pile, encircled by a zareba, wisely preserved in this way to prove they'd died a natural death, and hadn't been stolen, sold, or killed for food. This exonerated Selon of any blame for their loss. Three of the dead camels came originally from Bulhar, where there'd been an outbreak of rinderpest. Elliot quarantined the sick animal, hoping to contain the disease. The emaciated camels would've fared better on Toyo Plain, where grazing was better, the curator regretted.

Barclay made camp on the opposite side of the dry riverbed. The tents of both parties were pitched in the welcome shade of large trees lining both banks. It was "a great relief to be again where water is plenty . . . & to be able to indulge in the luxury of a bath," Elliot wrote.[12] As soon as the expedition's specimens could be made ready for shipment, the curator would forward them to Berbera under Dualla's care, with instructions to bring back additional supplies and camels. He needed many things, and this would be his last opportunity to communicate with the coast until his return from the far interior. He also planned to send back both cooks and one camel man, all of whom were too sick to continue.

In the afternoon, Elliot sent two men to a nearby village to retrieve a captive young leopard. They soon returned, bearing a flimsy cage of twigs housing five cheetah kittens. The curator was "much surprised to see them instead of the ordinary leopard." Cheetahs were rare, and he "never expected to obtain specimens of it, much less a litter of young." The month-old kittens were "lively & frolicsome . . . and made a most peculiar sound, more like the call of a bird than the voice of a mammal, & their little growl was like a purr of contentment," he wrote. Tame and fearless, they permitted anyone to handle them. "If I could only get them home alive they would make excellent pets," he claimed, "for the Chee-tah is very dog-like . . . & not having retractile claws is unable to inflict the terrible scratches that other cats are capable of."[13] Some of the kittens had bald patches on their heads and necks. Elliot intended to keep them alive until their hair grew back.

Later that evening, Barclay and Sartoris arrived for dinner. Elliot was pleased with his menu of "tartine a la humbug," or "mock turtle" soup, kippered herring, mutton, plums, cherry pie, Dutch cheese, and crackers. After the meal, they shared scotch and coffee, and smoked cigars, ciga-rettes, and pipes. "We enjoyed ourselves very much," Elliot wrote, "& as there was no move to be made in the morning we were in no hurry to separate."[14] They stayed up talking for many festive hours, going to bed at last very late.

Elliot spent the next day writing letters. Among them was a letter to Higinbotham to thank him personally for augmenting the expedition's finances. Akeley and Dodson, meanwhile, put the specimens in order for shipment to Berbera. Akeley opened some of the barrels that'd been washed away in the flood and found that all the skins were in excellent

FIGURE 16.2.

The dinner party at Adadleh, with (*from left*) Elliot, Dodson, Sartoris, Akeley, and Barclay. Photograph courtesy of the Field Museum, CSZ5888.

condition. Selon and his men had taken good care of them. Elliot saw "no reason why these skins should not go to Chicago & arrive in perfect condition."[15]

Elliot ordered his private water casks to be filled with boiled water. While this was ongoing, Dodson happened to visit the mess tent, where he saw three open tins of a commercial preservative for zoological specimens manufactured and sold by Rowland Ward of London—probably Taxidermine. He asked Musa, the cook, where he'd gotten them. Musa told him that Barclay's cook gave them to him to treat the water, that it was alum, and that he'd used it for this purpose frequently. Elliot, when he was informed, immediately ordered all the water to be poured out. Was Musa trying to poison them?

Elliot explained what'd happened to Barclay, who then "made a great row in his camp." When Musa learned that he'd made trouble for Barclay's cook, he changed his story, swearing that he got the preservative from Norman B. Smith's camp at Gelalo. "This was too stupid a lie to stand a moment," Elliot wrote, "as our camps were fifteen miles apart & Musa had held no communication with [Smith's cook]." Elliot concluded—perhaps naively—that neither man "meant any harm. But it only shows how careful one should be in leaving any articles about a camp where these ignorant savages are, for they pick up anything that is handy" and immediately put it to use. Elliot didn't know what Rowland Ward's compound consisted of, "but skin preservatives are manifestly not good things to be mixed with drinking water," he concluded.[16]

That evening, Dodson set some traps, baiting them with the stomach of a sheep butchered that afternoon. Elliot and his hunters were savoring their dinner just after sundown when a camel man approached and announced that a hyena had been caught. They went to investigate, accompanied by a crowd of men. They found that a great spotted hyena, *Hyena crocuta*, was caught by its hind leg. The curator shot it in the ear, killing it instantly. "It was a cowardly brute," he wrote, showing "no fight whatever, only growling once or twice, but trying all the time to get away." Elliot despised the spotted hyena. Of all the animals he hunted in Somaliland, this was "the most loathsome and repulsive. It is usually covered with scabs or sores, caused either by fighting with each other, or from the impure condition of its blood, or perhaps both." They smelled very badly, he thought. "Some individuals," he later wrote, "were so offensive that it was very disagreeable, not only to go near them, but even to stand to leeward of their carcasses."[17]

A careless jackal had also "succeeded in getting all four of his legs in as many traps." Dodson pressed his boot heel to the jackal's neck, released it from the traps, and carried it back to camp by the throat and legs. "It was a very fine specimen of the saddle-back kind & had a really beautiful skin," Elliot noted.[18]

Barclay and Sartoris visited one last time that evening, to pay their respects before setting out for Berbera in the morning. Elliot was sorry to see them go, as they were "very agreeable & pleasant fellows." On the other hand, he was fortunate in having acquired two of their guns with ammunition. Akeley was short on cartridges and had only the heavy Winchester for hunting, which left him "practically out of the shooting." Meanwhile, Elliot's Lee-Metford had malfunctioned several times, and he was reluctant to start for the interior with "a possibly defective rifle." From his English friends, he obtained a rifle made by John Rigby & Company, similar to the Lee-Metford, "but superior on its manufacture in every way." Akeley acquired a Lee-Metford made by William Moore & Grey. Elliot was pleased to think that the expedition was now "thoroughly equipped with fire-arms, thanks to . . . friends . . . who loaned me expensive 500 & 577 rifles for the dangerous game, & so saved the Museum a large outlay." Heavy-caliber rifles were "absolutely necessary," he wrote, when larger game, like rhinos, giraffes, and elephants, were to be "met & conquered."[19]

LEAVING FOR LEGUD

Both caravans—Dualla's and Barclay's—departed early the following morning, Dualla taking thirty-one of the expedition's camels. Elliot instructed him to buy another twenty, which should be "quite sufficient to bring everything back with him." Despite the episode with Rowland Ward's preservative, Elliot engaged Barclay's cook to return with Dualla. He was "an excellent cook & he cannot get access to any poisons in this camp," the curator reasoned. The camels were lightly loaded, most of them carrying nothing at all, so Elliot expected them to travel swiftly, certainly getting to Berbera in advance of Barclay's heavily laden caravan.

Elliot kept half the expedition's camels, including one sick animal that was then sprawling on the ground at the edge of camp. Selon let him know that it would never get up again. The curator decided that to leave it behind, ill and defenseless, would be cruel. It'd be more merciful to shoot it. The camel then rolled over quietly and died without a struggle. The remaining animals were loaded by two in the afternoon. Unfortunately, there was excess baggage, which had to be left behind. Elliot instructed Selon to send to the villages to hire more camels to carry the surplus to Hargeisa.

While loading, Elliot discovered that his Winchester was missing. Musa had carried it on the march, but now he'd gone to Berbera. Asking among the other men, he could learn nothing definite about the gun, and concluded that it'd been left behind at one of their camps. He sent a rider to follow after Dualla to see whether anything could be learned from Musa. He sent another rider to revisit their previous camps as far as Toyo Plain. Elliot, as was his habit, accepted none of the blame. "Such things as this are liable to happen at any time," he wrote, "for these people are not responsible for anything & careless as Savages always are. They break everything they get in their hand belonging to Europeans . . . nothing not even an iron crowbar I believe being safe from their utter carelessness."[20] Apparently, Elliot never suspected sabotage.

They departed Adadleh, riding over rough terrain, partly shaded by trees. The hunters traced a different route from that of their camels, which they didn't see until they reached camp late in the day. The new camp was pitched on a clearing of hard-packed sand, spread with wiry tufts of grass. There were villages nearby, with sheep and goats grazing around them. It was dark by the time dinner was served. With both cooks gone, their mess was a shambles. The bland fare was a foretaste of camp life without a competent cook.

They made an early start the next morning. The nearest reliable water source was an eight-hour march from their present campsite, and—since the camels were heavily loaded—Elliot divided it into two shorter marches for the morning and afternoon. The route was generally westward, heading toward a gap in a low range of hills at the edge of the plain. Hunting continued during the march. Using Sartoris's powerful rifle, Elliot killed three dik-diks, a warthog, and a jackal, this last sporting a fine coat. He'd spotted four young boars trotting along diagonally in his direction as he was stalking. In an instant, he was running to head them off. He picked out the last of the group as possibly the largest. His first shot bowled the pig over dead. His men, of course, refused to touch the carcass, but Dodson, who'd been blasting birds nearby, arrived shortly and began to skin it. Elliot sent a man for Akeley, who soon appeared. The skin was quickly removed, and the severed head placed in a sack and tied behind Akeley's saddle. The three dik-diks were packed on Elliot's mount. The jackal, meanwhile, had already been sent back to the caravan. Elliot was amazed by the superior power and accuracy of his new rifle.

After they had been marching and hunting for some five and a half hours, the heat became oppressive. They halted beneath some pleasantly shady thorn trees just beyond the edge of the plain. What Elliot called "the bush" began at this point, and they could see what appeared to be a dense mass of forest stretching away to distant mountains in the south. Beyond these mountains was a waterless tract of many miles lying between Hargeisa and Milmil, which the expedition eventually had to traverse. "Two lofty peaks, conical in shape, rose high in air in the west."[21] Near these peaks was Hargeisa, about two marches beyond their planned evening camp.

After a simple meal and a long rest in the shade, they set out again. They continued through another parklike country, much more open than it had appeared. The ground was sandy but peppered here and there by small stones strewn about "as if they had been sown broadcast." Akeley, using Elliot's Lee-Metford, shot a dik-dik, but brought back only the head as a trophy. Nothing else was left of the animal, the bullet from the powerful rifle "having scattered the body in every direction."[22]

By late afternoon, they reached another dry riverbed, this one called Legud. The men dug deep holes in its sandy bottom, and clear water seeped into them. They pitched their camp on the bank under a clump of large trees. Elliot was pleased with the setting, writing that this was "one of the prettiest situations we have had for a camp since leaving Berbera."[23] As oryx abounded nearby, and food for the camels in the shape of thorn bushes was abundant, Elliot decided to remain at this spot through the following day, at least.

The next morning Elliot went out. Akeley stayed in camp to focus on his camera, while Dodson hunted birds near camp. The country was so open that Elliot found it impossible to get close enough to shoot the oryx, gerenuks, and gazelles he saw in abundance. He returned to camp with nothing. "If all the country was like this," he wrote in frustration, "it would take us years to make a collection." Elliot's salvage mission was so urgent, though, that even one year's delay would have rendered it moot— or so he'd written to Director Skiff. Sensing the curator's ire, the shikari offered that there was better hunting near Hargeisa. Dodson, meanwhile, shot some twenty birds, including pretty parrots, sunbirds, shrikes, and tits.[24] He and Akeley had been busy preparing the skins.

A heavy traffic of thirsty sheep and goats visited the wells all day long. Shepherds climbed down into the deep holes and dipped water into a kind of basket woven from roots and twigs. A few sheep or goats at a time would then be allowed to drink from the basket. Women kept the various flocks in order and prevented them from comingling. Elliot was amused to see one little girl—not three years old—dressed elaborately with a bead necklace and armed with a long spear, standing guard over a watering hole and preventing the animals from crowding it.

Elliot's Winchester was brought back that afternoon by the man he'd sent to find it, accompanied by the man who'd had it, as well as the guide who led Elliot's man to the place where it'd been left. Each merited a small reward. The guide was a member of the same tribe as the looters who'd stampeded sheep belonging to the Eidegalla on the other side of Toyo Plain. He was afraid to return for fear the Eidegalla would kill him. Elliot gave him a single bullet and a piece of paper bearing a few lines "ordering all people to permit him to pass." The guide then went away contented. The curator remarked that this episode demonstrated "how much prestige" a white man had in Somaliland, and "what influence he exerts."[25] The bullet, which was symbolic of the tool of this prestige, spoke a clearer message than Elliot's English note, however.

The next day, Elliot and Akeley went out early for game, while Dodson continued to blast birds. Elliot covered a great distance through the same open country he'd hunted the day before. It resembled an immense, abandoned orchard, the sandy ground strewn with small, sharp stones. It was a "most unfavorable district in which to get near my wild game," he wrote with exasperation. He saw several oryx during the day and wounded two of them. Although he and his men followed the bloody trails, they were unable to find any remains. "Oryx," he wrote, "are very difficult beasts to kill & the distance at which we are compelled to shoot them causes me to think that the Lee-Metford is not . . . powerful enough." Elliot returned to camp empty handed. Akeley, who'd just returned, had killed a female Speke's gazelle. Dodson, meanwhile, had secured "a nice lot of birds." He'd also succeeded in killing a pig, which approached him unaware while he was hunting. Although the camp was "a very agreeable one & the water plenty & good," Elliot regretted that "the vicinity is not favorable for hunting." He determined to move on the next day, "as it seems useless to remain."[26]

In the early evening, the remnants of Dodson's pig were placed in a makeshift, open-ended zareba. Traps were then set for the spotted hyenas that'd been heard cackling near camp. Dinner had only just ended when an uproar alerted the hunters that something had been caught. They went to look, bringing lanterns. They found that a spotted hyena had dragged a trap along with the log to which it was attached "for quite a distance." The injured animal had entered a dense thicket of aloes and creepers. "He was very much out of temper & growled savagely, jumping about in his efforts to get loose," Elliot wrote. The curator shot it with the Winchester, and the men carried the carcass to camp. The hyena turned out to be "a good sized beast," with "a good coat."[27] A generous layer of fat showed it had lived well.

Akeley had hardly finished skinning the hyena when a furor at the zareba indicated another animal caught. Seizing their rifles, Dodson and Elliot went to investigate, finding that a trap was missing. Taking up an obvious trail in the sand, the hunters followed to a dense thicket of shrubs, where another hyena had hidden. Unfortunately, the wounded animal—still caught in the trap—had gone well into the thicket and gotten itself "thoroughly entangled" in a mass of branches and vines. It was still flailing about "in a great state of rage & fear." The hunters found it difficult to determine the animal's position. At length, Elliot "made him out in the darkness & shot him." The carcass was then retrieved with difficulty and brought back to camp. While the hyena was growling and screaming, another answered from a thicket nearby. The hunters, at first, thought that they'd caught a third hyena, but after counting the traps they found none were missing. The dead hyena, on closer inspection, turned out to be a juvenile. Elliot concluded that an adult hyena was answering the young animal's distress call. "It was careful not to show itself, & although we searched for it, we could not see it in the darkness."[28] Even so, the two spotted hyenas they'd secured would make good exhibit specimens. Elliot was now planning to build a group of them back at the museum.

THE WELLS OF MERODIJAH

The following day was a Sunday, July 12. Having stayed up late chasing and skinning hyenas, the hunters all slept in. The sun had long been up, and strong gusts of wind were swirling through camp by the time Elliot rose from his cot, calling enthusiastically for his rubber tub. Alas, even as he disrobed, the wind strengthened, raising choking dust and hurling bits of sand that stung his naked hide, spoiling the pleasure of his weekly bath.

The wells were again the "scene of great activity" that afternoon. A crowd of Eidegalla arrived for water, which they carried away in large chatties, hanging six from every camel. Elliot watched them slowly dipping water from the deep holes, letting them use two of his pails to expedite the work. Several unmarried young women were among them, their long hair hanging down their necks, as was customary. Elliot noted that they worked "as hard as the men, & quite as effectively." The women's tobes were of many colors, from the commonplace terra-cotta—made by washing them in red sand—to pure white. Some had tobes of bright red. One

FIGURE 16.3.

Somalis loading
water-filled
chatties onto a
camel. Photograph
courtesy of the
Field Museum,
CSZ6078.

had a red skirt, "a kind of crushed strawberry over skirt, & a yellow strip over her head." She was "very gorgeous," Elliot opined.[29]

Elliot continued to enjoy the company of the wildcat kittens he'd carried from Adadleh. They were "interesting little creatures . . . very tame, do not mind in the least any handling they receive, & delight in being let out of their cage for a run or play."[30] He regretted that they'd have to be sacrificed for science, as it'd be impractical to carry them alive much longer.

The caravan set out after lunch. Elliot and his hunters stayed behind, for the afternoon was hot and the march to Nagal—their next camp—wasn't long. Resting in the shade of a tree, Elliot watched the curious movements of a gaggle of eagles, crows, and vultures that were "trying to summon courage to approach" two hyena carcasses left lying on the bank. The timid birds would perch in the trees nearby and "utter all kinds of harsh notes as though daring each other to go on." The crows proved boldest when a few alighted on the bank—their bills splayed open—and hobbled cautiously toward the dead hyenas, eyeing Elliot closely, "ready to fly on the instant." Next, the vultures landed in the dry riverbed and lurched awkwardly toward the bank to partake. At long last, the eagles swept right up to the carcasses, glancing suspiciously at Elliot, and then went "right to business, by tearing the tough meat apart & swallowing it as rapidly as possible." The eagles "were undisputed masters of the field," no other bird "daring to touch any of the portions they had selected, though occasionally a crow would glide up sideways to the big bird, & view him in an inquisitive kind of way as if debating with himself how near he could go with safety." The vultures, meanwhile, maintained a respectful distance. Someone fired a gun, and "the whole assembly rose in a body, & the air was filled with waving wings & the rapidly moving . . . birds as they flew . . . in graceful circles hardly knowing whether to depart, or return to their interrupted feast." Elliot had netted many African birds already, yet not in the numbers or variety that the acquisitive curator would've preferred. Raptors and scavenging birds were so numerous and so tame in Somaliland that it would be "a very easy matter," he thought, "to get a fine series . . . in various plumages."[31]

Elliot's party departed, at last, in midafternoon, riding for three hours through "a sandy country covered with thorn trees without any under-growth save clumps of aloes." They saw no game, nor any tracks. Elliot's impression was that the nearer they approached Hargeisa, the less large game there was, hunting and other human activity having driven the an-imals away. They marched most of the way with difficulty through a dry, rocky riverbed. The sun was so hot that it reminded the curator of the ex-treme discomforts of the coastal plain. Various species of large thorn trees lined both banks of the river. Some patches of earth were "carpeted with a vivid green from the fresh growing grass." These spots were so inviting that they dismounted to give their horses a chance to graze. They spied many birds, "doves being especially numerous," before setting off again.[32]

Arriving at an island in the middle of the dry river, they found their men busy pitching tents. There were many birds, including flocks of weavers, *Quelea aethiopica*, "darting from tree to tree." Dodson took a shotgun for a stroll around camp, soon returning with two dozen specimens. From these, he selected the finest eight examples to put in the collection. The rest were given to the men to be cleaned and plucked for the next day's breakfast. Akeley, meanwhile, nearly filled another barrel with the skins they'd accumulated since leaving Adadleh, along with three leopard skins Elliot had purchased. "This makes the fifteenth barrel of skins obtained since we began to collect," Elliot noted proudly.[33]

They set out early the next morning, Monday, July 13, although their march was to be a short one. The wind had blown fiercely through the night, rattling the canvas and whistling through the guylines of their tents. This continued during the march, the blowing sand making for an uncom-fortable ride. It was especially hard on the camels, which had to balance their heavy loads against a furious crosswind. Several camels gave out but were persuaded to continue by brute force. They passed through more of the same country, with no large game visible. Yet birds were everywhere abundant. It seemed to Elliot as though he was "riding through a great aviary, so many were the feathered forms flitting in every direction belonging to many species."[34] They shot some for the collection.

Akeley killed a male dik-dik, which appeared to be different from any specimens they'd yet obtained. This led Elliot to muse about the differences between species. "I think the Dik-diks require careful examination," he wrote. "We have so far obtained what seem to represent four species." Yet only three had been described from Somaliland. "The one shot today," he continued, "is larger than any we have, with the red of the flanks extending high into the shoulders & far under the belly, having but a narrow line of grey between the shoulders, & a still narrower one of white on the belly."[35] Elliot thought Akeley's specimen was the expedition's finest dik-dik. But did it represent something new?

Around half past eight that morning, they arrived at the wells of Merodijah, where they found several deep holes sunken in the sand of a dry riverbed. This was where Elliot proposed to remain with the caravan until he heard from Dualla, who was expected to return from Berbera in a few days.

The little wildcats were sacrificed that afternoon. Elliot feared that their coats would be utterly ruined if the animals were kept any longer in their cage. He'd also learned from a villager that there were two young lions being kept near Hargeisa. He sent word to their keeper that he wanted to examine them. If he could secure these for the collection, "they with the one I already have would go far to form a litter for a group."[36]

LOCUST STORM

The next morning, Elliot and Akeley went out together to search for game, the curator having heard rumors of oryx nearby. They trampled through aloes and low shrubs into an open country very much like Legud, the ground sown with loose stones. Ascending a high hill, they spied one oryx standing below. But the animal spotted them at the same moment and sped away. Elliot fired, but this only inspired the oryx to run faster. He watched it go a great distance, never stopping so long as it remained in sight.

They descended the hill, Akeley and Elliot splitting up. The curator and his attendants were wandering again among the aloes when they caught sight of another beast sprinting through the brush. Elliot dismounted to follow, with Elmi stalking beside him. After walking a long distance without seeing anything, Elliot stopped for his horse, which didn't come up. They whistled and called repeatedly but heard no answer. Evidently, Ibrahim had lost them, so they returned to camp. It was a long way back, maybe five or six miles, and the sun was very hot. They saw nothing but birds the entire way. There seemed to be "a great scarcity of game of the larger kind in the vicinity." Elliot was spent when he finally reached camp. He'd been walking nearly the entire time since leaving his tent that morning, and he felt discouraged at finding no game. He'd heard Akeley fire several times in the distance and hoped that his partner had been more fortunate. Dodson, meanwhile, had hunted birds, including a beautiful striped kingfisher, *Halcyon chelicutensis*, the only one seen in Somaliland.[37]

After they'd finished lunch, Akeley arrived, bringing the skins of a male and a female gerenuk. These were killed at three hundred yards with two well-aimed shots as they stood looking him over with fatal curiosity. The female had the same stripes on the face that Elliot had noticed on the specimens obtained in the dibatag country, and he concluded that the variation didn't signify anything important. Akeley's two specimens were the largest of any yet obtained, and were the last ones needed to complete a group of that species.

The wind continued to howl, and the blowing sand was unpleasant. Signs of a possible thunderstorm showed on the horizon, which could signal a change for the better. Even so, Elliot decided to move camp the following morning nearer to Hargeisa. To remain any longer in their current gameless camp was pointless.

The next morning, however, they made a late start. Elliot had decided to stop on the near side of Hargeisa to await Dualla, so they planned to march for only a few hours at most. Arriving shortly at their next camp, Elliot immediately made a round with his men and noticed evidence of

FIGURE 16.4.

A shady camp at Hullier, Elliot's favorite campsite. Photograph courtesy of the Field Museum, CSZ5938.

considerable game. He saw lesser kudus, gerenuks, and wild donkeys, as well as the tracks of the greater kudu and oryx. Unfortunately, he was unable to attempt any shots, as the creatures were too wary—probably from having been hunted so often by sportsmen from Hargeisa. Dodson, meanwhile, managed to shoot some birds that were new to the collection. They had by now obtained about 130 birds representing sixty-eight species.

They were camped on the bank of another dry riverbed near a well called Hullier. The tents were pitched in the daylong shade of a clump of large trees at the river's edge. The opposite bank was a high bluff perforated in many places by deep, water-carved cavities. Green trees and bushes covered all the country near camp. Elliot described this as "one of the pleasantest sites we have yet found."[38]

Soon after they arrived, a man came to Elliot's tent to report that a leopard had been stealing sheep from a village "every night for some time past." The curator determined to have it for his collection, sending Dodson to the village to make the kill. Another villager reported that he had two live leopards in his possession that he wanted to sell. Elliot told him to bring the animals for examination the following day. The man soon returned, however, bringing another pair of cheetahs, a male and a female, each a little larger than the five kittens bought previously. Elliot was puzzled by the fact that there were so many young of this species available, when the adults were "so rarely seen much less procured." The young cheetahs "were in fairly good coat," so he bought them. But, with the long journey ahead, it would be too much trouble to keep the animals alive. They would soon "have to go into the brine barrel with the other skins."[39] Elliot hoped to bag a few adult cheetahs before beginning his long trek across the Haud.

A great swarm of locusts stormed through camp while Elliot was scribbling in his journal. They "exactly resemble snowflakes driven before the wind," he penned. Akeley assembled his camera to photograph the swarm. The tent boys kept busy driving the insects off whenever they alighted near camp. The insects would fly in a dense, blizzard-like mass. "Every green thing disappears wherever they settle, and a country is cursed indeed that is visited by such a scourge," Elliot observed.[40]

17

SOMEBODY MUST PAY

July 16–24

At sunrise the next morning, Elliot and Akeley set out in opposite directions. Twenty minutes from camp, the curator spotted two gerenuks. He stalked them briefly and shot one, sending the carcass to camp on a camel. He went on. Shortly after, he startled a small herd of lesser kudus, which sprinted up the stony slope of a hill. There were four cows with a big bull behind. Elliot shot the bull at three hundred yards, at which point it quit the herd and wandered away alone. The curator followed, but, after a long pursuit, he lost its track on stony ground.

Back in camp, Elliot found that Dodson had returned from the village, bringing only a dead goat. In the dark of night, a leopard had pounced on the bait in front of the zareba where Dodson and a gun bearer were sleeping. Suddenly waking when the doomed goat bleated, they made a noise that startled the leopard, which sprang away and didn't return. The goat, meanwhile, suffered a single, mortal bite to the head.

Another growling leopard close behind Elliot's tent had disturbed his sleep, so he set traps to catch it. Leopards were a terrible nuisance to the human population of Africa. They were "very destructive to the flocks of the natives & sometimes even carry away children & women," Elliot wrote. Yet the villagers always demanded high prices for any livestock he wanted to use as bait to catch them, he griped. Two men had even refused to tell him the name of a village where a leopard had killed a sheep unless he gave them a gift. Sliding easily into the role of the Great White Hunter who heroically clears the jungle of dangerous, man-eating beasts, he complained bitterly that "such actions make one feel that [the villagers] deserve all the losses they incur when they are so unwilling to render the least assistance."[1] He simply couldn't understand what made these inscrutable Somalis tick.

After lunch, Elliot was resting serenely in a camp chair when one of his men ran up suddenly, panting an urgent summons: a villager was trying to kill the curator's horse! "I immediately went out among a crowd of natives," he wrote, "& found that one fellow had speared my mare in the flank. I ordered him to be seized & brought within our zareba." As compensation, Elliot claimed a horse—"a fine grey stallion"—from the crowd of villagers that'd gathered, and had it hitched near his tent. He then wrote a letter of complaint to Captain Merewether, had the culprit's hands tied, and sent him on foot under armed guard of two of his own men to Berbera. The captain would decide what punishment the man should receive. Some in the crowd were unhappy about the rough treatment of their fellow villager. "There was a great outcry when the fellow was led away at the end of a rope, but no effort at a rescue. I think they saw I did not fear [them]," Elliot wrote grandiosely. The curator had charged into the crowd without a gun. He believed that any hesitancy on his part would've been seen as weakness, which might've invited more trouble. In a sensational letter to Skiff, Elliot related that "natives" had "crowded about the zareba" in a frenzy of excitement. "No doubt," he added, "there were enough savages about to wipe us all out if they tried."[2]

Later that day, a penitent villager came to Elliot and meekly explained that the seized stallion belonged to him. Naturally, he wanted it back, saying he would provide another—probably, the curator suspected, a broken-down, old nag of no value—in its place. Elliot refused. "I did not care who the pony belonged to, not even if it was Sultan Nur," he explained testily. He "had laid the whole matter before the Governor & they must wait his decision & abide by it whatever it was. If I was to keep the pony for the injury done to my mare, I certainly would do so & he might make up his mind to that fact." The curator rationalized, rather feebly, that the stallion might not belong to the man who now claimed it, and that he might've concocted his story to get the animal released for the real owner—possibly the man he'd sent, bound, to Berbera. "It is very difficult at times to get at the bottom of the motives of these wily savages," he wrote. "At all events I have the horse & shall keep it until I hear from [Captain Merewether]." Elliot wasn't bothered by questions of guilt or innocence; his only concern was reparation. "If my mare is permanently disabled," he declared, "somebody must pay."[3]

After the hullabaloo subsided, Akeley arrived with a camel bearing a magnificent greater kudu bull, its horns even longer than the specimen Dodson had shot in the Golis Range. "Its equal I fancy is rarely seen in Somaliland," Elliot wrote triumphantly. Akeley had headed to a pair of distinctive peaks, known as Nasr Hablod, following a leopard's growl, when he suddenly came across a group of greater kudu, including the bull, two cows, and a calf, only 125 yards distant. His first shot went directly through the bull's heart, and it dropped immediately. He also managed to wound one of the others, which escaped. "We have now three fine males

FIGURE 17.1.

Akeley poses
with a greater
kudu head and
feet. Photograph
courtesy of the
Field Museum,
CSZ6167.

of this grand species of Antelope," Elliot gloated, "one of the handsomest creatures in the world."[4] In his mind's eye, he envisioned a one-of-a-kind habitat group featuring meticulously taxidermied kudus substituting for these beautiful animals once they'd gone extinct in the wild.

Shortly after dark, a jackal's cry was heard from the direction of the traps. Elliot and his hunters went to investigate and found that two of the little carnivores had been caught and were now scrapping tenaciously for their freedom. A gun bearer tried to bring one of the animals to camp by dragging the trap's anchor. The feisty jackal quickly scaled the log to which the trap was fixed and bit him savagely on the arm. Elliot went to fetch his .22 rifle and shot both animals. He then cauterized the gun bearer's wound. Later, Dodson set the remains of a gerenuk out and hid under a nearby tree, watching for hyenas. The bait could be seen distinctly by the bright light of the moon. Three hyenas approached, and Dodson wounded two of them before killing a large male with a fine coat.

Elliot rode a long way the next morning, to the far side of the twin peaks where Akeley had killed his kudu. "The view from the other side of the two mountains was very fine," he wrote. There was "a great expanse of country covered with thorn trees their flat tops looking like a green carpet spread over the landscape. Hargeisa . . . was seen in the distance & a large village called Herrir lay spread out on a stoney plain, surrounded by a zareba."[5] He and his attendants saw "practically nothing in the way of game," save two distant gerenuks. He returned to camp without having fired a shot.

He found Akeley, who'd remained in camp to process the kudu. Dodson was just returning from tracking one of his wounded hyenas, which he'd spotted but couldn't capture. The beast had stopped at one point, and Arden—a gun bearer—immediately began to dig in the ground at that spot, searching for a root. "Don't you see," said Arden, "the blood has stopped & he has made himself well. He has cured himself with the root." No root was found, however, and "a great loss to medical science was incurred," Elliot noted ironically.[6]

The Somalis had many curious ideas about hyenas, according to Elliot. "They believe that the beasts can assume any sex at will, changing from male to female & vice versa," Elliot recorded. They also held that hyenas were blessed with great wisdom. "On important occasions, such as a looting expedition, the chiefs go out into the jungle & pretend to talk with them, as if consulting an oracle." Of course, as with any fortune teller, the animal's prognostications were reported in hopelessly ambiguous language, "so whatever happens the hyena is always right." Elliot found it odd that an animal he considered "such an unclean beast which the natives particularly hate" should be, at the same time, so respected.[7]

The following morning, Elliot and Akeley set out along opposite sides of the dry riverbed. A short distance from camp, a female lesser kudu "burst out of a clump of bushes and went away in splendid style . . . making a beautiful picture. She was in sight too short a time to get a shot, & we did not see her again." An hour later, Elliot and his attendants spotted an oryx descending the slope of a hill, making its way toward a screen of trees and shrubs. A successful stalk brought them within range, although the animal was difficult to see. A bullet from Elliot's Winchester "hamstrung him, & he started away up the hill but making poor headway." Elliot fired again, striking it in the hip. The oryx, however, disappeared over the top of the hill. The hunters followed as rapidly as possible. On gaining the high ground, they spotted the animal limping away some distance to the left. Elliot shot it a third time, hitting it "close to where the second ball had entered. This knocked him down, but when we drew near he was full of fight, shaking his formidable horns at us." Elliot then took the Lee-Metford and killed the oryx with a bullet behind the shoulder. "The vitality and ability of an Oryx to carry away lead is wonderful," Elliot marveled. This specimen was a large bull with thick, heavy horns. Elliot and his companions dragged the carcass to the shade of a tree, an "effort that required the unified strength of all four of us." Leaving the shikaris to guard the animal, Elliot returned to camp for a camel. It was still early in the day, and the shikaris wanted to continue the hunt. But there was a hyena skeleton to prepare back at camp, and Akeley might've gotten something, so Elliot decided that "there was enough on hand for the day."[8]

The oryx arrived on the back of a camel two hours later, and "all hands" began processing the specimen. The oryx skin was thick, "fully two inches on the neck & shoulders & also on the hips," so it took a long time and a great deal of effort to remove and prepare it for salting. Indeed, the entire afternoon was taken up with the oryx and hyena. A jackal of the saddleback species had also been caught in the traps the previous evening. Unfortunately, the trap had snapped shut on the animal's mandible, breaking it and tearing the lips. Elliot judged it "useless as a specimen." He nevertheless had one of the men skin it, "for it might be useful sometime if it was necessary to patch some specimen for mounting."[9]

The day was an unusually warm one, with a gentle wind, and thus the shade of the trees in camp "was very acceptable."[10]

MONEY TROUBLES

The next day was Sunday, July 19. Neither Elliot nor Akeley went out from camp. Dodson, however, took a short walk and shot a few birds, including his namesake bulbul, *Pycnonotus dodsoni*, and the Abyssinian scimitarbill, *Rhinopomastus minor*. According to Elliot, the latter bird, while "handsome and graceful," was the "noisiest creature among the feathered tribes of Somali-land."[11] Akeley spent some morning hours packing another barrel of skins—the expedition's sixteenth.

One man who'd accompanied Dualla to the coast returned to camp, bringing mail. He reported the arrival of ten camels with their loads at Hargeisa. The headman was still at Berbera, finalizing various purchases. Elliot ordered the new camels brought to camp. The new head cook had also arrived. This was appetizing news, as a tent boy had been serving ineptly in this capacity since Musa's departure. Elliot had hoped that the entire caravan would've arrived by now, as he was anxious to proceed into the interior. The messenger explained that there was some kind of money issue in Berbera, which accounted for Dualla's prolonged absence. Even so, Elliot expected the headman to arrive in a few days.

All the hunters set out the next morning. Dodson sought birds, while Akeley and Elliot went after larger game. Only a short distance from camp, Elliot sighted a small herd of gerenuks among the trees. He proceeded to stalk them carefully. The animals had spotted the hunters and were wary. But, with effort, Elliot managed to get within 150 yards and shot one, which proved to be a female. At that distance, and with so many trees and shrubs around, he'd been unable to tell whether the animal had horns. The rest of the herd immediately bolted. Elliot sent for a camel and had the gerenuk taken to camp. He then went on, hoping to find oryx or lesser kudus. Instead, he found a drove of camels, which normally indicated an absence of game. A woman approached and told Elliot that she'd seen oryx in the vicinity the day before. The curator found tracks, "but it is often a long way between tracks & the animal which made them," he reflected.[12] He returned to camp without seeing any more game. Akeley had also returned, unsuccessful. Elliot suspected that too many people in the area were frightening away the game. He considered moving camp.

Dodson, meanwhile, brought in some birds to skin for specimens, as well as some jungle fowl and francolins "for the pot." He skinned the birds, then hung the meat from a line near the kitchen tent. They typically suspended their birds in the shade for a day or more to tenderize. Even so, Elliot found that francolins were "not especially tender or well flavored, even to a hunter's appetite."[13]

Some expedition men were digging a hole in the sandy riverbed nearby, searching for water, when they found, some three feet down, a bulgy-eyed, toad-like frog, *Rana delalandii*, which they cheerfully captured for the collection.[14]

Later that afternoon, the new camels arrived, bringing fresh supplies. Tired from their long trek from Berbera, the camels would require a rest of several days before they could be expected to begin a long journey across the Haud.

The wells were crowded all day with villagers watering their livestock. Akeley attempted to photograph them, but they recoiled from his camera. The women went so far as to cover their heads with their tobes. They sent a man to ask Akeley to desist. The taxidermist nevertheless managed to sneak some snapshots by hiding the camera under his tent.

A NEW SPECIES OF ANTELOPE?

Akeley went back to the twin peaks of Nasr Hablod, where he'd bagged his greater kudu. Elliot and Dodson had just finished lunch when he returned, announcing that he'd killed a possible klipspringer—a small antelope they hadn't yet obtained for the collection. Akeley related his experience that morning: leaving a shikari to guard an oryx he'd just shot, he took his rifle for a ramble through the brush. The peaks were only a few hundred yards to his front, and he was "irresistibly drawn towards them, influenced probably by the memory of the Big Bull Koodoo I had killed at their base." He crept cautiously forward, crouching as much as possible in the gullies and ravines, until he approached the base of the smaller peak. He remained hidden behind some enormous boulders. He couldn't go any farther without exposing himself to any animal to the front of his position. From there, he made a careful scan over the ground, soon spotting a band of antelopes carelessly clamoring up the slope. The unsuspecting animals stopped in the shade of a spreading thorn tree and looked back in Akeley's direction. Based on the pattern of their gait and the size and peculiar pitch of their ears, Akeley concluded that these were kudu cows and calves. To his right was a great jumble of rocks, which—assuming he could reach them—would bring him within two hundred yards of his prey. He crawled slowly on his hands and knees, making a wide detour, keeping to the gullies. At last, he gained the desired ground. Peering cautiously over the rocks, he saw the animals still standing peacefully in the shade of the tree. He fired two or three times, the bullets all going high. It was then he realized that instead of shooting at kudu some two hundred yards away, he was firing at a much smaller antelope less than a hundred yards distant. Their kudu-like movements, together with their large ears, had duped him. Stricken with terror, the animals scrambled around in confusion. Akeley now took careful aim and knocked one down, but it immediately rose and scurried under a bush. With his next shot, he killed one in its tracks, and

FIGURE 17.2.

Nasr Hablod, the twin peaks where Akeley collected his small, mystery antelopes. Photograph courtesy of the Field Museum, CSZ5948.

the remaining animals made a mad dash to escape, running past their tormentor at about forty yards. Akeley succeeded in stopping one with another well-aimed bullet, but it struggled to its feet once or twice before falling for the last time—or so Akeley thought.[15]

The hunter then went to collect his prizes. He approached within ten feet of the first animal he'd wounded and stood gazing in wonder, surprised to find such a diminutive beast. "It was a hard thing to come down from the idea of a Big Koodoo, one of the grandest of African antelope, to a little thing not over twenty inches high at the shoulder," he later wrote. Just then, the wounded animal leaped up and dashed away to safety. "I have never seen any animal move so quickly," he wrote. "I had a fleeting glimpse of him a few moments later going up the side of the opposite peak on three legs, and from the way he traveled I think he could have got on fairly well with only one." He then approached the dead animal and found it was an adult female. Next, he went to look for the other wounded animal that he'd seen painfully kicking the gravel, but he found only a few drops of blood. The midday sun was burning fiercely, so he picked up the little antelope and carried it back to the oryx, only to find that a horde of vultures had already begun to peck at the carcass, their yellow beaks dripping red with blood. The shikari, having heard shots, disobeyed his orders and left the specimen to find the cause of the commotion. They returned to camp "with such booty as was left."[16]

When the new specimen arrived in camp, Elliot recognized at once that Akeley's antelope was no klipspringer. Examining the carcass carefully, he soon suspected that this was something new to science—a fact he could only determine with certainty by comparing it to similar specimens at the British Museum. Did this animal differ enough to be named and described as something new? "It is a beautiful little creature," he wrote, "smaller than any gazelle, with large ears shaped like those of a koodoo, & its color is a sort of mauve with purple intermingled & a white belly." Akeley had seen four animals: a buck, two does, and a juvenile. He shot the buck, a doe, and the young one. All but the doe, however, managed to get away. It'd been "almost impossible to see them they were so like the rocks, & they ran & hid themselves at once. They did not jump like klipspringers but ran more like koodoos. The hoof is peculiar, the bottom & hind part covered with a thick pad, but the front ends are pointed," Elliot wrote.[17] Akeley made a cast of the mystery antelope's head. Elliot decided to keep the skeleton too. He wanted to save as much of the animal's remains as possible, in case they were unable to acquire more specimens. Skin and skeleton would both be useful in trying—if possible—to differentiate this animal from other, similar species.

The following day, Akeley went back to Nasr Hablod. He returned to camp for lunch, bringing another of the mysterious antelopes—the one he'd wounded the day before. He'd hardly reached the base of the peak when the small antelope "jumped from almost under our feet and ran swiftly, but only for a short distance, as it was very weak." He quickly captured it, and after hunting the mountains thoroughly without finding a trace of the others, he abandoned the search. The new specimen was a young doe, exactly like the older one obtained previously. Elliot thought immediately of the animal's value as an exhibit object. "If we can only get a buck," Elliot wrote hopefully, "the group will be all right."[18]

Twice that day Elliot had to attend to medical patients. One of these was a woman bringing a baby with "the usual sore eyes." The other was a boy "with a dreadful leg ulcerated in a fearful manner." The boy had been afflicted for four years, the sores spreading from his foot to his calf. Elliot considered amputating the lower limb, but the boy was an orphan and there was no one to look after him during his recovery. The curator advised him to go to the surgeon at Berbera for long-term care. He found this case especially discouraging. "In certain ways these savages are perfectly callous to human suffering," he wrote disparagingly, "& if a person is no relation . . . & sometimes even if he is, they will let them die rather than take the trouble to care for them."[19]

Dodson brought in a striped hyena that he'd shot as it was speeding past him in the bushes. The locals called this animal *werra*, and the striped hyenas that they'd gotten before *deda*, considering them different animals altogether. The *werra*, they informed Elliot, hunted by day, and getting among a flock of sheep, it would kill them all before eating anything. Elliot was skeptical, although he conceded that the present specimen seemed to differ from others they'd killed. The markings were distinct and differently distributed, while the ears were a different shape. Again, it would require "careful comparison . . . before the question can be settled but it would seem strange to have two distinct species in similar localities so closely related."[20] The animal appeared to be full grown but still young, and much smaller than the ordinary striped hyena. While local knowledge suggested that there could be two separate species of striped hyena in Somaliland based in part on behavior, Elliot preferred to decide the issue for himself in London using the physical attributes of his specimens.

Elliot returned the stallion he'd seized to its rightful owner, his own mare's spear wound having nearly healed. For reasons he didn't bother to record, the curator had somehow concluded that the horse's owner was innocent of any part in the bizarre crime, "his horse having only been in bad company." Elliot must have felt a pang of guilt for his spiteful behavior. He gave the man a tobe to compensate for the anxiety he'd suffered when he thought his horse was gone for good.[21]

A heavy thunderstorm materialized one afternoon. The rain was accompanied by hail, and small bits of ice were gathered gleefully by the men. Although this was the rainy season, heavy showers were exceptional, and in Ogaden, where the expedition was headed, there'd been no rain at all—the grass was said to be withered and brown. Elliot thought that the outlook was rather bad for pasture for the livestock. He sent two scouts forward to Havale, a point on the march across the Haud, to determine whether there was any water to be had. If so, Elliot intended to move camp, for their present vicinity was denuded of game.

The next morning, Elliot made "a very long tour," setting out soon after sunrise. He saw nothing but a band of four gerenuks. One of these, a male, was standing broadside, looking in Elliot's direction, and started to run just as he pulled the trigger. The bullet, however, caught it on the back, breaking it. The animal fell with a single shot. It proved to be "a very fine specimen" with horns eighteen inches long.[22]

Akeley remained in camp to pack. There were enough skins on hand to fill another barrel and a half. This would make the fourth barrel of specimens since leaving Adadleh. Dodson, meanwhile, shot some birds new to the collection, as well as a common dwarf mongoose, *Helogale undulata*—the smallest carnivore in Africa—of which both skin and skeleton were preserved. Elliot couldn't identify this shy, retiring species in the field, calling it only a small mammal in his journal. This was so scarce a species in museum collections that the curator deeply regretted that he was unable to bring back more specimens.[23]

The scouts returned from Havale in the afternoon and reported no water, so Elliot decided to forego that place. He was now resigned to await Dualla's return at their present location. He sent a man to Berbera with a strongly worded message for his headman, expecting that it would "probably hasten his departure from that place."[24]

Just before dark, "quite an excitement was occasioned in camp" by three wild donkeys ambling down the riverbed, probably to drink from the wells. The men were making such noise in camp, however, that the donkeys stopped and rushed up the opposite bank. Elliot, Akeley, and Dodson all seized their rifles and immediately gave chase. They found tracks, but no donkeys. The spooked animals, Elliot thought, "probably ran out of the country."[25]

A PITIABLE SIGHT

The next morning, Elliot crossed the dry riverbed west of camp. Soon after leaving, he spotted a male gerenuk with a half-grown juvenile. He couldn't get a shot at the latter but managed instead to make a successful stalk on the male, shooting it in the breast. With the crash of the bullet, the animal reared on its hind legs and swung around, disappearing behind a clump of brush. Elliot knew the animal was wounded badly, and soon found it lying dead. It was young, with short horns. Elliot loaded the carcass on his mule and took it back to camp. He set out again, heading in another direction. Soon he found fresh oryx tracks. These he followed for a long distance, expecting to see the animal at any moment. But the tracks led to stony ground where he couldn't follow. Crossing a hill over which the oryx probably passed, Elliot spied a herd of seven gerenuks. The ground was open, with only a few thorn trees and no undergrowth, so the animals soon spotted Elliot and sprinted away. He followed them for nearly two hours over various terrain until, at last, he found them again among some aloes and thick shrubs. There, he managed to get within a hundred yards and shot one, which fell, rose up again, and moved gingerly away. The remaining animals ran a short distance and stopped to look, an old male presenting Elliot with its ample profile. He fired, and the gerenuk fell in its tracks. Elliot raced up, and Elmi cut the wounded animal's windpipe before it died. The first wounded gerenuk was nowhere to be found.

Worn out from a vigorous chase in the hot sun, Elliot sank exhausted beneath a tree. Elmi, meanwhile, went to the top of the hill to look for the wounded animal. Elliot heard him whistle and saw him beckon urgently, so he reluctantly trundled up the slope. The injured gerenuk was there, cowering under a bush. Elliot fired and killed it. His men carried the carcass—a female—down to the spot where the old male was lying.

FIGURE 17.3.

Akeley's shikari heading up the slope of one of the peaks at Nasr Hablod. Photograph courtesy of the Field Museum, CSZ5891.

CHAPTER 17

Both carcasses were then put on a mule, which made a heavy load. Fortunately, camp wasn't far.

Akeley, meanwhile, had returned again to Nasr Hablod, looking for male specimens of his mysterious antelope in order to complete a group. After hunting for a long time with a shikari, finding nothing, he sat down to rest on a boulder. The shikari suddenly made an excited gesture. Looking in the direction he pointed, Akeley saw a group of the small antelopes moving swiftly up the side of the peak, stopping now and then to look back. So perfectly were they camouflaged that they became nearly invisible when standing still. The hunters watched until the animals disappeared over the peak of the hill, then raced as fast as they could up the slope, hoping to head them off. But, reaching the top, they peered cautiously over and spotted the nimble creatures already heading out onto the plain on the far side. Akeley took a few hasty shots at them, but "only made the stones fly."[26] The hunters followed the antelopes out onto the plain for an hour or more without so much as seeing them again.

When Elliot returned to camp, Dodson was already there, having shot some birds and two more ground squirrels. After lunch, Akeley returned, bringing a male greater kudu about two years old and a female of the lesser kudu. He first spotted the greater kudu when he was too far away to shoot, and it rapidly ran out of sight through hilly country scored by ravines and tangled with scrub. He tracked the animal before losing its trail completely. Remarkably, he then decided to try to follow his prey by instinct. So, "constituting myself an escaping koodoo, I went where I thought such an animal should," Akeley later wrote. He knew he couldn't be on the kudu's true trail, because he could find no telltale tracks. Then, his progress was stopped by a thick, golden strand of web, "spun by a handsome yellow spider with black legs. Twisted together, it was substantial enough to be wound around . . . my watch chain where I wore it for several years. Had my koodoo passed between those bushes, the web would, I knew, have been his necklace." After following this instinctive track for some two or three miles, he gained the top of a ridge, which looked down across a ravine. He crept carefully to the edge and peered over, hoping to spot the kudu. Seeing nothing, he decided to stand, with the idea of scaring his prey out of hiding or giving up the chase. As he suddenly stood, he saw the kudu fleeing across the ravine, some three hundred yards away. Akeley knew he had little chance of hitting the animal, so he fired at the loose rocks on the far side of the ravine. Hearing the clatter of rocks at the spot where the bullet struck, the kudu stopped abruptly, listening. It was partly obscured by a magnificent euphorbia, the shrub's spiny branches spread like a candelabra. Akeley dropped down, resting his rifle on a stone, and, taking careful aim, he fired. A fraction of a second later, once he'd recovered from the recoil of his rifle, Akeley looked for the kudu, but it was gone. Puzzled, he raced for the euphorbia. There, he found the kudu, dead, but with no obvious wounds. Making a careful examination, he discovered that the bullet had entered behind the kudu's ear. As it listened to the falling rocks, the ear had been pitched forward. When the animal fell, the ear had swung back to its normal position, covering the tiny entry wound. The bullet had exited through the kudu's eye, but when Akeley arrived the dead animal's eyes were closed.[27]

A relatively inexperienced tenderfoot when he arrived in Africa, Akeley had by now developed into a skilled big-game hunter.

Camp that afternoon resembled a butcher shop, as all the animals were being skinned and the choicest cuts of meat were distributed among the men. "It is . . . pityable," Elliot wrote, "to see the miserable wretches [at] the zareba in hope of getting a piece of meat. . . . The old & feeble, crippled & sick are all represented, no body apparently to take care of them or to care whether they die or not." He reflected that it was "a great misfortune to be old in a savage land." It was better, he thought, "to die young or in one's prime, for no one has any use for old people."[28]

Elliot would learn years later that it was a misfortune to be old in his own country, too.

18

A PRETTY STATE OF AFFAIRS

July 25–30

Elliot's ideas about the Somali people he encountered were informed by books he read during the expedition. Harald Swayne's recently published hunting narrative *Seventeen Trips through Somáliland* was especially useful. Swayne wrote that "the whole country has been from time immemorial in a chronic state of petty warfare and blood feuds." Richard Burton recorded similar impressions of the Somalis. "They are a people of most susceptible character," he wrote. "Soft, merry, and affectionate souls," he called them, yet, "they pass without any apparent transition into a state of fury, when they are capable of terrible atrocities." Burton, who wore a terrible scar on his face from a Somali spear, emphasized that his attackers were prone to sudden violence. "When the passions of rival tribes, between whom there has been a blood feud for ages, are violently excited," he wrote, the Somalis "will use with asperity the dagger and spear."[1]

Elliot was puzzling over the enigmatic Somali character when he went out to hunt the next morning. Why were they so indifferent to the suffering of their fellows? Why so volatile? Why the perpetual tribal vendettas? He felt great admiration for many of them individually, but, like Burton, he tended to see the worst in his Somali comrades when he thought of them collectively. Ideas like these fostered a sense of cultural superiority, which was used to justify the colonization schemes of the European powers. This, in turn justified Elliot's salvage zoology mission. Lost in thought, Elliot tramped a long way over and beyond the mountains where Akeley had killed his big kudu, hoping to find one for himself. He found no game at all.

Walking back, Elliot observed that the hills around camp were paved with small stones. He'd never seen stones so "impartially distributed, & walking over them [was] frightful." The oryx seemed to prefer such ground, which made it difficult to stalk them, as there was no undergrowth and the thorn trees were too few to offer sufficient cover. Tracking over such ground was likewise difficult, as the animals left few prints. The valleys between the hills, on the other hand, were filled with sand, most likely washed down from above, and this was covered with a dense scrub of aloes, thorn trees, and bushes. In this habitat, lesser kudu and gerenuk could be found, though when startled or wounded the animals would generally leave their cover and flee to the stony hills. It wasn't easy country to hunt, Elliot concluded. Yet they'd done "very well" salvaging animals for their specimen barrels.[2]

Akeley and Dodson, who had specimens to process, remained in camp all morning. They filled another barrel, the fifth since leaving Adadleh. Elliot joined them for lunch at midday. In the afternoon, a heavy thunderstorm crept through the area, pinning the hunters close to camp. From the vantage of his tent, Elliot could see blue curtains of rain falling in various directions. In the evening, it struck their campsite, bringing a hard, steady downpour for three hours. The dry riverbed filled with a fast-moving brown slurry flowing irresistibly down from the surrounding mountains. It was easy to imagine his specimen barrels bobbing like toys on such a flood. The rain, Elliot hoped, would provide water on the Haud for his caravan.

Elliot played with his captive cheetahs, which were beginning to scratch and claw their way into the museum of his heart. They were "most interesting creatures," he confided in his journal. The male would eat scraps of meat from the mess, while the female took only milk from a saucer. He freed them from their cage several times a day for "a romp on the grass," where they played like tame kittens. He was strangely troubled by thoughts of their eventual fate. "I do not know that I can make up my mind to order their execution," he wrote, "for now there will be little trouble in raising them. The difficulty is in carrying them, for although they weigh but little now, before I get back to the coast they will require a large cage, & every pound of weight has to be considered."[3] Freeing them, of course, was out of the question. In Elliot's view, purchasing the cheetahs had transformed them from animals to specimens, and specimens belonged rightly to the museum.

The young male cheetah showed "great pluck" one afternoon. It was out of its cage, lapping milk, when the expedition's goats were brought in from grazing. When the cheetah spotted a half-grown kid, a vigorous chase ensued. The kid made immediately for its mother, which was then being milked. In a panic, the mother goat started on a run, upsetting the milk jug. The cheetah then went after it. "I do not know what he would have done if he had caught the goat," Elliot wrote, "but it was comical to see such a mite so earnest in the chase." Ahmet seized the troublemaking cheetah, which offered no resistance, and put it back in its cage. The same cheetah later chased some sheep the local villagers were leading to water. Sheep and women all panicked and scattered in terror, although the expedition's pet was no larger than a good-sized kitten. Obviously, as the captive cats grew in size and strength, the curator would have to take greater care with his livestock. For now, however, both cheetahs were relatively easy to handle. The male liked to sit in Elliot's lap and watch the movements of the men. Twice it managed to get beyond the zareba to scale trees nearby. The birds, recognizing a predator, "kept up a tremendous chattering" until the cat retreated to camp.[4]

A villager brought another beautiful, young leopard to camp, which Elliot bought. It was maybe four months old, and very spirited, spitting and growling with comical menace whenever anyone approached. The villager had captured it a few days previously. He'd found two siblings sleeping, but when one of them sprang at him, he speared it. The other tried to run but was caught. It was "as thin as a rail," Elliot wrote, "& looks as if it had not had a square meal since it was born."[5] Its coat, however, was rich in color and beautifully marked. Akeley took its photograph before Elliot dispatched it.

The next day was another Sunday, July 26. It was a beautiful day, bright and cool after the rain, but with a steady wind. The riverbed was a quagmire of mud and flotsam. Elliot remained in camp. Dodson visited Bodeleh—about three hours away—to look for water. A rumor had reached them that lions were killing livestock in that vicinity. The curator hoped to salvage more of these charismatic and fast-disappearing carnivores for his collection. Akeley started for the mountains for greater kudu but happened across two female lesser kudus nearby. He killed one and brought it back on a camel.

There was turmoil in camp shortly before lunch. Their work finished, some of the men wanted to socialize at a nearby village. A camel man insisted on joining them, although they told him "very emphatically" that his company wasn't welcome. When the camel man wouldn't relent, the others tied his hands behind his back with his own tobe and strapped him to a small tree with his cartridge belt. The captive managed to break free after a brief struggle. Humiliated and angry, he returned to camp to complain to Elliot, his hands still tied tightly behind him. The curator "could not help smiling at the fellow, & merely told him that next time he wanted to go to a village he had better go alone & take a different path from those who did not want his company."[6]

That should've been the end of the trouble. Unfortunately, there was another member of the aggrieved man's tribe in camp. This was a new man who'd recently arrived from Berbera with the new camels. This new man took the matter up and was planning to make it a tribal affair. He had "a hot discussion" with Ahmet, one of the offenders.[7] When Ahmet informed Elliot that the new man intended to complain to him formally, the curator told him to advise the man that he should stay out of it. He heard nothing further of the controversy for the time being. Yet it was far from over.

That afternoon, Elliot's informal medical practice was again very busy.

Men and women—some with babies—arrived in a seemingly endless procession for myriad treatments. Elliot prescribed basic remedies for most, but some he recommended to the hospital at Berbera, as their cases required more serious medical attention. One man reported that he'd had chronic pain in his chest for ten years, and now he wanted it cured. Another very old man wanted something to make him see better. Elliot referred him "to Allah[,] for I could not make him a new pair of eyes." Elliot regretted not having cough medicine or iodine, "for both would be most useful in many cases." Another sufferer was brought on a camel from a distant village. Contrary to Elliot's earlier complaint, the man had several devoted caretakers, including his wife, a brother, and his father—an old man. They explained that he'd been suffering for a year with severe pain in his abdomen. "He was a perfect human skeleton," Elliot wrote, "with absolutely nothing on his bones & the skin of his stomach drawn as tightly as the head of a drum. The pulsations of his heart could be distinctly seen on the skin of his breast, & altogether he was a woe-begone object." Examining the sick man, Elliot concluded "without much doubt" that his pain was caused by an enormous stomach tumor. He told the anxious family that he was powerless; he had no medicine that would do the patient "any permanent good." Elliot could see that the old man, in particular, was disappointed. If one does nothing, he remarked in his journal, the villagers attribute it to "unwillingness to help [rather] than . . . inability." The family begged the curator for some medicine, so he gave them "a simple drug that could do no harm if it did no good," and they carted the unfortunate sufferer away. "The land is full of such miserables," he lamented, "they just simply die, the fortunate ones quickly, the less so by degrees."[8]

Elliot, in a morbid mood, learned of a curious Somali belief about life and death that he noted forebodingly in his journal. The Somalis maintain, he wrote, "that when Allah made them he took up a handful of sand, & as he carried it along he dropped a little occasionally." A man would then and there be formed from the sediment. If this man should then pass over the same spot again in life, he would die. "It is not stated that if the individual in question never visited the place he was to live forever," Elliot explained, "but probably . . . it is fated that everyone in the course of his life must pass over the place where his particular sand was dropped, and depart his life without further bother."[9]

A GREAT STATE OF EXCITEMENT

Akeley and Elliot left camp early the next morning, the former for the twin peaks, the latter in the opposite direction. Dodson went on another bird hunt near camp. Elliot found the fresh tracks of a greater kudu and followed for several miles. He was certain, at times, that he must've been close to his prey, but he never saw it. He finally lost the track among those of two herds of cattle. He eventually came across these herds with their drovers, after which he made a great circuit, seeing four gerenuks, one of which was a very large male with beautiful horns, which he shot dead. The gerenuk had been standing, looking straight at the curator from about 250 yards. Though Elmi ran as fast as possible to cut the animal's windpipe before it died, Elliot was certain that he hadn't arrived in time. However, "he must have closed his eyes to the fact & made the necessary cut all the same."[10] The curator hefted the carcass onto his mare and headed back to camp.

There he found "a pretty state of affairs." An agitated Dodson was already back. He explained that when he'd returned to camp the men had been in "a great state of excitement & if he had not arrived on the scene in all probability they would have begun to shoot each other." Elliot soon learned that the man who'd been tied to a tree the previous day had tried to murder his tent boy, Ahmet. The culprit had stayed up all night fashioning a crude club. The next morning, while Ahmet was target shooting, the man snuck up behind him and bludgeoned him on the head, knocking him sprawling and unconscious. "The camp was in an uproar at once, & the men of the different tribes arrayed themselves against each other," the curator wrote. Thinking quickly, Selon "took the would-be murderer out of the crowd and led him away," thus helping defuse the crisis.[11] By the time Elliot arrived, Ahmet had recovered his senses and was lying under a tree with cold water bandages pressed to the side of his throbbing head.

Dodson had applied Pond's Extract—a topical cure-all—to Ahmet's injury, which helped reduce the swelling. After a thorough examination, Elliot determined that there'd been no fracture of the skull. He expected Ahmet to make a full and speedy recovery. "It is wonderful what these men can stand in the way of injuries of all kinds," he remarked in his journal. "Ahmet received a blow that would probably have instantly killed most white men, but thanks to the thickness of his skull he has escaped without injury." Elliot was glad for his own sake as much as for Ahmet's, "for I have become so accustomed to [his] services that I should miss him very much," he wrote. "It would have been a horrible thing," he added, "for him to have died at the hands of a cowardly assassin."[12]

The men lingered, curious to see what punitive action the curator would take. Elliot summoned Selon and had the accused tied (again) to a tree. He interviewed witnesses. He then wrote a summary letter to Captain Merewether and sent the captive to Berbera—his arms bound behind him— under guard of two men. The captain would determine his punishment. Elliot hoped this would be the end of the trouble, but he feared the worst. "From what I hear," he wrote, "there are some more men connected with the affair, & who probably incited the man to do the deed, & I have some whom I am watching pretty closely & who I shall probably dismiss from my service before I leave Hargeisa." Elliot attributed the conflict to "tribal difficulty, one man fancying himself insulted & the others of his tribe siding with him." He vainly hoped to end this ancient practice, and to get the members of different tribes to treat each other with dignity and equality.[13]

Akeley, meanwhile, completely unaware of the chaos in camp, sent for a camel to retrieve a greater kudu cow he'd shot. The camel arrived in short order. A cow was exactly what Elliot most wanted, as the one they already had was an inferior specimen, too ordinary to represent a disappearing species in a habitat group. Akeley himself came in much later, having hunted again in vain for more specimens of the mystery antelope.

Dodson went out alone the following morning. Akeley had his kudu and Elliot's gerenuk skin to process. Elliot decided that they now had plenty of gerenuks for their purposes, indeed, more than he thought he could use or exchange advantageously with other museums. So, he kept only the head and neck of the big buck he killed the day before: not a specimen but another personal trophy. The cautious curator thought it prudent to remain in camp after the trouble of the day before. Fortunately, peace and quiet prevailed all day. The "excitement" had apparently subsided, "but no one can tell when it may break out again, for it takes very little to arouse these people," he wrote.[14]

After Elliot went to bed that night, a heavy rain moved in and lingered over camp. His tent then filled with men and animals. When the weather was bad at night, he always allowed Ahmet to sleep on the floor alongside his cot. Likewise, his two shikaris took shelter just inside the flap. Additionally, the two kittens were brought in for fear that a soaking would be their death.

The next morning, Akeley started out at half past four for the twin peaks in yet another attempt to get more specimens of his mysterious antelope. Elliot went in a similar direction but angling more to the east. He rode several hours, somehow losing Elmi. Suddenly, a warthog sow and three little piglets zoomed across his path. He jumped off his horse and grasped for his Winchester, which Ibrahim was carrying wrapped in a linen cover. The curious pigs stopped for a quick look, then fled out of sight before Ibrahim could uncover the rifle. Seizing the weapon, Elliot ran at his fastest clip for more than a hundred yards and caught sight of the pigs running at top speed quite a distance ahead. Taking a quick sight, he shot the sow fatally through both shoulders, bowling it over on its bristling back. The piglets scattered in all directions. Elliot fired quickly at one of these but missed. Elmi hadn't yet come up, so Elliot told Ibrahim to guard the sow while he rode back to camp. Soon, however, he found Dodson, who went out immediately to skin the warthog, "for of course none of our men would touch it." This was "a good specimen . . . with nice tusks & a fine mane." Elliot was pleased, as they now had enough specimens to constitute "a very good group" for display.[15]

Dodson had gotten some good birds, including several new to the collection. Two of these he'd obtained once before, on his previous trip with Arthur Donaldson Smith into the far interior. Elliot was glad to record that this discovery extended the known range of these birds greatly, "as they have never before been taken in this part of Somaliland." These included an Old World babbler, *Argya aylmeri*, a small and rare woodpecker, *Dendropicus hemprichi*, and a sparrow hawk—possibly the African pygmy falcon, *Poliohierax semitorquatus*, which Elliot called "the most beautiful little hawk in the country."[16]

Akeley returned to camp empty handed at noon. Tired from stalking all morning in the heat, the taxidermist had been heading back when—turning to take a last look at the twin peaks—he saw at the top, clearly silhouetted, three pairs of ears forming a mocking row. "I imagined the animals had seen me before that morning," he wrote later, "and had become reckless when they supposed they were taking a farewell look at me." He started carefully to stalk them, "when they wheeled and disappeared." Doggedly, he climbed to the top of the peak, where—though he had a commanding view on every side—he caught no glimpse of the small antelopes. As he was stalking up the mountain, however, two hyenas sprang up from behind some boulders. He shot them both, killing one and wounding the other, which got away. The dead hyena proved to have no tail and was "a poor specimen altogether," so he abandoned it to other scavengers.[17]

FIGURE 18.1.

Drinking first from a freshet to gain poetical powers. Dodson is visible on the left. Photograph courtesy of the Field Museum, CSZ6057.

In the afternoon, they enjoyed another booming thunderstorm. Water began to flow copiously in the dry riverbed beside camp. Some of the men rushed in front of the torrent as it first raced down the rocky channel and attempted to drink directly from it. Elliot recorded that the Somalis believed that "he who drinks first will have especial poetical powers." The Somalis, with their strong oral tradition, considered this a great virtue. Selon, who was among the contestants, was renowned in camp for his storytelling ability. "Many times at night he has been surrounded by a crowd of listeners as he chanted long tales of wondrous deeds, most of which probably never happened," the curator noted disdainfully.[18]

DUALLA'S RETURN

A man who passed by camp on his way from Berbera brought word that Dualla had succeeded in buying all the camels Elliot wanted; he would be arriving with this caravan in about three days. But, with game so scarce, Elliot decided to move camp the following day to Bodeleh, where Dodson had gone the previous Sunday, and to await the headman's arrival there. He was irked by Dualla's dawdling.

To Elliot's surprise, the headman arrived that same evening, explaining that his long absence was the result of an error in the remitting of the money Elliot had asked to be forwarded from Aden to the government treasury at Berbera. Dualla had been obliged to wait for the transaction to be corrected. He'd ridden ahead of his caravan, which wasn't expected to arrive until the following day. Elliot had mixed feelings about the delay. He'd planned to be across the Haud by this time, yet they'd "done exceedingly well" collecting specimens. He was especially pleased about Akeley's diminutive antelope, "which may prove to be new. . . . It cannot be said," he remarked in his journal, "that the unforeseen detention has been any waste of time."[19]

Later that night, just as Elliot was getting into bed, "a great commotion arose in the camp." The curator distinctly heard the crashing of branches. Fearing more tribal conflict, he laid on his cot and braced himself for bad news. Instead, Ahmet came to his tent to inform him that a camel that'd been sick for several days had just died, collapsing into the zareba. He asked whether the men could cut its throat so that it could be eaten. Relieved that the news wasn't worse, Elliot was nevertheless horrified that the men were willing to eat an animal that'd died of disease. But he gave his permission anyway. The dead camel was another one bought at Gelalo from Norman B. Smith. It'd apparently died from an illness it contracted following a fly bite in Bulhar, "the second of the lot bought to die from the same cause."[20] Quiet was soon restored, and Elliot slept soundly.

Dualla's caravan arrived the next morning with the camels he'd purchased in Berbera. These were "a fine lot of animals," and Elliot expected they'd "last through the journey." The caravan also brought welcome mail from home. The hunters spent the day writing letters and arranging matters for moving the next day. The collections they'd made needed to be readied for shipment to Hargeisa for safekeeping until the expedition's return. Dualla's accounts also had to be audited, which proved to be "a long & tedious business." But after much wrangling and many debates, the curator and his headman "agreed in our figures at last."[21]

The batch of mail from Berbera included three outdated English newspapers that the hunters passed among themselves, eagerly devouring the weeks- and months-old news of Europe and America. The English news cycle in the spring of 1896 was awash in stories out of Africa. In March, for example, relenting to national outrage over his army's defeat by Abyssinian forces at the Battle of Adwa, the Italian prime minister, Francesco Crispi, resigned. Italy's loss destabilized the border region between Abyssinia and British Somaliland—exactly where Elliot's expedition was heading next. In June, British-led Egyptian forces defeated Mahdist rebels at the Battle of Ferkeh in their reconquest of Sudan. In its coverage of this momentous battle, the press touted the benefits of British imperialism. "The English occupancy of Egypt," wrote the *Sheffield Independent*, "has brought many blessings to an oppressed people, but none greater than the stiffening of their courage and the restoration of their good opinion of themselves, till they can defend their river basin like men." The Mahdist uprising, incidentally, had forced the withdrawal of Egyptian troops from Berbera in 1884, thus opening the way to British occupation. Meanwhile, in British East Africa, construction was underway on the Uganda Railway, connecting the port of Mombasa with Kisumu, on Lake Victoria. The press acknowledged that the expanding British imperial network along multiple fronts was putting increasing pressure on Africa's wildlife. A story in the *Huddersfield Chronicle*, for example, warned of the ongoing extirpation of elephants from British Somaliland. It recommended limiting elephant hunting to British military personnel. Such an act, "while tending to preserve one of the greatest animals from extermination, would at the same time secure for shooting purposes a fair slice of the Somali country for the exclusive use of British officers stationed at Aden and Berbera," it argued.[22] Elliot might have favored this act, so long as he'd had the opportunity to get his share of Somali elephants first.

Elliot dismissed a camel man, Hashee, who had attempted to instigate a tribal quarrel from the simmering animosity that'd culminated in Ahmet's attempted assassination. He ordered Dualla to explain in no uncertain terms to the men that anyone who attempted to raise any tribal strife in camp would be bound and sent ignominiously to Berbera.

19

AN ORANGE TO BE SQUEEZED

July 31–August 8

Camp was "in a bustle" before daylight broke on the last day of July. Elliot now had eighty camels at his command. It was "a wonder," he thought, "how rapidly so large a number are loaded. As each company finished loading," the camels were "led outside the camp" to "wait for the rest so that all may start together." Dualla had ridden hastily back to Hargeisa, bringing the expedition's bounty of five barrels of skins and several awkward sacks bulging with bones. These were stored for safekeeping in a sturdy, stone house belonging to a gray-bearded resident, Sheik Matar, called by some "the chief of Hargeisa."[1]

Elliot and his hunters left before the caravan was ready "so as to have a chance at any game which might be seen." They headed south, soon entering the Haud, an elevated plateau overlooking the country to the north up to the Golis Range. The landscape mirrored Toyo—open, parklike country, covered sparsely with a variety of thorn bushes. The ground, though, was stonier. They saw no game except a few gerenuks, which Elliot ignored, that is, until he spied one with "a very fine pair of horns." This one he proceeded to stalk. The animal spotted the hunters, scampered a short distance, then "stopped to have another look."[2] Elliot fired once, breaking its neck. Placing the carcass behind his saddle, they went on, arriving at their first well a little more than two hours from where they set out that morning. Though there was plenty of water and grass, Elliot wanted to push on.

They rode for another hour and came to a village with many sheep, but little grass. Worse, the water there was "so vile that even the mules refused to drink it." They immediately turned back to the first well, meeting the advance animals of the caravan on the way. Elliot transferred his gerenuk to a camel and galloped quickly back, selecting a campsite under the shade of some large trees and with "a long stretch of beautiful green grass smooth as a lawn."[3] This, he reflected, was an even prettier site than Hullier. Using his boiling thermometer, he measured the elevation at 4,458 feet. The caravan arrived in staggered detachments from both directions. The men soon pitched a comfortable camp.

An overcast sky threatened rain when the hunters set out the next morning at sunrise. Elliot's expectations were low, for there appeared to be nothing but gerenuks nearby, which he didn't need—unless a half-grown juvenile. Nevertheless, he went east and soon came to a village, where some men marched out to meet him. They advised him that Elliot would find nothing but more villages in the direction he was traveling, suggesting he should head north, where he might find oryx. Akeley, however, had already gone north, and Elliot didn't want to crowd him. He deviated slightly from his former course, though, as he hoped to avoid villages, where game was likely to be scarce.

About half an hour beyond the village, Elliot spotted a solitary bull oryx near a clump of bushes. Evidently, the bull hadn't seen him. He stalked within two hundred yards, took careful aim, and fired. The bull "kicked up his heels as if trying to throw something off his back & ran a short distance and stopped." Elliot fired again. The bull sprinted briefly, then stopped to have another look at its tormentor, this time exposing its other flank. Elliot fired again and "heard the bullet tell loudly." The wounded bull ran a few yards and dropped. On examining the carcass, Elliot discovered that all three bullets had struck just behind the shoulder, two on one side and one on the other. Yet the bull had shown no sign of duress until it fell dead. "The Oryx certainly takes a good deal of lead to finish him," he remarked with admiration.[4] Could such an animal as this—with so much vitality—really be flirting with extinction?

Elliot was using the Lee-Metford obtained from Sartoris. "I think it one of the very best weapons I ever had," he wrote, "very accurate but apt . . . to carry high at short distances." The bull oryx, he found, was old, and the men didn't want its meat. As camp was a long way, Elliot simply took the skin and gave the rest of the carcass away to some Midgans who arrived on the scene. Then he returned to camp. Akeley and Dodson arrived soon after Elliot, the former with a male Speke's gazelle, the latter with nothing. There was enough work on hand to keep both men busy for the afternoon, however, as the thickness of a bull oryx skin required "a good deal of paring before the brine can penetrate."[5]

On Sunday, Elliot slept late, then spent most of the beautiful day quietly loitering in camp. He freed his pet kittens from their cage, and they scampered off to the grass and "had a great frolic." Once he thought they'd been out long enough, he went after them, and they were so tame that they made "no objections whatever." The male was growing fast and "promises to be a fine fellow." His sister was smaller, "but makes up for size in her amount of spunk, & is at times irascible with him, but never shows any temper to people."[6] They were so interesting to watch that Elliot

FIGURE 19.1.

At its largest, Elliot's expedition encompassed some sixty men. Photograph courtesy of the Field Museum, CSZ6089.

wished to keep them alive as long as possible. The color of their coats was beginning to change, and Elliot wanted to observe these changes to see when they took on their adult pelage. He was beginning to suspect that they might be leopards.

In the afternoon, he bought thirty sheep for the long journey across the Haud, "for as we can get no water the men cannot cook their rice & must have meat instead." Elliot now had some sixty men in his party, and it took "a good deal of mutton" to feed such a large company. Elliot was again beset by "poor suffering creatures bringing every form of malady."[7] Some he could help; others were beyond human aid.

Later that night, a devout subset of the hired men held a prolonged meeting with fervent prayer and singing, at Dualla's tent. The meeting, which was kept up until quite late, was undertaken to ensure a favorable journey across the Haud. At the same time, there was a large and festive dance held on the opposite side of the camp, "by the unregenerate fellows," as Elliot called them. The songs and shouts of the latter group "mingled queerly" with the more solemn chants of the headman's congregation.[8]

Dualla served as something of a spiritual leader for the men. He'd been to Mecca three times and was regarded as quite a holy man. Elliot was nonetheless skeptical of Dualla's sincerity. "I have noticed," he judged rather harshly, "that all his religion has not taught him any self-denial nor to consider that he must not look out for himself & his own interests before any others." The curator, however, badly needed Dualla's leadership to make his expedition a success. He conceded that his headman was "an intelligent fellow—far beyond the majority of Somalis, & he does exert . . . much influence among [the men]." On the other hand, Elliot suspected that Dualla regarded his employer "as a financial sort of orange to be squeezed for his benefit whenever an opportunity presents & he creates as many of them as he can." The curator found this practice annoying, as it put him in a position of nearly constant vigilance. Despite all the headman's shortcomings, however, Elliot considered Dualla "without doubt the most competent Headman for an expedition there is in Africa today."[9]

BEDRAGGLED AND SOAKED

They all went out early the next morning, Akeley and Elliot heading in the same direction for a short way before separating. Soon, Elliot saw a fox darting among the aloes. A hasty shot missed its mark. Moving on, he spied a Speke's gazelle and proceeded to stalk it, almost running right into a herd of oryx, composed of three old individuals and eight young of various ages. Elliot crept "quite near" and was picking out the best one to shoot, when his "stupid" second shikari came "sauntering along with his tobe flying in the wind," and the oryx ran off at a full gallop. Elliot continued, seeing a small herd of eight or ten gazelles. He stalked these for a long time. They were wild, and the country was open, so it was almost impossible to get near them unnoticed. At length, he managed to get within fifty yards of "a fine buck standing broadside" to him and "felt sure" of an easy kill. Elmi handed him a rifle, and he took careful aim and fired. Away went the buck at the crack of Elliot's shot. How could he have missed? Checking his sights, he noticed they were set for three hundred yards. His bullet must have sailed a foot or more over the gazelle's back. The gun and gun bearer were to blame, Elliot concluded: "The sights move much too easily on this rifle & they must have rubbed against Elmi's tobe when he handed it to me."[10] He soldiered on. Shortly thereafter, he found a large herd of gerenuks. Among these he spotted some juveniles. He fired at these, hitting one, he thought. It ran off, however, and, reaching stony ground, he couldn't follow its tracks.

Convinced by now that this wasn't his day, Elliot started back for camp. Just then, a pair of dik-diks sprang up in his front. He leaped dexterously from his mount and shot one, putting the small carcass on the rump of his horse. Thinking his luck had turned, he kept on toward camp, gun at the ready. He soon saw three gazelles walking stealthily. He crept within two hundred yards, but with no bush or tree in the intervening space, there was no way to approach any nearer. He was too far away to tell buck from doe, so he took aim at the middle animal, sending a bullet crashing through its body. The others quickly sprinted away. The carcass proved to be a female, which was a disappointment. Placing it with the dik-dik on his horse, he returned to camp. He found both Akeley and Dodson had gotten in before him, the former having brought back a male Speke's gazelle, the latter nothing. Dodson, who was in a hunter's slump, had seen nothing save a lone jackal. In his journal, he wrote: "Shot nil."[11]

Elliot learned that the men who'd escorted the culprit who'd attempted to murder Ahmet had returned from Berbera. They reported that their prisoner had been sentenced to six months in jail and fined twenty rupees. Elliot assumed that this sentence carried with it the usual provision that the guilty party is forever banned from going to Berbera or Aden, which, the curator thought, was a severe enough penalty to serve as a deterrent for the other men.

Shortly after lunch, a heavy thunderstorm moved through and "freshened up every thing greatly." A caravan then arrived from Ogaden. These travelers reported that the country ahead was entirely desiccated by drought, and that all the wandering people with their flocks and herds were moving on to Milmil. Elliot expected to find a great crowd of villagers there when he arrived. The wells, therefore, would likely be "in a terrible state of nastiness." He hoped that his expedition might "carry the rain with us & change the condition of affairs." On the other hand, the distant Shebelle River, which the expedition would eventually have to cross, was reported to be "running full as its banks will hold."[12] He was now determined to begin crossing the Haud the following day. Dualla had hired eight extra camels to carry water. These, with his original six, made fourteen camels for water alone.

With a long journey looming, no one went hunting the next day. The men spent the morning preparing loads of supplies and equipment for the camels. The son of Sheik Matar arrived from Hargeisa, bringing four unburdened camels. The half-filled barrel of skins and sack of bones obtained at this camp were later sent back with him to be put with the rest of the expedition's specimens.

The caravan set out in the afternoon with the usual shouting and growling. They marched steadily across open country until they struck the Milmil road—an oft-trod, meandering camel track across the Haud. They gradually ascended until, looking back, they could see the Haud to its northern limit, stretched out behind them like a sea of green.

Clouds had been gathering in the west for hours, and the rumble of thunder portended a storm. Elliot hurried back to the road from a fruitless stalk, intending to halt the caravan and make camp before the rain commenced in earnest. First, he went after Akeley and Dodson, who'd passed on ahead. But, before he reached them, the rain began to fall in "perfect torrents," driving Elliot under a sheltering clump of trees. While he was taking refuge there, some of the caravan slipped by unnoticed. The ground all around was soon flooded with several inches of standing water, and Elliot's refuge "became a sort of quagmire." When the rain let up slightly, he went on, soon finding Akeley and Dodson huddled with their men under some nearby trees. They told Elliot that a part of the caravan had gone on ahead, and that they'd seen Dualla pass by, riding his horse while holding a tent over his head. Indeed, he was nearly smothered by the soaking-wet weight of it. Elliot went on after the camels, the rest of the men following reluctantly. The footing in the mud was treacherous, even for the horses. The camels with their loads must've been "having an awful time of it," Elliot thought, "slipping & sliding about & in great danger of falling & breaking their legs." In fact, the caravan had split into several detachments, each one making its way as well as possible. Elliot passed two stubborn camels that had plunked down in the mud and refused to go any farther. He struggled through this "slough of despond, the rain still falling & night coming on rapidly."[13] After a while, the ground became somewhat dryer and the footing surer as he ascended ever so slightly.

At long last, Elliot found Dualla with a few of the expedition's camels milling about on an open space, mostly free of water, where they could stop in relative comfort for the night. The camels kept arriving a few at a time, their loads absolutely soaked through and weighing a great deal more than when they'd set out that afternoon. Darkness fell rapidly, but most of the camels got in safely, save five of Elliot's animals and the eight hired ones carrying more than half of their water supply. Together with their handlers, these camels would have to fend for themselves for the night. Their vast flock of sheep came in "bedraggled and soaked, & looked a thoroughly miserable lot."[14]

The cage with the kittens had been put atop one of the loads, which came in among the last of the camels. The captive cats were drenched, of course. An uncharacteristically sentimental Elliot worried about their health, yet they were energetic and ate as voraciously as ever when let out of their cage for supper.

The men soon built cheerful fires. "A wetting to a Somali is of no consequence," Elliot remarked, "& the few clothes they wear are soon dried or exchanged for another tobe, or left on just as they are." The curator anticipated that they wouldn't be able to march out early the next morning, as the wet gear would be too heavy for the camels. "For a waterless tract as the Haud is always reported to be," he wrote, "we have found it a pretty damp sort of a desert."[15]

THE ROAD LESS TRAVELED

Morning broke clear and beautiful after the storm. The wayward camels had straggled into camp in the night, and all were now accounted for. Due to the thorough wetting everyone and everything had received, however, they didn't start out again until after ten o'clock. They had to give the camel matting, especially, an opportunity to dry. Elliot started out ahead of the caravan, after bidding goodbye to the son of Sheik Matar, who left the expedition there, wishing Elliot success and promising to be an excellent steward of the specimens left in his father's care.

Elliot described the road across the Haud as "only a well-beaten path, winding in and out among the trees, never straight for any distance, simply as the first camel laid it out so it has remained[,] every subsequent beast with its load following precisely in the footsteps of its predecessor."[16] How many camels had trod this path, he wondered, and for how many millennia? They rode in single file for several monotonous hours. No game was seen. They passed a woman who was tending a large flock of sheep and who fled at the sight of them. She was afraid of their guns, the men speculated. By midafternoon, as they approached Silo Plain, Elliot spotted three oryx and dismounted to stalk them. Firing from a hundred yards, he broke the back of an adult female. Akeley and Dodson arrived, so they all unsaddled their horses to await the caravan, the men in the meantime roasting the savory liver of the oryx over hot coals. Half an hour later, the camels arrived and camp was made.

Elliot estimated that Silo Plain was only ten or twelve miles across. He'd planned to be on the far side of it by then, but the storm had cost them half a day. Late in the evening, another spitting rain fell. Elliot feared more delay. Just as he was pondering the next day's march, Dualla announced that he'd just received news from a man who'd recently arrived from Milmil that—due to drought—there was no grass there at all. As this meant that there would be little or no grazing for the animals, Dualla suggested that they leave the Milmil road and strike out across the Haud, to a well southwest of Milmil called Higleleh. Dualla confessed that he'd never traveled that route, but he could hire a guide to lead the expedition there. Elliot gave his consent, although he was "loath to leave a beaten track."[17]

They awoke at half past three in the morning. A brisk wind was blowing, and the men shivered with cold. They stood huddled around several crackling fires, waiting impatiently for daylight and a hot breakfast. The caravan set out at five o'clock, quite a bit of time being needed to load the camels. Both Elliot's unhobbled horse and mule had started back on the road to Hargeisa. An attendant soon brought them in, and Elliot then forged ahead of the caravan. After riding for some five or six hours, they stopped for lunch, the main body of camels coming up shortly after they halted. Dodson came in soon thereafter, bringing a young female oryx he'd killed. Due to the scarcity of water, the men were treated to an agreeable meal of roasted sheep.

They set out again in late afternoon. After riding another hour, they came to a barren plain called Maredleh, which rose gradually to the southwest. They marched across this until early evening, arriving at a deserted village, the zareba of which furnished the expedition with abundant firewood. A large herd of oryx was grazing peacefully near the spot where they halted for the day. As they approached, a startled leopard bounded away into the twilight. Soon, the camels came up, having put in a hard day's march of more than twenty miles. The men pitched the tents in the dark, glad to be stopping for rest. One camel gave out and stubbornly laid

FIGURE 19.2.

A giant termite mound surrounded by aloes. Photograph courtesy of the Field Museum, CSZ5954.

down just short of camp. Its load was transferred to two willing donkeys and brought into camp.

The desolation of Maredleh was striking. This was evidently the work of a voracious species of termite—*Macrotermes bellicosus*—which had devoured all the trees and woody shrubs that once covered the plain. The destructive work of these insects could be seen everywhere. Another common insect was a large, black ant, *Megaponera analis*.[18] Elliot observed columns of these ants going on foraging expeditions and returning with struggling, white termites clutched in their formidable mandibles. When injured, the black ants emitted an overpowering stench. This odor was a great mystery until Elliot finally sniffed out its source.

A HEROIC ORYX

They made another early start the next morning. Elliot, who rose at half past three in the morning, led the march across the plain before dawn broke. They ambled forward for hours before reaching a dense clump of bushes that seemed to mark the edge of Maredleh. During the march, a shikari called Elliot's attention to an animal he called a "rhinostrich." Looking in the direction the shikari pointed, Elliot spied a distant ostrich— their first. After looking Elliot and his party over for a moment, the big bird strutted quickly away.

Not far from the edge of the plain, a grassy tract nearly concealed a herd of oryx. Elliot slipped from his mule and stalked them carefully, the tall grass hiding him completely. He crept within a hundred yards of a huddled pair and fired, the bullet striking loudly. The wounded oryx kicked up its heels and sprinted away with the rest of the herd. Elliot fired a second time, his bullet striking home. The oryx, however, kept on. Soon enough, the entire herd came to a sudden stop, looking back to see what the trouble was. Elliot fired again from two hundred yards, hitting another oryx. At this, the herd bolted at a full run. The animal Elliot had wounded first, however, split away from the rest of the herd and went alone into the partial cover of some grass. Elliot crept up and shot it twice more. The wounded oryx ran a short distance and stopped. Once again, Elliot moved in its direction, but the oryx had hidden itself so completely that he couldn't see it. Elmi wanted Elliot to press on, but, "mindful of what a wounded Oryx can do with its long horns . . . I advised caution, & it was well I did for suddenly the animal rushed from cover." Elliot shot again as it ran, and after going a long distance, it stopped. Elliot followed cautiously. He could only see its horns, as it had laid down in the grass. Guessing from the horns where its body must be, he shot it again. The oryx jumped up and was off on a full run. Elliot shot again as it ran, bringing it to another halt. The oryx looked back at its antagonist. Once more Elliot fired, knocking the wounded animal down. But it kept its head up and was more than capable of using its horns defensively if given the opportunity. So, Elliot fired one last time, killing it. He was using the Lee-Metford rifle with hollow bullets. The oryx had been struck in more than half a dozen places, several of which *should* have been fatal wounds. "I never saw any animal show such vitality & ability to carry lead," Elliot marveled.[19] The animal proved to be a large bull, which had probably survived many dangerous encounters with its own kind. Elliot sent for a camel, and his men skinned the heroic bull while they awaited its arrival.

Soon they could see the caravan passing, and so they made haste. Elliot had heard a shot before he started, and he stumbled on a bloody patch of grass where an animal had apparently been skinned. He overtook the caravan and, after riding rapidly, came up to Akeley and Dodson. The latter, he learned, had killed a half-grown gerenuk, which, together with an abundance of other specimens of all ages and sexes, more than satisfied their requirements for building a habitat group of that species. Late in the morning, they crossed the fresh paw prints of three lions. These they followed a long distance, but as there was no cover and, apparently, no end to the tracks, they finally gave up and returned to the caravan. The hot, midday sun made traveling odious. The camels were nearly spent. One, in fact, gave out completely and had to be shot, "as it would have been cruel to leave it to the mercy of wild beasts." They stopped for lunch at noon. Three hours later, they started again and made a relatively short march of two hours, when they stopped and made camp. Their guide expected that they would reach the wells the following afternoon. Elliot hoped he was right, "for this waterless country is not pleasant."[20]

Around midnight, camp was "thrown into an uproar" by a leopard that'd leaped over the zareba behind Elliot's tent, stole between it and Akeley's tent—despite a dense network of guylines—and seized and killed one of the sheep that had been huddled together protectively in the center of camp. That night was dark as ink, and the stealthy cat wasn't seen until it tried to bring the dead sheep out the way it had come. A sentry then spotted a shadowy form dragging the limp, pale carcass along. Of course, the first indication that something was amiss was a stampede of terrified sheep, which provoked the camels. Elliot at first feared that "the entire circus would go over my tent & mix me up in the ruins," but the men very quickly and efficiently rounded up the animals. Even so, there was "great shouting & rushing about."[21] In the confusion, the leopard escaped without the sheep, whose throat was quickly cut in order to salvage the meat for human consumption.

Camp was astir only a few hours later, and, by five that morning, the entire caravan was again on the move. As usual, Elliot went ahead with Akeley and Dodson. Unfortunately, only gerenuks were seen. The few trees, which grew far apart, offered no shelter for stalking. Elliot forged ahead until half past ten, when he halted to wait for the camels. These, however, didn't arrive until noon, thus making a march of seven exhausting hours. Once they were loosed, many camels stood in the shade of the trees and refused to eat. Seeing their fatigue, Dualla wanted to make camp there and continue the following morning. The guide, however, insisted that the wells were close, and Elliot thought best to finish the route before resting. Thus, by midafternoon, they were again on the march. In an hour, they began to see villages. Stopping at one of these, they purchased a sour and smoky portion of camel's milk. Many women fled from them. Elliot suspected that they'd been mistaken for Abyssinians, "who have from time to time raided different parts of Ogaden."[22]

In an hour, they reached a precipice that marked the limit of the Haud. From there, they gazed over "a wide extent of country lying below & beyond us, & the view was very fine."[23] Through the Jerer Valley below wound a dry riverbed where they hoped to find water. In the distance rose the tall peak of Sabatti, near which were the wells of Dagabur. This was the very edge of the range of rhinoceros, elephant, and zebra, each of which Elliot hoped to salvage for his specimen barrels. The place where they planned to make camp, Higleleh, was down in the valley below. Looking down from the height of the Haud, they could see a sharp descent to the plains. Akeley took a photograph before they began to march down the steep incline.

The route they had followed since leaving the Milmil road was one seldom taken by Europeans and was shorter than the one used by those who travel between Hargeisa and Milmil. In this way, they gained a day. The descent was rough, with a treacherous scree of huge boulders and loose cobbles covering the ground. They had to pick their way carefully among these hazards. When they finally arrived at the bottom, the ground was sandy and level and far easier to travel. They passed flocks of sheep and goats returning to the Haud after being watered at the wells. They reached their destination in late afternoon.

Elliot was disappointed to see that many deep holes had already been dug in the riverbed, and water was "not very plenty." Meanwhile, the surrounding country was "absolutely destitute of grass, a dismal prospect for our poor horses, mules & donkeys." Many villagers were already scurrying about the holes, dipping into the precious water for their livestock. Elliot rode a short way down the dry river, searching in vain for a good place to camp. He finally selected a "poor enough" spot with only a modicum of shade.[24] An hour later the caravan arrived, and the tents were quickly pitched. Many camels lay down immediately, not waiting for their handlers' command. The entire company was worn out from the ten-hour march. Elliot estimated that they'd covered nearly twenty-five miles.

His original intention had been to remain at this place for three days of much-needed rest, but with water so scarce and grass for the animals "practically not to be had," he began to reconsider. Perhaps it would be best to push on to Dagabur? They had, by now, left Somaliland and entered Ogaden, "among a different race of people although speaking the same language, and a different country presenting a different aspect."[25] A zareba was soon made, and the camp settled into an exhausted repose, the weary sentinels being the only men required to keep awake through the night.

20

FINISHING THE FIGHT

August 9–16

On Sunday, August 9, the hunters slept late. They passed a restful day in camp, writing letters. These would probably be their last for quite some time, as from now on their communication would depend entirely on chance encounters with caravans heading to Berbera. Elliot expected to hear nothing from the outside world until mid-September. Some men left to visit nearby villages, while the animals were led away to find grass.

The lack of water and grass near camp was a serious problem. If Elliot could learn with certainty that there were better resources at Dagabur, he would move there the following day, although the distance was long and the camels were by no means rested. They would recover more quickly with ample food and water, he reasoned. Selon, who'd gone to reconnoiter the hills about camp, returned and reported that there was better grazing there. The scarcity of water, however, had already forced Elliot's hand. He decided to push on. But, since the journey couldn't be completed in one march, he intended to start in the afternoon, as he didn't want to make two hard marches in a single day when his camels were already so fagged.

Early the next afternoon, they were off again, led by an Ogaden guide who was to bring them at least as far as Dagabur. The country resembled the Haud, with a smattering of thorn trees and no undergrowth of any kind. The ground was reddish sand. They saw many new birds. Dodson shot a shrike, *Dryoscopus rufinuchalis*. Elliot noticed a breeding pair of barbets, *Trachyphonus shellyi*, making loud, shrill calls, sounding "almost like laughter."[1] After riding three hours in the hot sun, Elliot stopped to rest in the shade of some trees. Shepherds approached, telling the curator of another marauding leopard that had taken two of their goats that very morning. They wanted him to kill it. He asked whether the leopard was still nearby, but they admitted that they'd scared it off. The curator knew it'd be foolish to chase after a skittish leopard while on the march, so he told them he couldn't delay.

He moved out again, continuing until nearly five that evening, when he stopped to await the camels. The guide insisted they'd reach a reliable water source after only a short, easy march the next morning. Akeley, Dodson, and Elliot took refuge in the shade of a large bush. Soon some villagers—many carrying spears—began to wander up to have a closer look at the strange-looking hunters. They offered traditional greetings—"salaam sikar" or "salaam sahib"—to which Elliot replied, "Aleikoum Salaam," touching his pith helmet politely to acknowledge their salutes. At this, they "squatted down in front of us for a good long stare." Elliot suspected the visitors then "made critical remarks upon our dress & personal appearance. We returned the stare & made remarks in our turn." One man seemed to have some authority, and when he felt that the others had seen enough, he ordered them away—and they obeyed. Elliot and his company endured this scrutiny until their tents were pitched. Then they entered the comfort of their own camp. Many curious locals watched, huddling in loose groups outside the zareba that had been hastily built by the expedition's men. The loiterers observed what was happening in camp with great curiosity. "We have evidently now come among a more primitive people not much accustomed to the presence of white men," Elliot concluded.[2]

During the long ride, Elliot discovered to his horror that his horse's tangled mane was crawling with bloodsucking ticks. Other horses and mules were likewise afflicted. "These terrible insects," he wrote with revulsion, "have been a regular plague ever since we entered the country."[3] The simplest solution, he decided, was to cut the hair off close. The loiterers watched the shearing in fascination. Their guide then asked for the hair, which he wanted to weave into a tassel to decorate his own horse.

They were once again underway shortly after daybreak, marching through "an uninteresting country of red sand covered with the usual low thorn trees & bushes." After a brief march of two hours, they reached the so-called singing wells of Dagabur. These wells were a remarkable fifty feet deep, cut through a layer of solid rock that lay some ten feet beneath the sand. Only the previous day, two men had died after falling down a well while fetching water. Elliot guessed that the wells were the work of the Galla. "Certainly no Somali of the present race would ever undertake such a work or even know how to go about it," he wrote contemptuously. Large numbers of camels were gathered on the high ground overlooking the wells, held there by a villager who permitted only a few to go down to the water at a time. It was "a comical sight," he wrote, "to witness the ungraceful beasts gambol down the hill when they were allowed to go."[4]

The method of extraction was simple and effective, "though it could hardly be termed a labor saving one," Elliot explained. Several men, "having laid aside their tobes," climbed down into the wells by means of a rickety ladder. Once the first to descend could reach the water, a double line of funnel-shaped vessels was shuttled up and down, "the descending one empty the ascending full." The water was then poured into a large, man-made trough from which several animals could drink at a time. The passing of water vessels was accompanied by a rhythmic chant that aided the timing of the work, one man uttering the first words, the rest answering in chorus. Between the camels coming and going, the men heaving water from the mouth of the well, and the singing and shouting, "the scene was a very animated one." Elliot and his men sat above the fracas and watched, surrounded by "a crowd of natives attracted by the unusual sight of the white man." After a long wait, the caravan finally arrived. They removed themselves to their new campsite, "a rather high shadeless plain" only a short distance from the wells.[5]

Once the camels were unloaded, they were taken to be watered and to find some shrubs to browse. Unfortunately, there was little if any grass for the horses. Worse, there was no game visible. "It is perfectly useless to remain in this spot," Elliot griped.[6] He determined to push on the following day to the next reliable watering hole. But, after talking the matter over with Dualla, he decided to remain another day to give the camels rest. Three of these animals had sore backs that warranted treatment.

From his tent, Elliot enjoyed "a splendid view of the country passed over during the last two days, bounded in the distance by the lofty plateau of the Haud which rises like a wall along the entire horizon." While they marched from their last camp, they had a mountain, Sabatti, constantly in view. Elliot described it as "a lofty mass of rock rising abruptly from the sea of green, . . . a stony land mark visible far and wide."[7]

Early the next morning, Elliot, Akeley, and Dodson fanned out from camp to try to find game. Their expectations were low, given the number of people about. They'd seen no tracks, save for some leopard prints just outside their zareba. Yet, less than an hour from camp, Elliot found a small herd of gerenuks, including two bucks and four does. The expedition didn't need more specimens of this animal, yet he "was getting pretty tired of mutton which was both tough & expensive." So, he dismounted and proceeded to stalk the herd. The ground was very open with little undergrowth. After following the herd a great distance, he finally fired at a buck that stopped for a look—striking it in the shoulder. The buck "reared & kicked & bucked" like a bronco, then ran a few yards and stopped. Elliot fired again, hitting it above the shoulder at the base of the neck. The buck then wheeled about and sprinted away. Elliot fired again, striking the fleeing antelope in his "hind quarter, the bullet breaking the hip & raking forward . . . shatter[ing] the front leg below the knee." Despite these injuries, the buck kept running, so Elliot shot again, this time fatally. "He was an old buck with good horns," Elliot remarked, "but so many bullets had spoiled him somewhat, for a specimen."[8] He didn't take the skin or skeleton for a specimen. He didn't take the head and horns for a trophy. He took some meat for the larder, and a patch of the animal's hide to re-upholster his worn-out camp chair. He saw no other game save two dik-diks, which maintained a respectful distance. He then returned to camp.

His companions were more productive. Akeley had come in before Elliot with three dik-diks. Soon Dodson returned, too, having killed another gerenuk, a dik-dik, and four birds, including two satin breasts and two larks, neither of which they could identify. Dodson, meanwhile, had suffered "a pretty bad toss" from his mule. The mule, "much frightened at the smell of blood," had begun to buck violently.[9] Dodson managed to stay in the saddle until his crupper broke, when the saddle shifted forward onto the mule's neck. He was then somersaulted over his mule's head onto a pile of stones. He was embarrassed and a little shaken up, but uninjured.

Some villagers had visited camp while the hunters were away. They'd come from another well, where lions were making trouble. Two men and several cattle had been killed, so the villagers dared stay no longer. This news changed Elliot's plans. He now proposed to visit the well the next day to see whether they could bag more lions. Despite his concern about the lion's diminishing numbers, Elliot had no difficulty justifying the hunt. "They would be much better in our pickle barrels," he reasoned, "than loose killing people." Fortunately, this place was not far out of their way, and apparently hadn't yet been visited by Europeans. They were now in terra incognita, so that to deviate from the path would be of "no consequence."[10] Rhinos, which Elliot was keen to get for his specimen barrel, were reported at this place also.

Large flocks of the glossy starling, *Dilophus carunculatus*, sang and fought in and around camp all day. These birds "enlivened the place very much by their beautiful plumage & sweet voices," Elliot opined. "In its lovely dress of metallic colors, with red under parts & white band on the breast it is one of the handsomest birds in Africa. It goes in flocks, flies straight & rather rapidly with quick movement of the wings."[11] It was utterly fearless, moving boldly throughout the camp and often very close to the men.

In the night, hyenas growled and scrapped noisily behind Elliot's tent, just outside the zareba. They were fighting over the carcass of a dik-dik Akeley had discarded. "They chased each other about & made a good deal of noise," Elliot wrote, "but made no attempt to enter the camp." As it was too dark to see well enough to risk a shot, the hyenas eventually withdrew "without a bullet being sent after them."[12]

PAIN LIKE A TOOTHACHE

In the morning, they started out early, the caravan moving out before it was even light enough to discern the path. Their guide took the lead, and Elliot rode ahead of his camels, with the three shikaris walking in front of his horse. Since their arrival in Ogaden, Elliot had changed the order of march. Instead of all the personal servants walking ahead, Elliot now permitted only the shikaris to lead. The rest, including gun bearers and syces, were to follow behind the animals. Also, from then on, talking was forbidden. This, Elliot remarked, was "a sore deprivation to a Somali."[13]

Their trail ascended "considerably," until they were west of Sabatti, and from the highest point reached they enjoyed "a fine view of the country ahead." Elliot described the scenery, which "lay beneath us broken up into gullies, with low hills, running in irregular lines, & covered sparsely with low thorn trees & bushes, & a mountain of considerable height, name unknown rising in the distance." They spotted their first wild dog—or so they thought—and both Akeley and Dodson went after it, unsuccessfully. They later concluded that it must've been a cheetah. After riding for three more hours, they reached another dry riverbed, probably the Fafan River. They rode along its sandy path for two hours. The sky was overcast, which made the journey pleasant. Great numbers of dik-diks continually zipped in front of them from one bank to the other, affording some target practice. Shortly after ten in the morning, they arrived at the well, which proved to be one shallow hole dug in the sand with the equivalent of a single washbasin of water in it. This was "certainly a very unpromising outlook," Elliot wrote.[14] They dismounted to await the caravan, which, to the curator's surprise, arrived in under an hour. They made camp on the bank of the river, and the men went to work digging more wells.

While the hunters were sitting down to lunch, Dualla approached and announced that no water could be found. He recommended that the expedition push on to another well called Hersi Barri. He'd already sent the guide and two men to confirm that there was water in the well. If not, the expedition would have to march as far as necessary the next day to find a reliable source. The trouble was that no one knew the routes, and they were dependent on "the uncertain information we can get from the native guides, who have no idea of distances or time & can only designate the former by 'far' & 'not far.'"[15] At this point, though, there was nothing to be done but to keep marching, as they couldn't remain even a day without water. So, after lunch, the camels were reloaded. Fortunately, before all had been made ready for travel, the guide returned with news that water had been found. Because he'd been gone only briefly, Elliot knew the water must be close. This proved to be the case, for after riding no more than fifteen minutes, they came to another well. No water was visible, but the bottom was damp. A stick penetrated the sandy bottom easily and came out wet. Camp was made on the bank close to the well, while some men worked at expanding the hole.

Elliot planned to remain there at least a full day to see what game was available in the vicinity. Promising signs abounded, including the tracks

FIGURE 20.1.

The order of
march, with Elliot
leading on his
horse, Dodson
following on a
mule, and the
attendants walking.
Photograph
courtesy of the
Field Museum,
CSZ6053.

of rhinos and oryx. The curator thought he saw an elusive Egyptian mongoose. He wanted a specimen, of course, but it managed to escape into the bushes, probably diving down into a burrow. Many different species of birds were flitting among the trees on both banks of the river. A beautiful, little bee-eater, *Melittophagus cyanostictus*, zipped through the air catching flying insects on the wing. Dodson shot a paradise flycatcher, *Terpsiphone cristata*, and a sunbird, *Cinnyris albiventris*. As he was moving through some bushes searching for one of his birds, he nearly stepped on a small black snake with a pearly belly and menacing fangs. He grabbed a stout branch and killed it. The Somalis knew that the snake, an asp, *Atractaspis microlepidota*, was venomous, and that its bite would kill a man very quickly.[16] Dodson coiled the deadly specimen tightly into a formalin-filled bottle. That night, the ominous roaring of lions could be heard distinctly in the distance.

The next morning, Elliot stayed in camp. He'd had a premonition the previous evening of another attack of fever. Akeley and Dodson went out, however, the former meeting with a flock of ostriches, including "two splendid males." He was stalking them quietly, "with every probability of getting within 100 yards," when Dodson blasted his shotgun in the distance and startled the ostriches, which sprinted away.[17] Akeley fired at a male, hitting it, but not fatally. It kept running with the flock and escaped.

On his way back to camp, he found fresh rhino tracks, but it was too late in the day to follow.

The country, it seemed, was full of game. Dodson brought back an aoul and another gerenuk. They found abundant lion tracks. Elliot had some traps baited with the offal from Dodson's aoul, thinking it might attract a leopard, whose tracks had also been seen. Remembering the roaring of the night before, Elliot ordered a zareba to be built, thinking that lions might be attracted to the smell of water. If the man-eaters were nearby, he intended to catch them. He proposed to tie up a donkey as bait. He'd planned to sit up that night himself, but a raging fever came on so strongly around midday that he went to bed instead. Dodson took Elliot's place, but no lion materialized.

Elliot suffered through "a miserable night, no sleep & temperature 101." The next morning, he was in no condition for hunting, and so remained helplessly prostrate in camp. In the traps, they found "a fine striped hyena, a Werra . . . with very distinct black stripes."[18] Akeley and Dodson went out again. Akeley went after ostriches, but was again unsuccessful, while Dodson killed two female aouls. They spent the afternoon preparing the skins.

Elliot then endured one of the longest and most uncomfortable nights of his entire life. "A peculiarity of this miserable fever," he wrote bitterly, feeling his own mortality, "is a sensation about the hips, as if one was carrying an enormous weight which threatened to pull him asunder." From his hips down to both feet he felt a continuous "pain like a toot[h]ache, so that it was impossible to remain a moment in any position." Sleep "was out of the question & I could only toss the night through & this morning I look as if I had been ill for a month instead of only two days."[19] He was grateful to see the dawn breaking through the open flap of his tent. His fever had broken. He took a warm Sunday bath that made him feel infinitely better. He then went back to bed and slept for several rejuvenating hours.

Rising later that afternoon, Elliot felt a strange new ambivalence about the expedition. Though he was on the very cusp of the range of rhinos, elephants, and giraffes—arguably the greatest zoological prizes in the world—he contemplated retreat for the first time. He'd already made an excellent collection, mostly antelopes, over the previous five months. It only remained to get a few more of the largest and most charismatic species. Yet, should his fever resume, he vowed to knuckle under. "I shall return to the coast," he wrote, "for it is not worthwhile to take any chances when the Expedition is practically a success, & only a few more species to be obtained."[20]

The traps that morning yielded two foxes, *Otocyon megalotis*, and another hyena. Two other traps were carried off by a pair of animals that left two diverging trails of prints. The trails were followed, and a hyena was soon overtaken and shot, while a leopard "made good its escape." Amazingly, a leopard had taken the baited body of a hyena and dragged it up some fifteen feet into the fork of a nearby tree, leaving it there for future provision. Elliot marveled at the "strength of the beast," which "must have been very great to accomplish this feat, for a spotted hyena is a large heavy animal & requires a considerable effort to lift it from the ground." While Elliot was resting in his tent, Ahmet approached and reported that a herd of aoul was passing close to camp. Elliot went out to have a look and possibly take a shot. After his recent bout with fever, however, he was "not in stalking trim," and the wary animals spotted him easily and sprang away unharmed.[21]

AKELEY'S BRUSH WITH DEATH

Akeley set out that morning to land an ostrich. He'd seen their "black-and-white plumed" bodies many times, but the big birds proved surprisingly elusive. Thinking that a smaller party would be advantageous, he brought only a mule and a single syce. Half a mile from camp, he spied an old spotted hyena "loafing along" after a night on the prowl. He shot it. On closer inspection, however, he found the hyena's skin to be "badly diseased." So, he left it. Shortly thereafter, he shot a warthog. He marked the spot, leaving the specimen where it fell.[22]

He shinnied to the top of an eight-foot termite mound to scan the ground with his binoculars. Adjusting his focus, he realized he was looking directly at the beaked heads and sinuous necks of a pair of grim-looking ostriches. "The birds remained perfectly motionless watching and I did likewise," he later wrote, "locating their position . . . by the termite hills which were nearly in line between us." Suddenly, their heads ducked in unison and disappeared behind some shrubs. Akeley slid down and sprinted to the spot but found only a trail of three-toed tracks in the sand. He followed this trail a long way, eventually reaching a large clearing. At its center was a dense green thorn bush a dozen feet in diameter. "A beautiful cock ostrich broke into the clearing at full speed," he wrote. Just as Akeley raised his rifle to fire, the bird disappeared behind the shrub. He aimed at the other side, waiting. Seconds passed. Soon, he felt foolish. He then ran quickly to the shrub, expecting to find his quarry hiding behind it. But the big bird had vanished. A divot and a swirl of sand told Akeley that the bird had pivoted ninety degrees and run the length of the clearing, always shielding itself behind the shrub. "I have not known many animals to do a more clever thing than this," he wrote in admiration.[23]

Much chagrined, he went back to camp for a late lunch and some rest. He set out again late in the afternoon for his warthog. But when he reached the spot where the carcass had been abandoned, there was nothing left but a gory string of vertebrae and some vulture feathers. Looking around, he spotted a hyena ambling away with the pig's head in its mouth. He took a quick shot, but "the brute" was already far away and he missed. He then

FIGURE 20.2.

Akeley had better luck shooting ostriches with his camera. Photograph courtesy of the Field Museum, CSZ5958.

remembered the diseased hyena he'd killed, and decided to retrieve its carcass, but he found that its remains had been carried a short distance away into some bushes. By now, twilight was setting in, and "objects were becoming indistinct." He spied a vague form moving through the bushes. Supposing it to be another hyena "making a feast on its late relative," he fired at it. The bullet struck home, and a low, menacing growl told him that the wounded animal was not a hyena but a leopard. He couldn't see his quarry, so he crossed the dry riverbed to the opposite bank in order to get a better vantage above it. While he was watching, the leopard rolled out from the bushes and lay writhing painfully in the dust. He shot at it again and, thinking the animal was finished, headed back to camp. But, glancing backward, he noticed that it was still struggling some twenty-five yards away. He fired again. At this, the leopard leaped up furiously and sprang directly at him. "For just a flash," he later wrote, "I was paralyzed with fear."[24]

There were no more bullets in his gun, and only one left in his hand, so he turned and ran as fast as he could, all the while fumbling frantically with his last remaining round. When the rifle was loaded, he turned back and fired hastily at the charging leopard, which was now only a few feet away. The shot must've missed, he thought, for in the next instant the enraged leopard was leaping dexterously through the air, aiming for Akeley's throat.

He clubbed at it desperately with his rifle, which had no effect. Yet the leopard missed Akeley's throat and bit down instead on his flailing right arm. Grabbing the leopard by the neck with his left hand, he hurled it to the ground, falling on it and planting his knees on its chest, pressing down with all his weight. He felt one of its ribs give way. Fortunately for Akeley, one of the leopard's forelimbs was pinned underneath its body, and one of its hind limbs was broken, so it couldn't claw him with these paws. Likewise, by means never known, he somehow kept out of reach of the two remaining paws. The leopard, meanwhile, kept crunching on his right arm. "I was conscious of no pain," he wrote, "only of the sound of the crushing of tense muscles and the choking, snarling grunts of the beast." He jammed his left hand into the leopard's mouth, pushing its cheek in with his fingers "so as to make a pad . . . to bite on." Once he got his hand wedged in the leopard's mouth, he was able to free his right arm. He then used all his remaining strength to suffocate the leopard, which struggled violently, "trying in every way to get her claws & fangs into him."[25]

The syce, meanwhile, "was scared out his senses & went dancing about shouting for help, but not attempting to render any." Akeley, calmer now, but still in grave danger, "began to think and hope" that he "had a chance to win this curious fight." He shouted for his knife. But when the leopard had charged, the attendant panicked, throwing both the knife and his own tobe in the leopard's direction and running a respectful distance away. Akeley called to him again, directing him to seize one of the struggling leopard's legs, but he was too scared to intervene. At length, the pressure on the leopard's chest subdued it. Its "struggles gradually lessened & finally ceased." Exhausted and badly injured, Akeley staggered to his feet. The leopard gasped and began to struggle weakly. The syce then summoned the courage to bring the knife, and Akeley "plunged it into the leopard's breast & finished the fight."[26]

BLOODY PANTALOONS AND BURNING ANTISEPTIC

Elliot and Dodson were already sitting down to dinner when they heard the first crack of Akeley's rifle. When this was followed by several more shots, "it denoted something more serious, either some dangerous animal or possibly natives lurking about the camp."[27] By then it was nearly dark, and nobody knew which direction the taxidermist had gone. Nor could they judge his position from the rifle reports. Once the firing stopped, they waited fatalistically in camp for Akeley's return rather than attempting to find him.

When he finally did come in, Akeley called first for antiseptic, explaining calmly that he'd been mauled by a leopard. He was a mess: "His pantaloons were all blood from the leopard when he had held it down, & I thought his legs had been badly lacerated, but on examining him found the bites were all on the right arm above & below the elbow, not touching the joint, and one below the thumb on the right hand," Elliot wrote. While Elliot and Dodson were rummaging through the medicine chest, the tent boys peeled the bloody, tattered clothing from Akeley's body and doused him repeatedly with cold water. Burning antiseptic was then squeezed into every puncture wound on his limbs, including fourteen on his right arm and two on his left hand. So much liquid was pumped into his arm that an injection in one wound would drive it out of another. This medical treatment, he later remembered, caused him more suffering than the mauling.[28]

Akeley had been "most fortunate," Elliot concluded, "in having the leopard seize him as it did, for the arm had gone so far into the creature's mouth that the fangs could only get a sliding bite & did not penetrate deeply, while his shirt sleeve having been rolled up, acted as a pad against the molars in their scissor like motions." Nevertheless, Akeley's wounds were "quite bad enough . . . and will probably keep him idle for some little time. After they were dressed he was quite comfortable with little or no pain mainly a numbness in the arm." Exhausted and exhilarated, he collapsed on his cot and slept like a dead man. Those few moments of terror—the leopard's athletic spring, the adrenaline rush, the metallic smell of blood—stayed with him for the rest of his extraordinary life.

The leopard, *Felis pardus*, was carried to camp by some of the men, who laid it respectfully alongside Akeley's cot. It proved to be a large female with one blind eye.[29]

21

NO ALTERNATIVE BUT TO TURN BACK

August 17–September 24

No one went hunting the morning after the mauling. Elliot seemed to be over his latest bout of fever, although another day's rest would certainly do no harm. Akeley, meanwhile, had passed a surprisingly comfortable night. The bloody bandages were peeled from his arm and hands and clean dressings applied. Elliot thought his wounds looked good, with little swelling. He was hopeful that his taxidermist would soon recover completely, although "for the present he can do nothing."[1] He thought it likely they'd have to remain in their present camp for several days, at least, or until Akeley felt strong enough to travel. Photographs were taken of the grim-looking taxidermist—his arm in a makeshift sling—posing with the leopard. With his thinning hair and scruffy beard, and his thousand-yard stare, he looked much older than his years.

There was "great excitement" among the men over the events of the previous day, and "they kept up an animated conversation." The leopard became "an object of much interest," and everyone crowded around to get a look at the strangled beast hanging by a rope from the ridgepole of Akeley's tent. Yet it was a specimen—not merely a curiosity—and it had to be preserved. So, Dodson later skinned and stowed it in brine.[2]

Elliot remarked that the leopard was perhaps the most dangerous animal they faced in Africa. "While not possessing the power of the lion," he wrote, "[the leopard] far exceeds him in agility boldness & cunning, & does not hesitate to enter any zareba in search of prey, & its movements are of such lightning rapidity that it generally manages to accomplish its purpose." It was probably responsible for more livestock losses than any other predator. It was also a threat to human life. "A less powerful man would have been killed," Elliot wrote of Akeley's mauling in a letter to Skiff. "A wounded leopard will fight to a finish practically every time, no matter how many chances it has to escape," Akeley later wrote. It was "vindictive."[3]

Akeley passed another comfortable night and was not nearly as stiff and sore as he'd been the day before. The pain of his wounds was much diminished, and his hand and arm now gave him no serious trouble. Most importantly, no side effects had developed. A much-relieved Elliot had "every reason to hope he will soon get over the effects of the mauling."[4]

After lunch, they all went to examine the scene of Akeley's desperate fight, only a short distance from camp. On the way, they passed the tree in which a leopard—possibly Akeley's vanquished adversary—had deposited the remains of a hyena. The carcass was still there, unmolested. They found a spot where the grass was "trampled down & stained with the leopard's blood." The other dead hyena—the "indirect cause of the trouble"— was still where the leopard had left it, this carcass likewise undisturbed.[5] They could find no fresh tracks of lions, which led Elliot to believe that they'd chased them off with all their shooting. The weather was mercifully cool with an overcast sky. Rain was expected.

To better prepare for possible hostilities as they marched farther into the interior, Elliot ordered four companies of his men to take target practice in the afternoon. Some did well, the curator conceded, "while others couldn't even hit the bank against which the target was placed, eighty yards away." Several of his men had accompanied Arthur Donaldson Smith's expedition, where they took part in a pitched battle that Elliot called "the Boran fight." These men, to Elliot's chagrin, were among the worst shooters of the lot. He decided it'd be prudent to give the men "considerable practice before they can be depended upon . . . in case we get into trouble ourselves."[6] Therefore, rifle drill was repeated several times over the next few days. The men enjoyed the novelty, and some showed marked improvement.

Elliot suffered an unfortunate relapse. Feeling weak that morning, he decided against making exertions of any kind. Nevertheless, he felt worse and worse as the day advanced. His fever returned with a vengeance, and his temperature spiked overnight. It raged unabated for days, while the curator remained fixed to his cot in abject misery. He considered moving to a higher elevation, thinking cooler weather might bring relief. He sent men on a short march to scout the wells at Berdadleh to see whether there was sufficient water and grass. They soon returned, reporting poor prospects. He'd heard that an entire village was converging on his camp, seeking to obtain the expedition's protection, and to "profit by the grass" that was so abundant there.[7] The curator, of course, would prefer not to be crowded. Tired of the scalding hot soups he'd been listlessly spooning, Elliot sent two men back to Dagabur to hire milk camels.

Akeley's wounds, meanwhile, were healing rapidly. Indeed, the bandage had already been removed from his right hand. He'd suffered no fever and no infection during his brief convalescence. Envious of his younger colleague's robust constitution, Elliot considered Akeley "very fortunate

FIGURE 21.1.

Akeley poses with the leopard he strangled. Photograph courtesy of the Field Museum, CSZ5974.

to get off so easily." His blood, the curator speculated, "must be in fine condition."[8]

Feeling considerably better than his American companions, Dodson hunted intermittently. One morning he shot two male aouls—both good specimens. Another day he bagged a single bird, a pipit, *Anthus sordidus*.[9] On still another, he killed an oryx. He'd seen a flock of ostriches, but he couldn't get close enough to shoot one. With game relatively scarce and a pair of ailing colleagues to monitor, he never stayed away from camp for long.

After six days of torment, Elliot had had enough. He gathered his comrades to discuss his predicament and weigh the options. Together they decided that unless the curator "wished to fight the fever out here to the bitter end," the wisest course of action was to start for Berbera—while Elliot was still strong enough to travel—so that he could get proper medical attention. The curator gave orders to move camp the next morning. "It is an awful disappointment to turn back when I am just on the verge of the zebra & rhino country," he wrote, "but there seems no alternative."[10]

FEVERED MARCHING

Elliot's fever ran high all night, and he slept fitfully. At daybreak, on Sunday, August 23, a sturdy camel was brought to his tent with a platform rigged high on its humped back. Elliot's mattress was arranged on the platform, and he was lifted bodily and strapped into the makeshift bed by half a dozen men. He set off at once with a few loyal attendants, leaving well ahead of the caravan, which wasn't yet ready to move. Their pace was as slow as a funeral procession.

The motion of the camel was "very trying," as, with each step, Elliot received a "dig in the back." He passed the time uncomfortably by calculating how many "digs" he would have to endure before reaching Berbera, 250 miles away. This was "not a promising outlook for my backbone," he lamented. They followed a dry riverbed for five painful hours, Elliot lying flat on his back, miserable, never moving. He couldn't sleep, but he kept his eyes shut tight against the brightening light. He heard birdsong, the clopping of the camel's feet, the crunch of gravel. He was fearful of the sound of his own racing pulse throbbing in his ears. They stopped at midmorning. The sun was climbing high, and the temperature soared. Once the caravan caught up, the men made camp. The curator "stood the journey well & was not especially tired." But, shortly after he took to his cot for rest, his temperature spiked again, to an alarming 103 degrees Fahrenheit. He rummaged through the medicine chest for a remedy, choosing a heavy dose of phenacetin, as quinine seemed inadequate with so high a temperature. As the afternoon wore on, his temperature dropped, and "there was a promise of a better night."[11]

Elliot learned that he was camped near the place where, in 1894, a young

Englishman, Lord Delamere, had been bitten in the foot by a wounded lion. Delamere was saved when his courageous Somali gun bearer leaped on the lion, giving the Englishman time to grab his rifle and fire. Badly wounded, he was detained at this spot for three months before he could hobble back to the coast. He limped for the rest of his life.[12] Elliot's fixation on Delamere betrays a niggling concern for his own health and safety. Would he outlast his fever? Would he suffer any long-term disability?

The curator's temperature the next morning was normal, but his pulse was racing. He took fifteen grains of quinine and set out again, stretched out lengthwise high atop his camel. He suffered a comparatively short march, reaching his destination around eight o'clock. The caravan arrived, and the tents were pitched among some trees. Dodson, shotgun in hand, watched a pair of owls, *Scops capensis*, glide gracefully into their spreading branches, then he shot them both.[13]

Elliot began to have second thoughts about abandoning the field. He decided to remain stationary for a few days, hoping to break his fever once and for all. Meanwhile, he thought of sending a fully recovered Akeley, with a portion of the caravan, on a brief, four-day reconnaissance to find zebra. He was reluctant to return to Chicago without any of these iconic animals. He would either await Akeley's return where he was or, if he felt well enough to travel, follow after him.

The men Elliot sent to hire milk camels caught up to the expedition that morning. They reported that they'd been roughly treated. Villagers had accosted them, seizing their guns and taunting, "You seem to think you are gods." Things might've gone badly for the expedition's men, but for the intervention of a wise, old bystander, who scolded his fellow villagers. They were acting very foolishly, he chided, and would surely get themselves into serious trouble. "The Ogaden," Elliot opined, "are wild savages & respect nothing but a force superior to their own."[14]

One of the expedition's camels, meanwhile, was unwell, which was a prelude to certain death, Elliot predicted, "for nobody seems to understand what to do for them." Sure enough, the sick camel died a few nights later. Elliot told Dualla to buy more, offering cloth, rice, or dates, as the expedition would no longer be needing so much of these bulky, heavy items. Dualla sent word to the nearby villagers that the expedition would buy livestock, yet he cautioned Elliot that it was nearly impossible to do business with them. "Everything they have they consider is better than anybody else's," Elliot groused, "& even when they may agree upon a price, they will back out at the last moment."[15]

No one seemed willing to part with a milk camel at any price. "The fact that it is needed for a sick man who probably cannot cross the Haud without milk has no effect upon them. If he cannot get across he can die & that is the end of it," the curator despaired. Dualla, however, was able to persuade the father of their guide to lend four milk camels to the expedition. He even agreed to let them go all the way to Berbera. A grateful Elliot sent two of the expedition's men to the man's village to retrieve the camels. Some of the men had spoken of taking milk or camels by force, but Elliot wouldn't hear of it.[16]

Around midday, Elliot's temperature again began to rise. "Nothing I could do seemed of any avail," he wrote in frustration. He could reduce the fever temporarily with phenacetin. Quinine, however, "had no more effect than water." It merely made his "ears ring" and his "head sore." Rifling through his medicines, Elliot decided to try arsenious acid, "to see if it would have any effect on the fever which still was having everything its own way." He'd never taken arsenic in any form and didn't know how it would affect his stomach—or his health in general—but he was determined to experiment with the highly toxic remedy. He began by taking thirty grains every three hours.[17]

Elliot's fever persisted the next day, his pulse surging to 110. He felt there was no immediate prospect for improvement. His symptoms were intermittent, a dramatic rise in temperature occurring regularly in the afternoon and often lasting up to eight delirious hours. His pulse rarely dipped below the century mark. The other hunters thought he looked visibly thinner and weaker.

Meanwhile, the idea of Akeley's zebra reconnaissance was given up. Instead, Akeley and Dodson both did some half-hearted hunting near camp. Dodson got some good birds, including a rare Somali courser, *Cursorius somalensis*, but neither he nor Akeley saw any mammals. The next day, Akeley went out toward Milmil to see whether there was any game to be found in that direction. He eventually came back with a round-nosed dik-dik, possibly Guenther's dik-dik, *Madoqua guentheri*, a larger species that they hadn't seen previously, and a male African fox, *Otocyon megalotis*. The latter was "a pretty creature with very large ears & a nice coat."[18] Dodson brought in some additional birds, one of which, Elliot suspected, might be a new species. He also brought back a female dik-dik of the large, red variety. This reminded Elliot that the dik-diks would require careful scrutiny and comparison, with the idea that there might be one or more undescribed species or varieties in his game bag.

THE STRETCHER

Riding atop the camel was so uncomfortable and so painfully slow that Elliot decided he couldn't possibly reach the coast by this means. He was simply too weak to withstand the relentless jostling. He asked Dualla to rig two poles and stretch a bed between them, harnessing a mule at each end. When poles of sufficient length couldn't be located nearby, Dualla sent men to search farther afield. These men returned with four poles, which needed to be trimmed and dried in a fire so that they wouldn't bend under the curator's weight. Given the work to be done, Elliot decided to remain at their present campsite for at least another full day.

Around midnight, Elliot awoke in a "dripping cold perspiration with everything wet through." His temperature was normal, but his pulse continued to race. He summoned his devoted tent boy, Ahmet, and had his linens changed. He sent word to Dualla that he wanted to start in the morning for the coast. Hours later, however, his fever was in full swing again. Dualla came to say that the new stretcher wasn't ready. They were obliged to wait another day in camp. Elliot's fever raged all day with varying intensity.[19]

Akeley shot a young gerenuk and another dik-dik. Dodson bagged more birds. A villager brought in a small, injured animal in a cage, which Elliot bought. Although it was too young to be determined with certainty, the curator identified it as a civet cat, *Genetta pardina*, an uncommon species.[20] It was beautiful, he wrote, with a long bushy tail ringed in black and white. Akeley made a skin of it. They now had four more barrels full of mammal skins.

Another day passed with no abatement of Elliot's fever. His temperature rose and fell with the regularity of a yo-yo. His pulse continued its runaway pace. The ordeal was exhausting. The curator was able to sleep only two or three fitful hours a night, as he had to take his medicine regularly. He also monitored his temperature every hour or so to make sure it didn't spike dangerously high. Ahmet helped in any way he could, "but of course he knows nothing about fever & is not a nurse." Akeley and Dodson were too tired with the day's work to sit up with Elliot. He expected that they would do so willingly, but he was too proud to ask. "It is a tough job," he wrote, "to be at the same time your own physician, nurse & patient, & sometimes I feel as if I would like to divide the labor with someone else." His fever was a mystery, "a very nasty obstinate thing. . . . I would much prefer . . . a regular attack of malarial fever for you know what . . . remedies to give."[21]

They set out again early the next morning, Saturday, August 29. Elliot was still laid out atop a camel, the poles for the new contrivance not having dried. They made ten miles—about half the distance back to Higleleh—before halting for the day. There was no change in Elliot's temperature or pulse. They made another early start on Sunday, reaching Higleleh around nine in the morning and making camp within the confines of their old zareba. A tormented Elliot endured the journey staring helplessly into a pitiless dome of blue sky. He urged Dualla to make sure the new stretcher was ready the next day, as he refused to go any farther perched uncomfortably atop a camel's humped back. The headman then spent the afternoon personally directing the men in shaping the poles. Under Dualla's watchful eye, the stretcher was finished by nightfall.

FIGURE 21.2.

Kittens at play in camp. Elliot's stretcher is in the foreground. Photograph courtesy of the Field Museum, CSZ6121.

On Monday, they broke camp early. Elliot found, to his infinite relief, that the stretcher was vastly more comfortable. There was far less jarring motion, no painful jabs to his spine. Moreover, the mules walked relatively quickly—easily surpassing the plodding camels. In this way, Elliot could make the same distance as the camels but march only a little over half the time these animals required. To cross the waterless Haud, the caravan needed to make about twenty miles a day, which meant a morning and an evening march. It required two men to lead the mules and six—three on each side—to keep the stretcher steady and to prevent Elliot from tumbling out. At times, the stretcher would tilt precipitously, and the curator wasn't strong enough to hold himself steady.

They reached the bottom of the pass leading up to the Haud. The way was steep and narrow with sharp, switchback turns, too tight for the stretcher. The men encouraged Elliot to ride up on a mule or be carried by one or more of the men. He chose to ride a mule and managed to hang on awkwardly until the summit was reached. "I do not think I rode in very good form," he admitted.[22]

On Tuesday they marched for six and a half hours, stopping to eat lunch and to let the camels feed. Since leaving Higleleh, Elliot was feeling better, his fever less severe. Indeed, his temperature had been normal each morning. He was not sure whether to attribute this to the arsenious acid regimen he'd self-prescribed or to the cooler, "purer air" of the Haud. "Perhaps both had something to do with it," he concluded.[23] In the afternoon, they made another long march. They traveled by a more direct route than they had some weeks previously, bypassing Maredleh to the east instead of crossing it. Dualla didn't know precisely where they were, but he expected to reach Silo Plain the following day.

After another early start, Elliot surged far ahead of the caravan. At eight in the morning, he arrived at a line of shrubs, which marked—he thought—the edge of Silo Plain. He'd come much farther on the journey than he dared hope. He stopped there to await the caravan, as he didn't know how far the camels could travel without rest. An hour later, they arrived, but they plodded on without stopping. Dualla explained that they had to go some distance farther to reach the true edge of Silo Plain. There they could stop for rest, and then cross in the afternoon. So, the poles were fixed to the mules once more, and Elliot forged ahead. The sun was very hot at midday, but the sky clouded over, promising rain before nightfall. Reaching the plain, at last, Elliot stopped and waited half an hour before the camels arrived in staggered bunches. Camp was quickly pitched, and he settled comfortably into his tent just as rain began to fall. They'd made the quickest passage of the Haud on record, he suspected. Only three marches remained to reach Hargeisa.

Elliot instructed Dualla to keep trading their excess cloth and provisions for livestock. Soon, the campsite was as raucous as a bazaar. Dualla bartered rice and cloth for twenty-eight sheep. Elliot later estimated that the bargain cost of these new sheep was a little over four rupees each. A new crop of dates was expected soon, so this commodity proved hardest to unload. As they were only six hours from a reliable water source, Elliot decided to remain where he was to try to regain some strength, while Akeley and Dodson continued to hunt. They sent camels every other day for water.

Over the next several days, a much-encouraged Elliot felt sure he was improving. His temperature was normal; his pulse slowed to ninety-six. He continued to take arsenious acid but stopped all other medications. By Sunday, September 6, he felt so much better that he indulged in a warm bath and began to eat meat and rice again. He was tired and weak, yet he was all but certain that his fever had finally broken.

Hunting continued whenever practical. Akeley shot a hartebeest while crossing the plain, killing it at four hundred yards. But, because they couldn't stop to process the skin, he kept only the head as a trophy. The following day he shot three aouls. Akeley and Dodson went hunting together one morning and brought back a male and a female aoul. They also shot a male aoul with horns eighteen and a half inches long, the finest specimen yet procured. In the following days, they got another pair, a male and a female aoul, and a bull oryx. After shaving down the thick oryx skin, Akeley took a short walk out on the plain and killed a young male aoul, its horns just beginning to sprout.

Dodson told Elliot of a "curiously colored bird" he'd seen. About the size of a small bustard, it sprinted through the concealing grass with a shy, stooping posture, head down and tail up. He tried but failed to get a specimen. Elliot suggested that he and Akeley join forces, one lying down in the grass with the shotgun, while the other drove their prey to the hunter. Dodson and Akeley went out together the next day, and, using Elliot's tactic, they secured the bird. It proved to be a beautiful bustard, *Lissotis hartlaubi*, which Elliot had never seen before, with "white wings, jet black innerparts & pearl colored neck, & mottled back." Elliot believed it'd never been taken in Somaliland. Akeley also shot a rare secretary bird, *Serpentarius secretaries*, at an impressive five hundred yards. Other birds acquired included a beautiful little hawk, *Poliohierax semitorquatus*, a warbler, *Cisticola cisticola*, and two larks. Someone also shot a specimen of Dodson's fantail warbler, *Cisticola dodsoni*.[24]

A mounted party of Eidegalla, numbering more than a hundred well-armed warriors, passed by camp that evening. They were going to raid an itinerant group from Ogaden, who'd established some villages at Maredleh. If they achieved surprise, Elliot expected they would "carry away considerable booty."[25]

Elliot's fever returned—although less virulently—and lingered for several uncomfortable days. His temperature climbed to a hundred degrees in the afternoon, then dropped back down in the evening. His pulse surged alarmingly high. When he wasn't feeling well, he remained in bed until nearly eight o'clock some mornings. Thursday, September 10, marked exactly four weeks since he had first developed a fever. He continued to regain his strength slowly, yet he was feeling increasingly frustrated and helpless. "I am getting pretty tired of it," he admitted in his journal.[26]

Akeley and Dodson both went out very early, returning to camp with a varied bag consisting of a young male hartebeest and three aouls, including one large buck, which would make an excellent skeleton. Elliot decided to relocate camp the next day, if Akeley could be ready to move. With four animals to skin, a skeleton to scrape clean, and two barrels of skins to pack, Elliot feared he wouldn't be able to finish that afternoon.

Dualla brought the expedition's flock of sheep to ninety-one. All their excess rice and dates were now gone, and they had only tobes left to trade. As the nights were beginning to get cooler, Elliot thought the demand for cloth would grow. "I expect to turn all my bales into mutton," he wrote hopefully. A courier arrived from Berbera and reported that sheep were bringing up to seven rupees. The curator expected to get the same price, hoping to "do a good piece of business."[27] The man brought welcome letters from home; Elliot's latest was dated August 9. He also brought medicine that Elliot had requested but no longer wanted. Akeley, working at his usual frenetic pace, managed to complete all his tasks in the afternoon.

Elliot felt unwell and slept poorly that night. In the morning, his temperature was normal, but his pulse raced like a rabbit's. Setting out at half past five, the caravan marched for five hours before pitching camp near a pool of rainwater. Elliot wanted to remain at this place for a day or more, with the idea of getting some additional oryx specimens. So, Akeley and Dodson both went out hunting. The former brought back a bull oryx; the latter bagged a pair of birds.

At this site, they learned that fighting between Somali tribes had begun. One Eidegalla had apparently been killed. Meanwhile, various groups of armed Somalis visited the expedition's camp throughout the day. Their chief object at present, Elliot deduced, was to find out whether rumored raids on their camels were true. The curator expected there would be reprisals in kind. "We may see a battle between the tribes," he wrote. Worried about the sanctity of his specimens and his few remaining possessions, he urged the men to make "a compact camp & put up a strong zareba, so that the fighting, if any occurs, may not take place among my things."[28]

Later that day, Elliot bought a young oryx from a villager. This animal was a female, maybe a few months old, and "as tame as a young calf." It suckled two goats in camp, "which seemed to offer no objections to this strange foster-child." Elliot's salvage mission must have seemed less pressing during his long convalescence, as he considered keeping the young animal alive. Ultimately, however, he decided it'd have to go into the pickle barrel with their other specimens. This, he lamented, was a real pity, "for it is a beautiful little creature, not unlike in coloring a Jersey calf, with beautiful large eyes & a well shaped head."[29]

MORE ENGLISHMEN

On Sunday, September 13, the caravan departed early. By a quarter to nine that morning, Elliot had reached Bodeleh and halted. The caravan came up about an hour later. They didn't set up shop at their old campsite, but instead chose a nearby hill, as Elliot preferred to pitch his tent away from the bustling water source. A villager came to report that he'd seen greater kudus nearby. Akeley went to investigate. Unfortunately, he fired at an oryx before getting to the place where the kudus had been spotted, and the report of his rifle chased them off. The next day, Akeley went hunting again, returning late in the morning with a bull oryx and a male Speke's gazelle. Elliot recorded that the oryx had the longest horns of any they'd yet obtained, at thirty two and three-quarters inches.

Dodson and Dualla set out to visit an Englishman's camp that was said to be near Hargeisa. They returned just before lunch, bringing Henry

S. H. Cavendish and Henry Andrew. Only twenty years old, Cavendish was an Eton graduate and a fellow of the Royal Geographical Society. He'd spent several years hunting in South Africa. His companion, Andrew, had recently resigned his commission as lieutenant in the Black Watch (Royal Highlanders) battalion. Cavendish met him in Aden and invited him to join his expedition. Together, these would-be big-game hunters aimed to reach Lake Rudolf. Their headman, however, had no experience beyond the Shebelle River. Worse, they were "not at all properly fitted out for [their] journey." They needed many things that Elliot could supply, which was "a good thing for us both," the opportunistic curator observed.[30] Cavendish sent for his tent and had it pitched inside Elliot's zareba, where he and Andrew spent the night. He expected his headman to arrive soon from Berbera, bringing the rest of his threadbare caravan.

The next morning, Akeley and Andrew went hunting together. Akeley saw nothing of interest, while Andrew shot four gerenuks with the Lee-Metford rifle he had bought from Elliot. Dodson and Cavendish also went out together, and the former killed another gerenuk. Dodson was so charmed by his new hunting companion that he agreed to accompany him to Lake Rudolf. Cavendish's camels arrived in short order, and they pitched their camp close to Elliot's. In addition to the Lee-Metford, Cavendish purchased Elliot's four Winchester rifles and all his ammunition, as well as ten provision boxes.

Elliot suffered another relapse that day. His temperature was normal in the morning, but his pulse was high, and he felt weak and sluggish all day. He was keen to get to Aden to confer with a competent physician, "for this thing has hung about me long enough." By the evening his pulse had soared to 115. He felt "very uncomfortable."[31]

Feeling stronger the following morning, Elliot spent the entire day in camp, disposing of the expedition's surplus to Cavendish. The young Englishman bought fifty of the curator's camels and "practically everything I have to sell, including two of my tents, all the beds & bedding, blankets, boots & shoes & even some of my clothes." Nearly all of Elliot's guns were sold, save his .22 rifle. Elliot sold nearly £600 ($3,000) worth of goods, including most of the expedition's livestock. This welcome transaction with Cavendish was a boon for Elliot's bottom line, but it put a dent in his larder. "I have just about enough provisions left to last to Berbera," he estimated. Cavendish later wrote: "We have very pleasant memories of encountering Prof. Elliot and his caravan. In addition to giving us valuable information, Prof. Elliot very kindly supplied us with a large quantity of trading goods, baggage animals, and live stock."[32]

They spent the next day delivering their goods. "My camp," Elliot wrote, "[now] looks as if . . . savages had looted it." He'd planned to march to Hullier that afternoon, but it proved impossible to get his dwindling things in order. To carry what remained, including the precious barrels of specimens left at Hargeisa, Elliot was obliged to hire sixteen camels. His own camels were scrawny and covered in sores. "I expect to get very little for those I have left," he wrote.[33]

Elliot anticipated five more days of trekking to reach Berbera. Dodson, meanwhile, who was showing signs of serious illness, gave up any idea of joining Cavendish's expedition. "He has been very miserable with indications of dysentery," Elliot wrote, "& I think it is well for him we are drawing near the coast & civilization."[34]

The following day, Friday, September 18, they set out around five in the morning. Elliot continued to ride his stretcher, although he was feeling significantly better and had benefited from a good night's sleep. It took the caravan three hours to reach Hullier, where they camped at the spot where Elliot had awaited Dualla's return from Berbera in July. There must've been a great deal of rain since they left, as the grass had grown considerably. This was "a very pretty place, the pleasantest camping ground I think we have had on the trip," thought Elliot. Their camp was much reduced in size. Elliot missed the great herd of camels that used to come in each night. But, at the same time, he was relieved that he'd sold so many of them at such a good price. "I do not think the Somalis at Berbera will make much out of the remnant of my possessions," he wrote.[35]

One of the expedition's few remaining camels died soon after reaching camp, and another appeared to be weak and sickly. Elliot worried about keeping enough camels alive to reach Berbera. Five of his men had joined Cavendish's caravan to Lake Rudolf, and he sent fourteen more ahead to Berbera as there was nothing for them to do anymore for the expedition. He knew that any idlers hanging around camp would be bad for discipline.

Akeley ventured into the nearby mountains to find a male of his mystery antelope. It was "rather a forlorn chance," Elliot suspected, but "worth taking." The taxidermist returned empty handed late that afternoon. He reported "no game whatever."[36] Meanwhile, some hired camels arrived in camp, bringing all the items the expedition had left in Hargeisa.

The next morning, "quite a row" took place. The men who owned the hired camels refused to take the heavy loads assigned to them. "One fellow," according to Elliot, "was particularly obstreperous." Selon brought the angry camel man to Elliot, who ordered him to take the load, otherwise he'd bring him before the governor at Berbera on charges of disobedience. This threat succeeded, and the man returned to his camel and loaded up. They then made a late start from Hullier, at half past five in the morning. The march to the next reliable water source was uninteresting, save for the occasional sweeping views it afforded of the coastal plain beyond. It was to those "intensely heated plains" that the expedition was gradually descending. Elliot rode mainly over flat terrain studded with stones, small aloes, and a few low thorn bushes. The heat intensified as the morning advanced. It was not until half past nine that he reached Robeleh, still riding his stretcher. The rest of the caravan staggered in about an hour later. The weary camels looked emaciated. A few had "pretty sore backs."[37] Elliot expected that some would give out during the final miles to Berbera.

They executed another short march that afternoon, as Elliot wanted to finish the journey in four days or less—otherwise they would run out of food. Dodson was feeling better, and Elliot hoped he would soon be back to full strength. As for the curator, he was also feeling stronger. The fever, he thought for the third time in as many weeks, had finally run its course.

THE FINAL PUSH FOR BERBERA

They started out again the next day, a Sunday, at five in the morning. Soon they transitioned to a hilly terrain, populated by groves of trees, and featuring occasional views of the hazy country to the north, lying below and to the front of the expedition's path. Elliot intended to push the camels as far as possible in order to shorten the next two marches. The curator's mules, as usual, outpaced the caravan, even forging ahead of Akeley and Dodson on their horses. He bypassed a watering hole and arrived at his destination just before nine o'clock.

An hour later, Akeley and Dodson arrived, the latter bringing some birds he'd bagged along the way, while Akeley brought an adult and a juvenile dik-dik. There they met Cavendish's headman, marching in the opposite direction to link up with his client, bringing several camels loaded with provisions. The headman had already lost two camels since leaving Berbera, and he was relieved to learn that Cavendish had acquired some healthy animals from Elliot. He was carrying letters from Berbera and a message from Captain Merewether. The captain belatedly advised Elliot not to sell his camels in the interior, but to bring them all to Berbera. Some Europeans, he reported, were then trying to acquire camels for an expedition, but there were none to be had at any price.

Elliot passed the day comfortably in camp. His fever had entirely left him, he thought hopefully, although his pulse "still keeps on the rampage & seems to have no idea of reducing its pace." Dodson went out in the afternoon to hunt birds and returned with a nine-foot, green tree snake. Elliot thought it was venomous, as it had "two long fangs hidden away in a sheath."[38]

The following day, they marched to Adadleh, which Elliot reached first in about four hours. He sheltered under a large, shady tree, as the sun was very hot. It took the caravan another two hours to catch up. Unfortunately, there seemed to be no level space anywhere to set up camp, save for the riverbed, which was damp. Akeley and Dodson arrived, the latter bringing more birds, including a migrating oriole, *Oriolus larvatus*, and a pair of small hawks. While Elliot, Akeley, and Dodson were all sitting in the shade, waiting for camp to take shape, they heard the singing of a pair of yellow-breasted barbets, *Trachyphonus margaritatus*. The birds "were sitting close together on a limb," making "loud, shrill cries [and] the most absurd contortions of the neck, wings and tail." This performance was their undoing: "So busily engaged were they with their own affairs, that they paid no attention whatever to Mr. Dodson as he approached them, and both birds were secured at one shot."[39]

Because Adadleh to Laferug was too far for a single march, Elliot decided to make a short jaunt of two hours in the afternoon. The camels set out at three o'clock. The hunters followed an hour later. The sky was overcast, and the ride comparatively pleasant. They made camp on the edge of a small, dry riverbed. Soon, the sun set and the waxing moon—nearly full—brightened the landscape into "one of the most perfect nights" Elliot had ever known. Hyenas serenaded the camp "with their curious mournful howl, but although we had no zareba to protect us, the cowardly brutes did not dare come near the fires."[40]

They set out the next morning around half past four, their earliest start in several weeks. The route became even more hilly. As they neared the Golis Range, the scenery improved, "the mountains with their sugar loaf peaks standing in sharp profile in front & on the side of us." It was a long and tedious four-hour march for Elliot. The camels, meanwhile, needed an extra two hours to catch him. These "poor brutes" were "pretty well played out & are very thin." Arriving at Laferug, they made camp "nearly on our old ground." The day was "exceedingly hot, a foretaste of what we may expect when we reach the maritime plain & Berbera."[41]

Shortly before reaching their destination, they passed the grisly scene of a recent battle between tribes. Elliot counted the remains of seventeen bodies—many evidently scavenged by hyenas and vultures—lying a short distance from the camel track. "If we had only arrived a day earlier we would have been able to witness the fight," Elliot remarked morbidly in his journal.[42]

Elliot departed the following morning at four o'clock, arriving at Daragodleh in about three hours. Dodson and Akeley, who continued to hunt en route, arrived an hour later. The overburdened camels, however, needed nearly six hours to make the journey. They made camp along a dry river, pitching two of the expedition's tents on a bluff overlooking the gravelly bed. When a powerful gale blew through, it was all they could do to keep the tattered canvases standing. At a well nearby, one man killed a five-foot-long, broad-hooded snake. The venom-spitting cobra, *Naja nigricollis*, had been coiled in ambush near the water, probably awaiting the doves that drank there. Dualla, meanwhile, killed five sandgrouses, *Pteroclurus exustus*, to feed the men. Elliot saved the skin of one female as a specimen.[43]

They were now ten hours from Berbera. Elliot planned to reach his destination as early as possible by marching through the dark. By early afternoon, the heat reached its zenith. The wind ceased to blow. Elliot sat uncomfortably in the shade of his tent, fanning himself vigorously with his hat. Late in the afternoon, he set out again, the caravan having departed an hour earlier. They marched until half past seven that evening, when they made camp to eat and rest before the final leg of their journey.

On Thursday, September 24, they departed again at three in the morning. Elliot rode his mare upright for the first time in six weeks, as he preferred to reach his destination triumphant in the saddle, rather than feet-first on a stretcher. Anxious to get to town before the sun should get too high, he rode on ahead as fast as his men could manage. It was warm, and the air was still. Yet the ride was pleasant until daybreak, when it became suddenly very hot.

Elliot reached Berbera at half past six that morning and was glad indeed to dismount at his old headquarters. He found that the building was then partly occupied by Lieutenant Henri-Gustave Joly de Lotbinière, of Quebec, and his retinue, who were busy working on an engineering project in town. They divided the spacious house amicably between the two parties, and Elliot, his men, and all their belongings were soon comfortably installed.

22

BRINGING IT ALL BACK HOME

Captain Merewether came for a visit as soon as he learned of Elliot's return. The curator found him looking thinner, the "hot summer having told upon him." The handsome captain had a wound on his face, the result of an accident that could've been much more serious. He'd been taking target practice with Cavendish a few days earlier, when the breechblock blew on his Lee-Metford and the rifle burst in his hands. The accommodating captain invited Elliot to be a guest at his Berbera house. But Elliot had so much to do that he politely declined—it would be more convenient to remain where he'd be easily accessible, he explained.[1]

In the afternoon, Elliot paid off his camel men. "It was a noisy time while the money was being distributed," he wrote, "for although I only permitted one at a time to come in the room, yet the verandah was crowded & they kept up a continual talking." Elliot was relieved when the last man was finally sent away, "for not being very strong I found it a wearisome task."[2] He paid off and dismissed most of his remaining men the next morning. Only Dualla and the obliging tent boys were now left in Elliot's employ. The tent boys—some of the same young men who'd chased the scourge of locusts from the camp at Hullier—would escort the hunters from Africa.

Cavendish had purchased so many of Elliot's things that there was little left to sell. Twenty-two exhausted camels remained. They were "very thin and many of them had sore backs & it is a question whether some of them will live after the hard work of the journey just finished." Elliot sent Dualla to auction them off at the market, his expectations low. To his surprise, the animals were bought by some of the expedition's former camel men at fair prices. Elliot also had three threadbare tents remaining, which he hoped to sell at Aden. No one in Berbera wanted them. Meanwhile, Akeley kept busy overhauling the expedition's thirty barrels of skins. With only two or three exceptions, he found all the skins to be "in splendid condition."[3]

To beat the heat, Elliot had his cot placed on the veranda and slept in the open air, while Akeley and Dodson chose to sleep out in the street near the house, and "got along very well in such a novel situation."[4] The weather, still warm, was not as insufferable as when they were last in town, for the hot, southern monsoon had stopped blowing and a salty breeze wafted pleasantly from the sea.

A fellow curator, William Henry Holmes, had urged Elliot to make a collection of typical Somali objects for the museum's Department of Anthropology. So, Elliot held court one morning in his room, purchasing various articles brought for inspection by local merchants, including weapons, utensils, jewelry, garments, religious implements, and more. With an amateur's eye for the work, he bought anything that was interesting and inexpensive. In the afternoon, Captain Merewether sent a carriage to take Elliot to the crowded bazaar to make more purchases. But, after he had bought only a few things, the great crowd that followed Elliot around, pressing him closely, shouting and gesturing with their wares, made him so uncomfortable that he was "obliged to give it up and leave." Still, he'd acquired many objects, which "will give a very good idea of Somali work." He lamented that he'd been unable to purchase much from the Ogaden villagers in the far interior. He'd acquired spears and two ornamented bags in which the women carried their babies on their backs—nothing more. "They are not a pleasant people to deal with," he reflected sourly.[5]

There was an urgency to ethnological collecting in the nineteenth century. Just as the spread of Western civilization was driving animals to the brink of extinction, so was it pressuring Indigenous cultures in places where Europeans had settled. Museum anthropologists in Europe and the United States were scrambling to collect the artifacts of these cultures before they disappeared forever. No doubt Holmes stressed this urgency to his zoology colleague. Elliot apparently never recorded his own views about the probable future of the Somali people. Yet there is reason to suspect that he believed the Somalis must "civilize" in the face of European expansion. Those that could adapt themselves to the new British presence and to Western practices, in general, would survive. Those that couldn't would inevitably perish. "Verily the Field Columbian Museum will be known in the midst of darkest Africa," Elliot had written months earlier.[6] Perhaps this casual comment betrayed a future vision of an Africa reshaped by Western culture, including institutions, values, and more.

Western pastimes were already making inroads in Somaliland, Elliot was pleased to see. Returning to his quarters from his salvage mission for anthropology, he was disappointed to discover that the Merewethers and Lieutenant Joly de Lotbinière were just finishing a friendly game of golf in the sand. He regretted that he hadn't returned earlier to join them.

On Saturday morning, the reeking *Tuna* docked in Berbera. Her captain proposed to sail for Aden again at three o'clock that same afternoon. Despite his eagerness to embark, Elliot protested that he couldn't possibly be ready to leave by then. Therefore, the captain decided to go first to

Bulhar, return in the evening, and start for Aden the following day. Elliot and his men then raced to pack and make everything ready to ship. He found that he had altogether "80 barrels & boxes containing the results of the Expedition." One of these containers housed two live wildcats, for which he paid ten rupees in freight charges.[7]

Back in July, Elliot had learned of a modest export duty on animal skins shipped from Berbera. Although the charge per specimen was slight, he feared that the total amount would be considerable, given that he was planning to salvage skins by the hundreds. He had corresponded with Lieutenant Colonel Ferris, a British official in Aden, about getting an exemption for the expedition's scientific specimens. The colonel, to his surprise and mortification, denied his request, advising Elliot to pay the duties and then apply for reimbursement. Elliot, who'd been confident of approval, was outraged by this unenlightened policy. He refused to be cowed, asking the museum to intervene. "It seems to me," he wrote indignantly to Skiff, "that the Museum should make application either directly to Col Ferris at Aden, or to the India officer in London through our State Department . . . to have the duties waived." To Skiff he explained that "the specimens are purely for scientific purposes, & as such are not chargeable with duties by any Government that I know of, save here." Ferris, by contrast, saw little difference between Elliot's salvage collecting activities and those of the sportsmen who hunted Somaliland for trophies. The curator, of course, saw this equivalence as "absurd." In mid-September, Skiff wrote to inform Elliot that thanks to a lengthy chain of correspondence between the museum, the US State Department, and the British Foreign Service, the customs agent at Aden had been instructed to waive export duties. Unfortunately, these instructions arrived too late, and Elliot was required to pay.[8] It isn't known whether the museum was ever reimbursed for these costs.

The *Tuna* returned in the evening, as scheduled, and anchored near the wharf. The next morning, Sunday, September 27, Elliot and his men loaded their cargo, boarded, and bade farewell to Africa. Though they'd started the expedition with a shared purpose, Akeley and Elliot were, by now, beginning to drift in different directions. Akeley, having faced certain, violent death and survived, felt exhilarated just to be alive. A young man, he could begin to see the thread of a long and productive career unspooling before him. The possibilities for museum taxidermy in Chicago must have seemed almost limitless. Having drunk of Africa's fountains for the first time, he was thinking already of drinking again. Elliot, on the other hand, was still suffering from occasional bouts of fever and a runaway pulse. Homesick, exhausted, and feeling the weight of more than three-score years, he couldn't get away from Africa soon enough. The ailing curator was yearning for hearth and home. "I am anxious to get back," he wrote in a pitiful letter to Skiff, "for the last seven months have been anything but a holiday experience & I have had enough."[9] They arrived safely at Aden the following morning.

The expedition was fortunate in its timing. On the night of October 14, a cyclone struck the Somali coast. Twenty boats were wrecked, including the *Tuna*, which disappeared with all hands. A formal investigation headed by Governor Cuningham of Aden concluded that the ship was "quite seaworthy and fit for the service she had to perform." Nevertheless, some fifty-five passengers and twenty-five officers and crew were lost. Elliot's celebrated headman, Dualla Idris, was also on board and did not survive.[10]

SPLENDID COLLECTION TWELVE

Some sorting and repacking were done to the collections at Aden. Once the expedition's personal baggage was separated out, and the last few salable items were disposed of, the remaining collections filled thirty-one barrels, twenty-five boxes, and two shapeless bundles, or fifty-eight packages in total. These were consigned to Cowasjee Dinshaw & Brothers, which would ship them via the steamer *Edendale*, which was scheduled to sail directly from Aden to New York, departing October 11 and arriving about November 15. The shipment was to be forwarded from New York via railroad to Chicago. Following instructions from Skiff, Elliot mailed the invoice and bill of lading to the New York office of the Chicago retailer Marshall Field & Company, advising it of the expected time of arrival of the shipment. Staff in the foreign office at Marshall Field then took out an insurance policy for $10,000 on the consignment.[11]

In a cable to the museum sent the following day, Elliot reported the party safely arrived at Aden with a splendid collection. He provided no details as to the quantity or variety of specimens obtained, yet he led Skiff to believe that the expedition had been an unqualified success. Using some forgivable hyperbole, Elliot boasted in a letter a few days later that "a finer collection of mammal skins & skeletons was never before brought out of any country. I trust it may arrive at the Museum in good condition for it has cost much wear & tear of mind and body, not to mention money." By "finer," Elliot referred not only to the number of specimens salvaged, but also to the scarcity of the species both in the wild and in other museum collections. No doubt, "if the material reaches the Museum in good condition," Skiff later wrote confidently, "the results will exceed all expectations."[12]

Elliot telegraphed Skiff from Aden: "Splendid collection twelve," meaning that he valued the collection at $12,000 for insurance purposes. This, he belatedly realized, was a blunder. Since the museum had given Elliot $15,000 to fund the expedition in the first place, the $3,000 shortfall was likely to raise eyebrows on the executive committee. Elliot, therefore, sent a follow-up letter explaining that he'd greatly underestimated the true value of the collection: "In giving the figure 'twelve' in my telegram to guide you in effecting insurance of these packages," he wrote, "I placed their value much less than it really is. But having had considerable experience with Insurance Co's, which are always ready to insure for any amount but not so ready to pay a loss, I placed the amount of a figure that could certainly be recovered in case of disaster." Elliot insisted that the real value of the specimens was much higher than his estimate and was a factor of the rarity of the species collected. He also warned that should the collection arrive in Chicago before Akeley, then the barrels (containing specimens packed in brine) should be moved to the basement of the taxidermy building, and the cases (containing bones and casts) should be placed in the museum's warehouse. He stressed that "*on no account whatsoever must any of them be opened until Mr. Akeley arrives.*" This, he explained, was "a preventative measure, for as there are many skeletons & other Osteological specimens, and Africa supplies a breed of insects that will eat anything from dough to cast iron, there should be no opportunity given them to escape, & Mr. Akeley best understands how to deal with these."[13]

After a brief stay in Aden, and a requisite visit to the doctor for Elliot, he, Akeley, and Dodson departed again on October 1 on the steamship *Himalaya*. Elliot expected to arrive in England by October 20. Among his personal effects, he brought a few boxes containing certain birds and mammals. He planned to take these to the British Museum for comparison with other Somali specimens in that storied collection. This was the surest way to determine whether any of the animals collected were indeed new to science.

Arriving in London, Elliot checked into a hotel. When his local friends protested, he moved to a private residence in South Kensington, near the natural history museum. Akeley remained in the hotel. For the duration of his stay in London, Elliot felt "rather seedy." He consulted another physician, who noted that "while I was perfectly sound, I must go slow & that I could congratulate myself not only that I was *alive* but that I had such great recuperative forces." He'd lost fifteen pounds. Adding to his suffering, he caught a terrible cold—the worst he'd had in years. Akeley, too, was miserable with what Elliot diagnosed as sun prostration. And Dodson suffered a brief attack of fever. "Another month in Africa would I think have closed out the white men of the expedition & it was fortunate I headed back when I did," Elliot reasoned.[14] Though his salvage mission was urgent, it was hardly worthy of the ultimate sacrifice.

Although his doctor warned him to take it slow and easy, Elliot went nearly every day to the natural history museum, often working together with his friend Richard B. Sharpe, who provided privileged access to its incomparable zoological collections. Together, they worked out the identities of some of the expedition's mystery birds, including several completely new forms that Elliot later described in Chicago. As for the mammals, El-liot carefully compared some of the more interesting skins he'd acquired in Somaliland with known specimens in London. Here he concluded to his chagrin that there was nothing new in his collection. The small antelope that Akeley hunted over a period of several taxing days at Nasr Hablod—acquiring two female specimens, an adult and a juvenile—he identified as *Dorcotragus megalotis*, or the rare baira antelope, which had first been described by German zoologist Josef Menges only two years earlier. Re-garding the dik-diks, Elliot concluded somewhat conservatively that he had only three recognized species in his extensive collection: *Madoqua swaynei*, *M. phillipsi*, and *M. guentheri*. Elliot wrote nothing about it in his letters, but he must've been deeply disappointed by the lack of new mammals in his collection. His other great regret was that the expedition—thanks to the curator's devastating fever—had failed to collect the very largest and showiest African game, including rhinos, giraffes, elephants, and zebras.[15] These animals, especially, were all facing imminent extinction, and the window of opportunity to collect them in the field was closing. Still ur-gently wanted for the museum collection, they would have to be acquired by purchase or on a follow-up expedition.

In London, Elliot learned of several desirable skins of African ante-lopes and other species that were then available for sale. Although he had money remaining from his appropriation—several hundred pounds, in fact—he didn't feel authorized to use it on purchases. "If the Ex[ecu-tive] Com[mittee] are willing to trust to my discretion & judgment, you might telegraph me," he wrote to Skiff, seeking permission to spend. He also had £40 ($200) left from a private donation provided by Martin A. Ryerson, the museum's vice president. This he gladly spent on the skins of two rare antelopes and a zebra. "One thing is certain," he emphasized, "& it seems at length to have penetrated the heads of naturalists generally, that the Antelopes of the African continent are doomed, & in a few years most of them will be as extinct as the Mastodon." Before that happened, Elliot wanted to sweep up everything he could, placing the Field Colum-bian Museum, "in that respect at all events, far in advance of all its sister Institutions in America." Elliot stressed that the collection he'd already shipped would exceed all others in terms of Somali antelopes. "I want to be able to say the same of other parts of Africa as regards the species yet to be had," he continued, "& I strongly advise the Trustees never to permit an opportunity to acquire an Antelope not in the Museum, to pass." Elliot called it "merely a question of a little money." Only a few years hence, however, "the values of these animals cannot be estimated in money, any more than one could say today, what a perfect specimen of a Mastodon or a Dodo would be worth. . . . It is for this reason, being on the ground when I could see the skins myself, I should like to buy these specimens."

Skiff dutifully presented these arguments to the museum's executive committee, and then cabled their reply in a terse telegram dated November 2. "Committee objects purchase," it read.[16]

Elliot also made a short visit to Paris, to the Musée national d'Histoire naturelle, to visit with its director, a friend, zoologist Alphonse Milne-Edwards. There, he arranged to exchange eight of his African antelope duplicates for surplus specimens in that museum's collection that would be difficult or impossible to obtain elsewhere.[17]

Akeley, meanwhile, stayed busy in London developing the last of his photographic plates. He'd made about three hundred "splendid negatives, worth a large sum of money, a unique collection." Elliot chose a number of these to be made into colorized lantern slides in London. "It is a great thing to have these," he wrote, "for they . . . increase greatly the value of our collection & the interest of our journey."[18]

THE HAY EPISODE

Following his sudden termination in June, acting curator Oliver Perry Hay had written a letter to Elliot in Africa, protesting the injustice of his sudden termination. Weak with fever, Elliot was shocked and disappointed by the news. He wanted an explanation. "In a letter lately received from Dr. Hay," he wrote to Skiff, "I was surprised and grieved to learn that he had been summarily dismissed . . . no cause or reason assigned." This was "a grave and serious thing to do to an able scientific man," he stressed, adding that a scientist's reputation was his chief asset. "I am greatly surprised that in your late letters you have made no mention of an act, fraught as it is with such supreme importance to myself and my future work in the Museum."[19]

A letter providing Skiff's version of events reached Elliot soon thereafter. The curator's impolitic reply to Skiff's letter suggests skepticism and frustration: "I cannot understand why [Hay] should have acted in the manner you describe as he never exhibited such characteristics to me. . . . I am sorry for the whole business, and I have lost an able assistant whose place cannot be easily filled."[20]

A prickly man, Skiff could be extremely sensitive to criticism—especially from a subordinate. He'd gone to extraordinary lengths to ensure that Elliot's expedition was properly supported by the museum. He didn't take kindly to what he perceived as the "tone of complaint and censure" in Elliot's letters, which smacked of ingratitude. "I am unable to withhold from you," he fired back, "my feelings of discouragement . . . that after my deep and single-hearted interest in your welfare—far exceeding official concern—and the abiding, affectionate faith I have, in every possible manner, exhibited in you and your future, you should with apparent willingness—upon exparte and discredited statements—base expressions that would offend the most ordinary friendship."[21]

By the time he received Skiff's letter, Elliot was in transit back to Chicago. He was "astonished," he replied, writing immediately to "repudiate the idea of any intention of censuring anyone much less you, who have always behaved like a 'brick' . . . to me." Above all, he stressed that Skiff "must not take offense, when none was intended." Meanwhile, someone— almost certainly Hay—had written an anonymous letter to the museum's executive committee criticizing the director. Skiff must've reported this episode to Elliot, also. Elliot responded diplomatically: "Anonymous letters never did anyone any harm except the writers, and of course the Ex[ecutive] Com[mittee] treated the one on your management with the contempt it deserved. One must be an awful coward to write such epistles."[22] Apparently, Skiff was confident that the anonymous letter would do him no real harm, and Elliot agreed. But the director knew by September 1896 that his administration was under assault. What he didn't know was whether Elliot could be counted an ally or an adversary.

Soon, the attack on Skiff widened and went public. The following month, in his October editorial in the *American Naturalist*, Edward D. Cope renewed his criticism of Skiff. He began by recapitulating the arguments of his previous two editorials, suggesting that the first, more critical one, had hit much closer to the mark. "Other resignations have occurred," he wrote in reference to Hay's dismissal, "and the institution is evidently destined to be a failure unless a reorganization is effected." He concluded decisively that "no museum can become great unless its administration is in control of scientific experts, who should be responsible to each other and to the trustees only. With an organization of this kind . . . there is no reason why the Field Museum . . . should not rival the greatest museums of the world."[23]

Copies of Cope's critical editorial were later mailed anonymously to officers of the museum in a scheme to get Skiff sacked. One recipient, Robert McMurdy, warned Skiff that someone had sent him a copy of Cope's remarks. "I sincerely trust that the [trustees] will have the good sense to retain you in your present office. I do not understand the meaning of . . . the article sent me . . . but I believe in *you*." Skiff replied coolly the following day with an ad hominem counterattack on Cope and Hay, explaining: "I presume that all this eminated from a man who was refused a position in the Museum in conjunction with another who was discharged, both of whom are of unreliable character."[24]

COMING TO AMERICA

Akeley shipped from London the first week of November, bringing several specimens Elliot purchased there with the Ryerson fund. "He looks well," the curator noted, "but is not. Sun prostration is a nasty thing, worse I believe than fever." Elliot cautioned Skiff to say nothing to the taxidermist about his illness, as he was "sensitive on the subject."[25] This suggests that there was tension between Elliot and Akeley in London. Had they been traveling together too long, or was there another, larger problem between them?

Akeley arrived in Chicago in mid-November. But, since the African collection hadn't yet materialized, there was little for him to do at the museum. Having worked almost continuously since early March, the taxidermist had certainly earned himself a vacation. So, he left the city for an extended visit to Milwaukee, Wisconsin.[26]

Elliot booked passage for New York on the *St. Louis*, sailing on November 14 and arriving one week later. He was met at the pier by his wife and daughter. He expected to remain in New York for a few days to attend to some long-neglected personal affairs.[27] He celebrated Thanksgiving there with his family and friends on November 26.

Meanwhile, a story appeared in the *Chicago Daily Tribune* claiming that there was an ongoing feud between Elliot and Akeley. It was rumored that Akeley intended to resign his position as chief taxidermist at the museum. The rumor must have originated with Akeley—was he unhappy in some way with his museum position, or with Elliot, his mentor? In any case, an angry Skiff vehemently denounced this for gossip. Akeley, he stated confidently, "has the highest admiration for Mr. Elliot, and there is no possibility of any trouble between them. Neither is there any possibility of Mr. Ackley [*sic*] resigning his position before his work is finished. . . . Prof. Elliot is in the best of health, and will be in Chicago in a few days."[28]

23

BETTER THAN SEEING THE ANIMALS ALIVE

From the very beginning, Field Columbian Museum patrons regarded the Department of Zoology as the runt of the litter. The museum's annual report for 1895–96 candidly acknowledged that "in extent and character of material [Zoology] did not . . . rank with the other Departments." Elliot's African expedition was part of a deliberate campaign to bring zoology to "a higher standard of completeness" by aggressively acquiring specimens through purchase, donation, and especially fieldwork.[1] At the same time, dressing up the appearance of zoology's exhibit halls with a series of spectacular group displays of rare African mammals was expected to draw attention away from the building's shabby exterior, thus elevating the entire museum mission.

Once Elliot returned to Chicago, an accounting of the results of his expedition was in order. What had he collected? the trustees wondered. Had it been worth the exorbitant cost? Had the curator demonstrated that fieldwork was the best and most economical method of acquiring specimens? How soon could the new specimens be transformed into creditable exhibits?

In terms of vertebrate animals, Elliot's draft report detailed that the African expedition had accumulated 193 mammal skins, 22 mammal skeletons, approximately 300 birds, 115 fishes, and 30 reptiles and amphibians. Moreover, by exchanging duplicate specimens with natural history museums in New York, Paris, and London, additional rare mammals were acquired by trade. Elliot's final report was uniformly positive. According to the curator: "The collection obtained is very valuable, probably the most important, certainly so as regards quadrupeds, ever brought out of any country by one expedition."[2]

Elliot's report showed that collecting was done explicitly with the intention of mounting rare African specimens in elaborate group displays at the museum. Akeley, he noted, had made plaster casts of many animal specimens in the field. These would be "beyond price" when it came time to mount their skins for display, as the casts showed "every muscle, artery, and in the case of heads, the proper lay of the hair and contour of the face." The taxidermist had also taken photographs of the animals, living and dead. These would be useful for re-creating natural-looking poses. The curator suspected that such exacting representations of individual animal specimens for taxidermic purposes had never been taken in the field before. Nor—if the species went extinct, as Elliot expected—would there ever be an opportunity to take them again. Authentic accessory materials for habitat groups, including rocks and plants, had also been collected. None of this would have been possible had the specimens been acquired by purchase or exchange. Fieldwork, the curator stressed, was "the only proper way to secure collections for a Museum."[3]

Elliot took pride in having conducted such an ambitious salvage zoology expedition, and he wanted museum patrons to feel this pride too. Yet this was no time for the institution to rest on its laurels. African mammals, he noted, were disappearing most rapidly, "and although the Field Museum by its recent acquisitions is ahead of all its sister institutions in the United States as regards the large quadrupeds of Africa, yet there are large numbers not yet represented." Moreover, what remained to be accomplished in terms of collecting would have to be done expeditiously, "for the time is near at hand when, in certain lines of Zoology . . . it will be forever impossible to procure examples." Most of the world's large mammals, the curator insisted, were certain to become "as extinct as the Mastodon or Dodo are to-day."[4] In short, much more urgent fieldwork remained to be done to salvage zoology specimens for the museum before it was too late to collect them.

To appease the trustees, who'd so generously funded the expedition, Elliot addressed the monetary value of his collection. This was impossible to know precisely, he explained, as so much depended on the condition of the specimens and the rarity of the species obtained. Salvage zoology was an investment against expected losses of species that would pay mammoth dividends in the future. In a few years' time, the curator predicted, with "the gradual disappearance of the large animals," which had been going on in Africa for decades and had "already resulted in the complete extinction of some of the finest species, the collection brought to the Field Museum will be practically priceless." Meanwhile, Elliot estimated the present monetary value of the collection at $18,000.[5]

Perhaps as a way to smooth Akeley's ruffled feathers, Elliot made special note of his chief taxidermist's invaluable contributions to the expedition. "It gives me great pleasure to acknowledge my indebtedness," he wrote. The chief taxidermist had "devoted himself exclusively to [the expedition's] service, and strove to ensure its success." He'd worked constantly, "totally unmindful of himself, and accomplished results that would have been praiseworthy if attained by . . . half-a-dozen men, and much of the credit . . . is largely due to his untiring effort." The curator had come to regard Akeley as indispensable. "[Akeley] is an exceptional man," Elliot confided to Skiff, "one we cannot replace."[6]

THE VALUE OF FIELDWORK

Elliot labored to convert his field success into professional capital. Within months of his return, he delivered a pair of public lectures at the museum describing his African expedition. These popular lectures were beautifully illustrated with colorized lantern slides made from the best of Akeley's photographs. He consolidated these into one lecture that he delivered later at the American Museum of Natural History, where he "took pleasure in telling the people of the rare antelopes that our Museum . . . alone possessed in such numbers as to afford an adequate representation of them." By "adequate representation," Elliot meant sufficient to represent the species after its extinction. "Altogether," he assured Skiff, "it was a great night for the Field Museum." He explored the possibility of writing a popular book about the expedition but abandoned the project when the museum declined to provide financial support.[7]

Elliot did research on his African collections as well. Early in 1897, he published one article each on the birds and mammals collected by the expedition. In his ornithological paper, he described and named seven new species and one new subspecies of birds. One of these new birds was a kestrel, which he shrewdly named *Cerchneis fieldi*, after Marshall Field, the museum's "most liberal patron." Elliot also designated two new larks. One, *Ammomanes akeleyi*, he coined for his hardworking chief taxidermist. The other, *Minafra sharpii*, he named after his friend and collaborator Richard B. Sharpe, of the British Museum. In a letter to Sharpe announcing the eponymy, Elliot alluded to the illusion of immortality in naming new species: "Glad to put your name to a Lark . . . and it is a great satisfaction to know they have to carry it through all eternity whether they like it or not."[8]

The acquisitive curator continued to focus on augmenting zoology's collections. He did this through any means possible, including purchase, exchange, and donation. Mostly, however, he stressed the importance of collecting in the field. To build collections, work in the field was "the most expeditious as furnishing the greatest returns, and by far, in the end, the most economical," he maintained. To streamline the effort, he clamored for more resources and greater autonomy. "If Zoology is to take its proper position in the Museum," he argued, "much greater advances must be made. . . . At the best, the acquirement of a truly representative Zoological Collection is very slow and surmounted with difficulties." For best results, the curator wanted to be empowered to "to avail himself instantly of such opportunities for obtaining specimens as his knowledge and judgement tell him would be advantageous for the Museum."[9] Time was fleeting, he insisted, and nimble decision-making was imperative.

In 1898, Elliot recommended hiring a full-time collector. "At times . . . it is necessary for the Head of the Dept to take the field himself," he wrote, "but in the majority of instances his presence is not absolutely required, as he can . . . direct the movements of his assistants from . . . the Museum." It was impossible, he found, to be responsible for research and curation at the museum while also aggressively salvaging specimens in the field. For the latter, he needed full-time help: "I can, with great profit and advantage to the Museum," he argued, "keep a collector busily employed throughout the year in various parts of our country." He stressed that hiring a collector made good financial sense: "I think his services could easily be made to yield in return to the Museum four times the amount he would cost us." It would be, he offered, "a great stroke of business."[10]

Nevertheless, Elliot led Akeley on another collecting expedition, to Washington's Olympic Mountains, in 1898. They returned with 520 mammals and reptiles. This collection, Elliot claimed, was exceedingly valuable, coming as it did from a remote locality where few, if any, naturalists had collected systematically. He anticipated several undescribed species among the new specimens. Proving his point only a few months later, he named a new rodent, *Peromyscus akeleyi*. In 1899, Elliot accompanied the privately funded Harriman expedition to Alaska, bringing back several valuable specimens for the museum.[11]

Zoology hired its first full-time collector, Thaddeus Surber, in 1899, and the results were immediate. According to Elliot, "The value of field work has never been more thoroughly demonstrated." The curator suggested that several more collectors could be employed advantageously in different parts of the country. He again raised the specter of extinction and stressed that the museum must make haste. "Many of our mammals are annually becoming scarcer and more difficult to obtain," he asserted, "and of these the Museum has yet to acquire a large number." He wanted to build habitat groups for these diminishing species, and the best method of procuring the requisite specimens was through fieldwork. In a letter to Skiff, he argued: "I cannot too strongly emphasize the importance of field work, especially for a young institution that has everything to acquire, as it is the cheapest way to obtain material."[12]

Elliot grew increasingly impatient with the museum's sluggish bureaucracy. He knew patrons were pleased with the growth and development of zoology, and he wanted to parlay this into greater freedom to run his department as he saw fit. He particularly wanted to end the practice of constantly seeking permission from the director to spend small sums of money to acquire rare specimens or to send a collector afield. Elliot believed that

Skiff's oversight was unnecessary and inefficient, and that the inevitable delay was costing him specimens. "We cannot afford to lose time in this matter," Elliot stressed. "The animals are being killed off rapidly, & the game laws of the States are becoming yearly more strict."[13]

No doubt Skiff resented Elliot's machinations and what they implied about the director's role at the museum. Yet, if he did, he never recorded it anywhere.

IMPROVING THE LOOK OF ZOOLOGY

The museum established a new Section of Taxidermy in 1896, with Akeley—who remained answerable to Elliot—as its head. Elliot wanted to repopulate his department's exhibit halls with elaborate group displays featuring multiple specimens, mounted in active poses, and exhibiting the animals' typical behaviors and habitats. In Akeley, he found someone with the peerless talent and soaring ambition to implement this plan.

The construction of new habitat groups from the specimens salvaged in Somaliland began immediately. A group of six lesser kudus, of various ages and sexes, was first to be installed in the museum's West Court—in a large, custom-built, mahogany and plate glass case—in the summer of 1897, only seven months after the expedition's return. An alert, adult male with spectacular horns formed the center of attention for this group, while a scrawny, doe-eyed juvenile in the foreground returned the gaze of museum visitors. According to the *Guide to the Field Columbian Museum*, this group was "the largest and most complete in the world. In fact, it may be said to be the only one in existence that gives a full representation of the species." Akeley's second group, installed later in 1897, consisted of six Waller's gazelles (gerenuks) wandering through a patch of flowering aloes, one adult male posed gracefully on its hind legs to browse the leaves of a tree.[14]

The museum was immodestly proud of its new group displays. The annual report for 1896–97 boasted: "The striking manner in which these . . . animals are arranged and posed, the life action and naturalness of the picture presented, no less than the scientific fidelity and faithfulness of accessories, stamp them at once as of the very highest character of work." In the kudu group, a termite mound with an African owl perched on top made "a striking accessory." The plants were "true to life, and, . . . the impressions of the desert are forcibly conveyed to the spectator." The gerenuk group was likewise "most effective and dramatic . . . finished in every artistic detail, and complete in every requirement of the scientist and hunter. Attention to the work of Mr. Akeley . . . cannot be directed too flatteringly." Elliot reveled in a report to Skiff that he "believed that [Akeley's] groups and those that are to follow will surpass any similar exhibit in the World."[15]

In 1898, Akeley completed another "striking" group, this one of five oryx surrounding a distinctive termite mound, with one large adult grazing

FIGURE 23.1.

Akeley's Somali wild donkey group. Photograph courtesy of the Field Museum, CSZ12559.

the threadbare grass. The museum continued to gush about his results. The museum's annual report proclaimed: "The same appreciation of the high purpose of [Akeley's] work, the same skillful blending of nature and of art, the same conspicuous care in detail, distinguish [this] group."[16]

A problem emerged: the new taxidermy laboratory built in 1896 wasn't meeting Akeley's needs. The space was too small, its appliances inadequate. Elliot submitted a plan for expansion and improvement. In 1898, the museum added a second story to the lab. The lower floor was used for storage. The upper floor was converted into a new, more commodious working space. Taxidermy could now be pursued "with comfort and without the risks and inconvenience that obtained in the previous confined space."[17]

FIGURE 23.2.

Akeley's
cheetah group.
Photograph
courtesy of the
Field Museum,
CSZ76368.

Akeley's productivity then markedly increased. In 1899, he finished three new groups of African mammals, including Somali wild donkeys, cheetahs, and striped hyenas. The donkey group consisted of four immaculately mounted specimens in a barren landscape of stone and sand. The cheetah group featured a breeding pair of adults and their brood of five kittens—Elliot's onetime pets—feasting on a dead Soemmerring's gazelle. The adults have turned to stare back at the museum visitor as though they've been startled by the approach of another predator. Finally, the striped hyena group included four individuals scavenging the nearly bare bones of some large ruminant.[18]

The museum continued to boast of the superlative quality of Akeley's art: "The groups executed by Mr. Akeley keep fully up to [his] high standard of excellence . . . and place the Museum's possessions in this class of work not equalled by any other institution." They were less happy with the rate at which Akeley produced the work, however. Once Skiff authorized the hiring of additional assistants, including Charles K. Mason, the work proceeded "with much more satisfaction with regard to its quantity."[19]

The annual report for 1900–1901 proclaimed that "work in taxidermy has been unusually active, and results of the very highest character have been attained." More "fine groups" appeared, including a magnificent greater kudu group, in 1900, which featured four adults—one female and three males on a rocky slope. The form of the kudus was faithfully sculpted, and the realism of the individual mounts looked uncanny. Another group completed circa 1900 was the spotted hyena, three specimens of which were mounted in the act of plundering the piled stones on a Somali grave. A "striking" group of northern warthogs exhibited the following year immediately "attracted much attention." Elliot's magnificent boar with huge tusks and a menacing stare was the centerpiece of this ensemble of four specimens. Patrons were thrilled with Akeley's results: "A perfection of work never before attained has been secured."[20]

FIGURE 23.3.

Akeley's greater kudu group. Photograph courtesy of the Field Museum, CSZ6217.

In 1897, the museum had abolished several departments. Specimens in these departments were sold, donated, or returned to their original owners. By 1899, it'd been practically decided to restrict the scope of the museum to natural history and anthropology. The museum then made sweeping changes to its exhibits in 1900. In the shuffle of halls that followed, zoology gained the spacious South Court to provide much-needed room for growth. Elliot decided to use the West Court exclusively for African groups. Cases in the West Court containing other specimens were gradually shifted to the South Court and replaced by new African habitat groups as they were finished. Additional space was found by contracting and consolidating other, less-showy zoology exhibits. Hall 24, for example, displaying protozoa, sponges, and corals, was rearranged in 1902 into new, specially built cases. Not only did these new cases show specimens "in the best possible way," but they also effected "a great economy of space, the collection . . . now occupying about one-half the space it formerly did."[21]

The last two habitat groups made from Somali specimens included Clark's gazelle (dibatag) and Swayne's hartebeest. The dibatag group consisted of six specimens: two adult males, two adult females, and two juveniles of different ages. The foremost male was mounted with its black, bristly tail sticking "straight up in the air" as the dibatag would often do when fleeing from hunters. The specimen collected in utero was mounted

FIGURE 23.5.

The West Court, including several of Akeley's new habitat groups from Africa. The empty case on the middle right would soon contain a group of Somali wild donkeys. Photograph courtesy of the Field Museum, CSZ8466.

as though attempting to nurse. The habitat was bare, reddish sand with a single, leafless thornbush and two small patches of grass. The hartebeest group included seven individuals of varying ages and sexes, in an open, grassy plain. Elliot wrote of the hartebeest's "ungraceful" appearance, yet Akeley's mounted specimens belied that assessment.[22]

By 1903, Elliot's plan for exhibiting rare, one-of-a-kind African mammal groups had been largely implemented. A row of cases arrayed along the north side of the West Court housed cheetah, Clark's gazelle (dibatag), oryx, Swayne's hartebeest, and striped hyena. A large case at the far west end of the court contained greater kudu. A second row of cases on the court's south side contained spotted hyena, Somali wild ass, lesser kudu,

FIGURE 23.6.

Akeley's spotted hyena group. Photograph courtesy of the Field Museum, CSZ12697.

Waller's gazelle (gerenuk), and northern warthog.[23] Many of these groups were unique to the Field Columbian Museum. Should the species represented disappear at some future date, the groups could never be duplicated. Thus, the realization of Elliot's dream of creating a "Mecca for all naturalists" now hinged on extinction.

Much to the satisfaction of the museum's patrons, Akeley's habitat groups elicited very favorable attention for the museum from outsiders. In 1897, for example, the journal *Science* reported that the "mammal groups of Mr. Akeley, who is unrivaled in this work, are deserving of special notice." In 1901, William T. Hornaday—once an innovative taxidermist himself, but now director of the New York Zoological Park—wrote Akeley a letter praising his animal groups. "In thinking over your surprisingly fine groups of mammals," Hornaday wrote, "I am more and more strongly impressed by their artistic excellence. They are the finest examples of taxidermy in the world—so far as I have seen it." Hornaday then wrote an article about American taxidermy for the *New York Herald*, where he trumpeted Akeley's work: "When I saw Mr. Akeley's splendid array of groups of African mammals—close haired, every one of them, exquisitely modeled and posed, neither fat nor lean and not a seam visible anywhere—it gave me a thrill of pleasure to think that those masterpieces . . . had been produced in America." William A. Bryan, formerly an assistant curator at the Field Columbian Museum but now a curator at the Bishop Museum in Honolulu, wrote in a flattering letter that Akeley had "done for Taxidermy what Michelangelo did for sculpture and painting."[24]

LIKE FIRING INTO A MOUNTAIN

Museum visitors were equally impressed by Akeley's taxidermy. For example, a suburban Chicago teenager with two of his friends visited the Field Museum circa 1910, "embued with a desire for knowledge." Their first impression as they approached the dilapidated building was of its "massiveness." Next, they were struck by its "appearance of decay." Once inside, it was the realism of Akeley's taxidermy that wowed them most. In the building's central rotunda, they stood in front of "a replica of a battle between two mighty elephants and we gazed with awe at the huge beasts who did not look in any sense stuffed but as tho suddenly arrested in the most exciting moment of the fight." The teens were speechless: "We moved around this group too admiring to comment until one of the boys said 'and to think that a woman shot one of those, it must of been like firing into a mountain.'" They left the elephant group and wandered through the exhibit halls "looking at the wonderfully realistic groups of various wild animals and birds. The taxidermists . . . that mounted these various groups were artists and showed it by their attention to the little things," he noted admiringly. What was the educational value of seeing Akeley's habitat groups in the museum? It was "almost as good," the teenager wrote, "as seeing the animals themselves alive to see them in [their] native haunts in attitudes true to life. It is better in fact than the impressions most of us would get from seeing them alive for here we can study them and note new details."[25]

Whether he was inspired by a conservation message, young Ernest Hemingway didn't record.

24

UNPLANNED OBSOLESCENCE

By late 1902, the museum's home—the former Palace of Fine Arts—had reached the "limits of repair." This was especially true of the building's exterior, which was "gradually falling to pieces." Large sections of the museum's decorative plaster veneer had been spalling off the building for years, exposing ugly gashes of underlying brick. (Akeley's invention of a pump to spray concrete on the exterior would later slow—but not stop—the deterioration.) But the damage wasn't merely cosmetic. The building's slow decay was endangering collections, staff, even visitors. The glass-topped cases housing Akeley's revolutionary habitat groups were particularly vulnerable to falling debris. Winter storms wreaked havoc on the building. Each spring, a team of temporary craftsmen swarmed like ants over the building's roof, replacing the glass from cracked or broken skylights or hammering sheets of tin to the museum's crumbling dome. But these costly repairs were merely delaying the inevitable. In 1903, all attempts to improve the appearance of the exterior were temporarily abandoned as wasteful and ineffective.[1]

Patrons committed to build a vast, permanent, and absolutely fireproof structure to be the institution's new home. Marshall Field—who would be footing the bill—approved blueprints for the monumental new building in the spring of 1905. Curators inspected the plans for the first time in June. Extraordinary efforts would now be required from the staff to expand the museum's collections and create worthy exhibits with which to furnish the great, new edifice.[2] But were all the scientific departments up to the task?

Skiff raised concerns about a persistent lack of progress in ornithology. "I should be glad to be advised," he wrote to executive committee chair Harlow N. Higinbotham, "as to what steps can appropriately and effectively be taken to prosecute the work of developing the Department." Charles B. Cory, the curator, had visited Chicago, on average, once a year for a few days only since the start of his honorary appointment in 1894. The only person on the department's payroll was an assistant curator, who, although "excellent and capable," had been given insufficient initiative or authority to make any meaningful progress in placing new bird groups on display. Few of the museum's existing bird exhibits were suitable for installation in the new building. Skiff argued that the present condition of the department was "a detriment to the Museum." Moreover, it would be impossible, he feared, to bring ornithology up to museum standards without "the constant and personal direction" of a full-time curator. Knowing of Cory's personal relationships with several trustees, Skiff was reluctant to raise the issue. Yet something clearly had to be done to "give vitality to Ornithology."[3]

The Department of Zoology also came under scrutiny. At an executive committee meeting on July 24, 1905, museum patrons made two crucial decisions that would shape the future of zoology. First, they appropriated $5,000 to fund a collecting expedition to British East Africa. Elliot—now seventy and a recent widower, and still suffering from occasional recurrences of fever—declined to undertake another physically demanding expedition to Africa. Akeley, therefore, took charge. This was the museum's most important zoological undertaking of 1905.[4]

The second decision was more unexpected. A quorum consisting of Higinbotham, board of trustees president Edward E. Ayer, Ryerson, and Skiff decided to dispense with Elliot on the first of January 1906. Higinbotham was asked to pen a letter "as he might deem advisable," notifying Elliot of this fact. The chairman matter-of-factly wrote: "I was instructed to advise you that it would be [the executive committee's] pleasure to discontinue your services at such time as you can conveniently conclude such work as you may now have in hand." To soften the blow, he added: "The Committee also desires me to express to you their appreciation of the valuable service that you have rendered, both to the Museum and to the scientific world." Dumbfounded and humiliated, Elliot replied immediately, pleading for a reason. Higinbotham explained that in view of the work to be done building collections and installing quality zoology exhibits in a new and larger building, "a younger man, with larger powers of endurance, will be necessary."[5]

What had caused this shocking reversal of fortune? Elliot wondered. Perhaps he recollected hunting near Jerato Pass—Akeley and Dodson scaling the mountain while he tarried on the plain. Hunting greater kudu was a young man's job; even Elliot recognized this. He might've thought for a moment of the old, enfeebled Somalis he'd encountered on his expedition. They'd been cast out into the wilderness to fend for themselves or be devoured by lions. Was he really so useless? The curator was at a loss to understand why he'd been pushed out so suddenly and ignominiously.

Yet, Elliot's demise was neither as sudden nor as unpredictable as the guileless curator felt. Indeed, his fall from grace began in Africa. From the beginning of his tenure at the museum, Elliot had an excellent working relationship with Skiff. The director's rapport with other members of the scientific staff, on the other hand, was poor. Tension with the curators had flared when Skiff fired Elliot's assistant, Oliver Perry Hay, in June 1896. Edward D. Cope had fanned the flames of discontent with his pair of caustic editorials in the *American Naturalist.* Skiff was under siege. He still enjoyed the confidence of the trustees, yet he had few friends among the scientists. Elliot, he had believed, was one of them. So, he had felt especially hurt and betrayed when the zoology curator—his closest confidante on the staff, and the curator for whom he'd moved mountains to help make the African expedition a success—called his leadership into question over the Hay episode in a private letter. When Skiff had called him out for this, Elliot had been mortified. The curator knew that he owed a great deal to Skiff for the director's steadfast support. He had tried to make amends for his gaffe in a conciliatory reply, but the damage had already been done.

Meanwhile, discontent had spread among the other curators. Tension boiled over in January 1897, when they revolted. In general, the bone of contention was autonomy. The curators—Elliot included—wanted the freedom to do science exactly as they thought best, answering only to the trustees. They hoped a generous museum endowment would guarantee this freedom. Unfortunately, Skiff's autocratic leadership style kept the curators in check by limiting their spending. This ruffled many feathers. By early 1897, the curators had had enough. According to a brief article about the insurrection, "charges of incompetency were [proffered] against Director Skiff by . . . the curators." The *Chicago Daily Tribune* reported that "friction . . . has existed between the curators and Director F. J. V. Skiff, and, it is said, resulted from the policy which the director has followed in administering . . . the museum. The curators drew up their grievances . . . and this was brought to the attention of the . . . Trustees."[6]

Anthropology curator William Henry Holmes provided a firsthand account of the crisis, identifying Skiff's personality and his management style as the root problems. "The scientific staff of the Museum was gradually getting into a state of rebellion against Director Skiff as a result of his unappreciative and tyrannical attitude," he wrote. To Charles D. Walcott, a trusted colleague, he explained: "The trouble developed out of the Chicago idea that only a business man . . . can conduct the business of a . . . museum." This, he conceded, "would have been well enough had the man chosen as director been qualified. . . . [The] difficulty was due not only to the business direction which extended over the scientific work, but to the personality embodying more unfortunate elements than I have ever known assembled in one individual."[7]

The executive committee, chaired by Higinbotham, met privately with the disgruntled curators. No record of the meeting is known, and no major changes to the museum's management resulted from the rebellion.

Skiff would continue as director. Henceforth, however, no curator could be removed from his position without the consent of the executive committee. According to the *Chicago Daily Tribune*, the curators decided to let the matter drop: "The curators, it is claimed, are now satisfied with the existing condition of affairs at the museum." There's no record of Elliot's role in the rebellion, yet he almost certainly sided with his fellow curators. Skiff survived the insurrection, but he harbored long-lasting resentment for those who'd rebelled against him.[8]

SKIFF BIDES HIS TIME

Years passed without incident, and Skiff and Elliot seemed to get along fine. But things took a turn for the worse early in 1902. That spring, Elliot was so ill with typhoid fever that he couldn't work for several months. Speculation about his long-term health and his future with the museum became common conversation among the staff. In 1903, the museum began to scrutinize the money Elliot was spending to build his department's study collection (as opposed to specimens collected explicitly for display). The pace of Elliot's research also came into question, particularly the delay in issuing a publication on the mammals of Mexico. Higinbotham, who didn't normally involve himself with the day-to-day activities of the research scientists, wrote to Elliot in December, asking: "Will you kindly give me at your early convenience, a statement of the status of the work . . . and also please advise me what has occasioned the delay." In a lengthy, carefully worded reply written the following day, Elliot explained that his text had been finished for quite some time, but that there'd been much time lost in making the illustrations. He insisted that there was "no extra expense whatever" in the delay.[9]

But the delay provoked a personnel crisis in the department. In January 1904, Elliot complained to Higinbotham that assistant curator Seth E. Meek—Hay's replacement—had been "maligning" him and "using language that was so obnoxious that he did not feel like retaining him any longer in his position." According to Elliot, Meek had asked the museum photographer to "slur" the work he was doing on images for Elliot's publication. When given an opportunity to explain himself, Meek suggested that it was all a misunderstanding. He admitted that he'd asked the museum's artist, J. Howard Stebbins, to hurry the work he was doing for Elliot, but said that he hadn't meant to suggest "that it should be slighted or slurred." Meek simply had illustrations of his own that he wanted the artist to finish. He also admitted that he might've made some careless remarks about Elliot's work for the museum being "nearly concluded," but claimed that this had been when the curator was seriously ill. It wasn't his intention, he insisted, to injure Elliot in any way, and he was glad to issue an apology if that was the "right and proper" thing to do.[10]

Higinbotham met with Elliot to defuse the crisis. He recorded that the curator told him that his health was "much better than a year ago. Said also that his work on mammals had gone to the press and it now remained for them to get it out. His work had been completed and he did not think he would ever undertake such a thing again." More ominously, Higinbotham "judged from his manner and his conversation that aside from a general supervision . . . of his department he feels that his work is about done. He has, however, no idea of retiring."[11]

Skiff, too, weighed in on the controversy. To Higinbotham, he opined that the antipathy between Elliot and his assistant curator "began long before Mr. Meek thought of making any slighting remarks of Mr. Elliot. You must know that Mr. Elliot is of a temperament . . . that invites levity." This was a reference to Elliot's rigid formality. "Mr. Meek and Mr. Akeley," Skiff continued, "are good friends and sympathize with each other. They are both interested in the Museum. . . . Elliot, . . . in my opinion, . . . cares nothing for the Museum as a physical presentation of scientific facts"—a remarkable charge to bring against a man who'd worked so hard and sacrificed so much to build the zoological collections—"but simply uses [the museum] as a means of propaganda of his own ideas on certain scientific deductions." Skiff's short-term solution was to dismiss the department's fish biologist, Howard W. Clark, "who is a tale bearer and a mischief maker by all accounts," and to ask Meek to apologize to Elliot.[12] Clark, who'd ratted Meek out to Elliot, was fired. Elliot lasted another year before he, too, was notified of his termination.

In a letter to Higinbotham written after he'd been fired, Elliot protested that he'd been working on a comprehensive catalog of the museum's mammals. The curator considered this work to be important both to the museum and to naturalists generally. As this project couldn't possibly be completed by January 1, 1906, it would have to be abandoned. Another important work on which the curator was engaged was the identification of specimens in a large collection recently acquired by the museum. "Until they are named those specimens have no value," Elliot argued.[13]

In October 1905, in his final annual report, Elliot touted the impressive growth of the museum's mammal collection. "The entire collection," he wrote, "now numbers something like 15,000 specimens, a creditable showing for ten years' work." With regard to the department's future, he argued that he still had a vital role to play. Again, he stressed that building collections was best accomplished through fieldwork done by trained men. But, he emphasized, this should be done "under the control and guidance of the head of the department. . . . Any other procedure will surely end in . . . unsatisfactory results, with both time and money wasted."[14]

By the new year, the museum had reconsidered. "The Executive Committee," Skiff explained to Elliot, "reluctant to contemplate the complete severance of your connection with the Museum . . . instructs me to ask you if you will accept the position . . . of Honorary, Supervisory, Curator." The position would pay a stipend of $1,500 per annum and would permit Elliot to live anywhere he chose. His duties would be "largely regulated" by himself. The director would be free to refer zoological questions to him, to ask him to determine specimens, and to direct expeditions. Elliot, of course, was flattered. "It will give me much gratification to accept the position," he replied. "I shall be at all times ready to do my utmost for [the museum's] welfare and prosperity."[15]

In the museum's next annual report, Skiff spun the reorganization of zoology as a mutual decision reached amicably. "The resignation of Curator D. G. Elliot," he wrote, "and his acceptance of the post of Honorary and Supervisory Curator . . . was an important change." The change was the result of a two-year deliberation, the director explained, and was intended to give Elliot "greater freedom of action . . . and at the same time to retain . . . the benefits of his scholarly attainments and wide Museum experience."[16]

If Skiff intended to canvass the scientific community for a successor to Elliot, those plans were scuttled when a personal financial crisis visited Charles B. Cory, curator of ornithology. "This was nothing less than the complete loss of his fortune," Cory's obituary explains, "reducing him in a few months from a man with the income of a millionaire to one required to earn the daily bread of himself and family." Consequently, Cory—at age fifty—uprooted his family from Boston and moved into a modest Chicago bungalow. He then settled down to a life of quotidian curatorial responsibilities. Ornithology was folded into the Department of Zoology. Cory, who had once insisted on the separation of ornithology from zoology, now had dominion over the entire animal kingdom. The museum began paying him a salary for the first time on July 13, 1906.[17]

But Cory's salvation created myriad problems in the new department, leading to an exodus of employees, including Akeley and others.[18] Akeley's resignation was a particularly bad loss for the museum. With his superlative skill as an artistic taxidermist, he'd revitalized his institution's exhibits, raising the global standard for museum zoology in the process. It's no exaggeration to claim that Akeley had revolutionized museum display during his decade in Chicago. Now, he'd be plying his trade at a rival institution. Indeed, the Akeley Hall of African Mammals at New York's American Museum of Natural History is considered by some to be his magnum opus.

Cory's salvation also proved costly for Elliot. Early in the new year, "in view of the necessity of the strictest economy," the executive committee decided, once again, to let Elliot go. Skiff rose to the occasion with a magnificent—if insincere—letter, explaining that the executive committee had expressed its "highest regard for you as an individual" and a scientist. "Personally," he added, "I must leave to some opportunity for a personal conference to express to you the extent of my regret that a relation that has been of such satisfaction and benefit to me is thus discontinued."[19]

Elliot was gone for good by May 1, 1907.

CONCLUSION

The zeitgeist of the nineteenth century was progress. One important hallmark of progress was increasing industrialization, with its dependence on natural resource exploitation. Another was European imperialism and the global spread of land-intensive practices like farming and ranching. An unfortunate consequence of these activities was environmental degradation, habitat loss, and—with increasing frequency—extinction. As Western civilization advanced around the world, wild things gradually receded. By the mid-nineteenth century, extinction was understood and accepted as the price of progress.

What's more, many nineteenth-century naturalists believed that extinction wasn't necessarily bad. Geologist Charles Lyell suggested that it was a natural, regular, law-abiding phenomenon. Charles Darwin's *On the Origin of Species* argued that its cause was the struggle for life. Species driven to extinction were those that failed to adapt to the challenges of their changing environments. According to Darwin, "The extinction of species . . . which has played so conspicuous a part in the history of the organic world, almost inevitably follows on the principle of natural selection; for old forms will be supplanted by new and improved forms." Rather than something to be prevented, extinction, according to historian David Sepkoski, "was understood to be nature's way of strengthening and improving itself by weeding out the unfit, and competition was celebrated as a source of natural progress." To Victorians, progress was paramount. Although a small and vocal conservation movement was already afoot by the final third of the nineteenth century, the phenomenon that we now call biodiversity was simply not valued at that time in the same way and for the same reasons that we value it today.[1]

Museums of the nineteenth century endorsed this worldview to a greater or lesser extent. Chicago's Field Columbian Museum, for example, was initially conceived not as a natural history museum, but as a kind of panorama of progress. The exhibits—many derived directly from the

World's Columbian Exposition of 1893—were intended to showcase humanity's greatest achievements in the arts and sciences. The idea was to create a self-evident, object-based narrative of universal progress. A reporter discerned the message of progress written into the exhibits on the museum's opening day: "The history of the race and the record of earth's evolution in all its forms are presented in the museum arranged with scientific exactness and system, so that the student may see at a glance and study the record of progress." Or, as Skiff wrote, "from one end of the museum to the other can be traced, almost without a break, the living and instructive story of nature and of man and his works."[2]

The arc of progress that Skiff described began in the museum's West Pavilion with geology and ended with locomotives in the East Pavilion. The Departments of Zoology and Ornithology, meanwhile, shared space in the Central Pavilion with botany, anthropology, and industrial arts. Bridging the conceptual gap between zoology and industrial arts were two halls devoted to animal industries and fishery industries, respectively. The former featured tanned animal skins, leather footwear, and models of Chicago slaughterhouses, among other exhibits. The latter included a veritable "cabinet of curiosities" from an authentic Arctic whaling ship. A text from the *Guide to the Field Columbian Museum* describing this exhibit and explaining its value was indicative of the museum's attitude toward extinction. It read: "The time may come during the life of the present generation when the sperm and the right whale on the high seas will be almost as much of a curiosity as the buffalo upon the prairie; the introduction of a modern harpoon fired from a gun, having taken the place of the hand lance, is devastating the sea in a manner . . . similar to the slaughter of the denizens of the prairie by the repeating rifle."[3]

The museum presented extinction matter-of-factly as a byproduct of modernization. Using a tone of triumph rather than regret, the *Guide to the Field Columbian Museum* implied that there was something inevitable about the fate of the buffalo when faced with the hunter and his repeating rifle. Ditto the sperm whale, which was doomed by the modern harpoon.

Incidentally, the name of the whaling ship in the exhibit was *Progress*.

SALVAGE ZOOLOGY

Salvage zoology was a common practice in the late nineteenth century. If certain species were doomed to extinction, including many large-bodied mammals, flightless birds, and island species, then it was incumbent on zoologists to collect the remnants of these animals for their museums while they still had the chance to get them. Alfred R. Wallace explained the rationale for this practice as early as 1863. "The extinction of the numerous forms of life," he wrote, which "progress . . . invariably entails will necessarily render obscure this invaluable record." It was, therefore, imperative "that in all tropical countries colonised by Europeans the most perfect collections possible in every branch of natural history should be made and deposited in national museums, where they may be available for study."[4]

Salvaging scientific specimens and data in the face of extinction was critically important to nineteenth-century naturalists. Hugh Strickland and Alexander Melville, for example, who acknowledged humanity's role in exterminating the dodo bird, made no general call for conservation. What good would this do, after all, when extinction was the inevitable consequence of human progress? Rather, they argued, it was the duty of the naturalist "to preserve to the stores of Science the knowledge of these extinct or expiring organisms . . . so that our acquaintance with the marvels of Animal and Vegetable existence may suffer no detriment by the losses which the organic creation seems destined to sustain."[5]

Naturalists around the world heeded these calls to salvage. Zoological collecting in the Galápagos Islands around the turn of the twentieth century, for example, was done explicitly for salvage purposes. English naturalist Osbert Salvin raised an alarm about the imminent extinction of giant tortoises in the Galápagos in 1880. "What is required," he wrote, "is the search in each island not only for the few lingering individuals that may still survive, but for any remains." No time should be lost, he insisted. Once these animals were extinct, "which must ensue in a short time, all traces of them must follow." George Baur collected in the islands in 1891 in order to salvage specimens for science. (Some of Baur's specimens were later exhibited briefly in the Field Columbian Museum.) "Such work ought to be done," he emphasized, "*before it is too late.*" In 1897, zoological collector Lord Walter Rothschild, certain that Galápagos tortoises were on the brink of extinction, instructed his fieldworkers to bring home every specimen they could find, dead or alive, to "save them for science." In 1905–6, scientists at the California Academy of Sciences swept the Galápagos Islands of some 78,000 specimens, including 266 tortoises. The Galápagos trip was part of a series of aggressive collecting expeditions designed to salvage island specimens before those species went extinct. An article in the *San Francisco Chronicle* explained that "science has awakened to the state of affairs, and strenuous efforts are being made in various parts of the world to secure specimens before it is too late."[6] Despite acknowledging the imminent threat to the Galápagos fauna, each of these expeditions was undertaken explicitly to get specimens while the getting was good. The reasoning was clear: since the species themselves cannot be saved, the specimens must be. Salvaging the material remains of these doomed species for museum collections, then describing and illustrating them, would allay the loss to science occasioned by their extinction from the wild.

Museum zoologists believed that the inevitable loss of species would be further mitigated by the construction of spectacular, lifelike habitat displays to exhibit their mortal remains. These would serve—must serve—as a reasonably worthy substitute for what was being lost in nature. Habitat groups created an illusion of immortality that blunted the impact of extinction.

The exhibition of rare and extinct species brought to the museum a measure of prestige and relevance in a rapidly changing modern world. Rarity implied value. The rarer the specimen, the greater its value to the museum. By the mid-nineteenth century, the rarity of the great auk—driven to the brink of extinction by overhunting—rendered it an invaluable museum commodity that paradoxically accelerated its demise. The last two known specimens of great auk were captured and strangled for the museum trade in 1844. Or consider the passenger pigeon, an animal that was once so abundant in North America that migrating flocks sometimes obscured the sun for days. Little attention was given to the bird's anatomy, behavior, or ecology while it was still common. In 1914, however, when the last known passenger pigeon, Martha, died in captivity at the Cincinnati Zoo, its body was frozen and shipped to the Smithsonian. "Now suddenly Martha's little carcass was precious," wrote David Quammen, "in accord with the scarcity theory of value." The specimen was later stuffed and put on display. Today, we read the fate of an extinct animal as a cautionary tale.[7] In the nineteenth century, on the other hand, it could just as easily have been seen as emblematic of progress.

Some repentant museum zoologists, including William T. Hornaday, viewed taxidermy as a kind of atonement for the sin of extinction. In 1886, when he set out for Montana to hunt some of America's last wild bison for a habitat display at the Smithsonian, he already believed the animal was doomed. He undertook his expedition explicitly to salvage specimens for display in the National Museum. In 1891, he made a clarion call for salvage zoology. "The rapid and alarming destruction of all forms of wild animal life which is now going on furiously throughout the entire world," he wrote, "renders it imperatively necessary for those who would build up great zoological collections to be up and doing before any more of the leading species are exterminated." There was a practical reason for urgency: "If the naturalist would gather representatives of all these forms for perpetual preservation, and future study," Hornaday argued, "he must set about it at once. . . . *Now* is the time to collect. A little later it will cost a great deal more, and the collector will get a great deal less."[8]

Daniel Giraud Elliot, who bought this argument hook, line, and sinker, practiced salvage zoology at the Field Columbian Museum. He believed that the museum that accumulated especially rare specimens was doing a great service to science. This belief was the impetus for his ambition to make his institution "the Mecca for all naturalists."[9] If he were to succeed in acquiring and exhibiting scores of animals that then went extinct in the wild—as the curator confidently predicted—then future naturalists would be obligated to visit the Chicago museum to see the last remnants of nature. Thus, extinction, to Elliot, was a self-fulfilling prophecy with the potential to benefit those museums of Europe and America that collected most aggressively.

THE SLOW DEATH OF SALVAGE

This idea about the inevitability of extinction gave way once museum zoologists, in ever-increasing numbers, began to see wildlife conservation as part of their professional mission. According to Harriet Ritvo, "the point of view that had accepted the elimination of wild animals . . . as an inevitable by-product of progress was replaced by one that viewed them as a valuable resource requiring protection." Often, the turn toward conservation followed a particularly harrowing hunting experience. American Museum ornithologist Frank M. Chapman, for example, became an advocate for bird protection after shooting fifteen rare Carolina parakeets in Florida in 1889, although he continued to collect specimens for the museum and to defend scientific collecting in print. Once an avid big-game hunter, Hornaday in 1896 became a zoo director and a fierce advocate for wild animals and wilderness. He battled the feathered hat industry, poachers, gun makers, and lax game laws for the last forty years of his life. He also railed against fellow naturalists for not doing more to advance the cause of conservation. He is credited with saving the Alaskan fur seal and the American bison from total extinction. According to a recent biographer, Hornaday experienced his ethical awakening while hunting bison.[10]

Other naturalists, too, advocated for a kinder, gentler approach to studying zoology—to shoot with a camera, for instance, rather than a gun. Ornithologist and nature writer Florence M. Bailey, for example, advocated birding with opera glasses in the late 1880s. William Finley favored photography over collecting at the turn of the twentieth century. The tide was turning against salvage. In fact, some ornithologists became so concerned for the future of scientific collecting that Joseph Grinnell—director of Berkeley's Museum of Vertebrate Zoology—wrote an impassioned paper called "Conserve the Collector," published in *Science*, which argued for the critical importance of this practice.[11]

Elliot, on the other hand, never made the transition from acquisitive museum zoologist to wildlife conservationist. A lifelong hunter, Elliot remained an unapologetic and unrepentant collector of animals. His experience on the African expedition, though, showed that the distinction he made between scientific collecting and sport hunting could sometimes be a fuzzy one. John F. Reiger has argued that it was "impossible . . . to make a clear distinction between the ornithologist and the sportsman. A better term would be 'sportsman-naturalist.'" Elliot is a case in point. Indeed, following the expedition of 1896, Elliot wrote several contributions for Henry A. Bryden's guidebook to Africa's sporting animals, *Great and Small Game of Africa*. "The great bulk of the book," Bryden noted in the preface, "has been contributed by well-known African sportsmen, who have had intimate personal experience of the animals." Another of Elliot's most popular publications from his Chicago years was a book about the game birds of North America. The subtitle described this as "a book written . . . for those who love to seek these birds afield with dog and gun."[12]

In April 1900, while the US Congress was debating the Lacey Act, which was designed to protect wild animals and plants from illegal trafficking, Elliot wrote a lengthy opinion piece for the *New York Times* about the world's best hunting grounds and the disappearance of big game. In this article, he advocated tepidly for the protection of game animals from "indiscriminate slaughter," yet he also noted that game laws were seldom enforced and largely ineffective. National parks, he argued, were the "surest method" of protecting "the most species for the longest time." Yet even this, he expected, was merely a stopgap measure. He concluded that the rapidly expanding human population—which was putting enormous pressure on wild animal habitats—the improvement of firearms, and the increasing numbers of sportsmen were "the infallible signs that foretell the speedy arrival of that day when North America will no longer be regarded as one of the world's great hunting fields for big game."[13] Elliot lamented the loss of sporting opportunities more than the loss of species.

In short, Elliot was no conservationist. Rather, he subscribed wholeheartedly to the nineteenth-century view that many species were doomed to extinction, and he often used that idea as a rationale and a justification for collecting specimens with abandon. He remained a committed salvage zoologist throughout his career.

As the conservation movement gained momentum in the twentieth century, some museums and museum zoologists were a little late in getting to the party. One reason for the lag was the conflict between the ideals of conservation and the practice of making scientific collections. Mark V. Barrow Jr. has pointed to controversies in the early twentieth century and argued that collecting rare animals showed a callous disregard by museum zoologists for the plight of endangered species. One could go even further and argue that there was no question that some museum zoologists at the turn of the century were more interested in pursuing their science than in protecting vanishing wildlife. As a matter of fact, systematic collecting continued at museums. The practice gradually diminished in significance, however, over the first several decades of the new century, as the focus of museum zoology broadened from taxonomy to include new questions about ecology and animal behavior. Field observations, experiments at research stations, and limited collecting for carefully circumscribed projects with more modest specimen needs became the norm. A period of reckless collecting by museums gradually gave way to one of more carefully considered acquisitions. This was partly a function of the diminishing returns of systematic collecting, and partly the result of changing priorities. Vertebrate life on the planet was far better known by the 1930s than it had been in the 1890s, and thus the need for, and the benefit of, systematic collecting dropped off considerably.[14] Museums were arguably then in a position to curtail their voracious zoological collecting because they finally had it all (or nearly all, certainly). What's more, being in possession of rare or endangered specimens put museum zoologists in a position where they could support conservation without sacrificing the completeness of their own collections. Indeed, having a rare specimen that could no longer be taken legally in the wild could even be considered a competitive advantage.

Another important consideration in the collecting slowdown was undoubtedly the realization by museum professionals that unlimited growth was unsustainable in the long term, when housing and caring for specimens in perpetuity was so resource intensive. Even Grinnell, who'd defended scientific collecting so passionately in 1915, soon changed his tune. By the early 1920s, he was poised to ratchet down the systematic collecting in favor of working on specific projects. "This sporadic type of field work," he wrote, "as immediate opportunity and problems in hand seem to demand, is now, perhaps, more justified . . . than consecutive all-around vertebrate collecting."[15]

The advent of conservation brought the slow death of salvage zoology in the twentieth century. Museum zoologists continued to make and maintain collections. Museums, likewise, continued to build habitat groups. But the museum's mission gradually evolved. Rather than collecting and exhibiting with the idea of salvage in mind, museums—following their zoologists—eventually made a fateful turn toward conservation. Some progressive museums led the charge. Indeed, some museum zoologists were banging the drum of conservation in the 1880s. Other museums lagged, but all—or nearly all—eventually made the shift.

Extinction would no longer be expected complacently; it would be actively combatted. The quixotic goal of creating a museum mecca to replace nature was abandoned in favor of a new purpose: the conservation of wild animals in their wild habitats. In the twenty-first century, conservation has become a central pillar of the mission of nearly all museums of natural history. Indeed, it is so ubiquitous now that the practice of salvage zoology has been forgotten. Yet salvage—once the working model of many museum zoologists—provides the proper context to understand why they once shot pandas and harpooned manatees, stuffed them, and put them on display.[16] Given that so many early conservationists began their careers as traditional hunting naturalists, one could argue that salvage zoology was a necessary first step in catalyzing the conservation movement. In any case, it will no longer be necessary to rationalize that museum zoologists slaughtered animals in the nineteenth century in order to save them. Nor will it be acceptable to claim that museums have *always* been at the forefront of zoological conservation.

The Field Museum, for example, made a modest beginning in zoological conservation in 1930, when Stanley Field—the museum's long-serving president and nephew of its founder—joined the first advisory board of the American Committee for International Wild Life Protection. This was a small, informal group whose members represented museums, zoos, and conservation societies, and whose object was to promote the global preservation of rare species. It was a start, but little is known of Field's involvement with the committee, and no evidence has been found that the committee's work had any impact on the museum's zoological collecting policy. It wasn't until the late 1990s—a century after Elliot's African expedition—that the Field Museum formalized its commitment to conservation, with the creation of the Office of Environmental and Conservation Programs.[17]

Regarding the diorama, its role in natural history museums has done an about-face. No longer intended as a memorial of nature, the diorama is today seen as a wake-up call. It "offers an altar for the worship and contemplation of nature that strikes each viewer with an individual chord of emotions," argued a recent book about Field Museum dioramas. "In some cases, if it were not for a diorama featuring an endangered species, the call to action for rescue might never have been heard."[18]

APOTHEOSIS

In 1921, Akeley led a gorilla-hunting expedition to Mount Mikeno, in the Belgian Congo, on behalf of the American Museum. The hunt was a traumatic and transformative experience. Seeing the expression of pain and fear on the face of a dying gorilla, Akeley felt like a murderer. For the rest of his life he advocated for a gorilla preserve in the Virunga Mountains. In 1925, having been influenced by Akeley, Belgian king Albert I established Africa's first national park. Opposed to hunting gorillas for sport, Akeley still maintained that collecting them for scientific purposes was acceptable. He led his fifth and final journey to Africa late in 1926. He died of dysentery on November 18 and was buried there, close to the spot where he'd had his apotheosis. The "true sportsman," Akeley once argued with uncanny prescience, wants "to remain [in Africa] as long as possible."[19]

The exquisite taxidermy Akeley crafted for the Field Museum at the turn of the century—including nearly every habitat group he made from specimens collected in Somaliland in 1896—is still on display in the Carl E. Akeley Memorial Hall.

Daniel Giraud Elliot, who died in December 1915 in his eighty-first year, would likely have been puzzled by Grinnell's spirited justification of scientific collecting, which, to his way of thinking, needed no defense. For Elliot, there was no question that building museum collections was the right way to pursue zoology. "Science has no use for a wobbly disciple," he told an enthralled audience of admirers—young and old—in an address given before the Linnaean Society of New York on March 24, 1914.[20] (Elliot was there to receive the Linnaean Medal of Honor for mammalogy and ornithology.) Science was a heroic pursuit, he suggested. It required masculine virtues like courage, steadfastness, and perseverance. He was taking a moment to look back on his scientific triumphs, to find a life lesson for future naturalists, to take stock.

"Stand by your guns," he told his audience.[21]

NOTES

Preface

1. For more on fieldwork in the late nineteenth and early twentieth century, see Kuklick and Kohler, "Science in the Field"; Kohler, *All Creatures*; Vetter, *Field Life*. On Jones, see Rideout, *William Jones*. On Stanley, see B. D. Patterson, "Dr. Bill Stanley."

2. On the vital role of museums in wildlife conservation, see Arengo et al., "The Essential Role of Museums in Biodiversity Conservation."

3. Moehlman, Kebede, and Yohannes, "African Wild Ass."

4. Moskovits, "1994."

Introduction

1. Daniel Giraud Elliot, manuscript journal on the Field Columbian Museum's zoological expedition to Africa (Elliot journal, hereafter), 1896, p. 120, Mary W. Runnells Rare Book Room, Field Museum.

2. Elliot journal, 121.

3. D. G. Elliot to F. J. V. Skiff, June 14, 1896, Director's general correspondence (DGC), Field Museum Archives (FMA), Field Museum.

4. Elliot to Skiff.

5. Elliot to Skiff. See also Elliot journal, 121–22. *Savage* was a relatively common nineteenth-century term for a group of people who had not developed agriculture (see, for example, Morgan, *Ancient Society*). Elliot used the term in this sense, but also with the more pejorative meaning, in his journal. Elliot used other offensive terms too. Often, these terms appear in passages unrelated to the main narrative of the expedition. These passages have usually been omitted. Where such passages remain, the terminology used is faithful to the original.

6. Elliot to Skiff, June 14, 1896. On the *dibaltig*, see Pease, *Travel and Sport in Africa*.

7. Elliot journal, 122; Elliot to Skiff, June 14, 1896.

8. Elliot to Skiff. See also Elliot journal, 123.

9. Elliot to Skiff, June 14, 1896. See also Elliot journal, 124.

10. Elliot to Skiff, June 14, 1896. See also Elliot journal, 123. That flattery was obviously aimed at Skiff, Elliot's friend and ally, who was pleased with his adopted city and very proud to be director of its young museum.

11. Elliot to Skiff, June 14, 1896. See also Elliot journal, 123.

12. Sepkoski, "Extinction and Biodiversity," 29–30. This is referred to as the "realist" position in Powell, *Vanishing America*, 15. Some naturalists believed that many species would survive progress through domestication; others felt that species could endure if only civilization was more careful and compassionate.

13. Farrington, "The Rise of Natural History Museums," 52; Field Columbian Museum, "Annual Report [. . .] for the Year 1896–1897," 186. Biologist David M. Armstrong was apparently first to use the term *salvage zoology*, doing so in his 1972 monograph *Distribution of Mammals in Colorado*. He defined salvage zoology as "an attempt to understand patterns of distribution and symbiotic relationships before they are destroyed or distorted by human intervention" (39). He noted that "inevitable" development was deeply disruptive to ecosystems (v). "Once disrupted by extinction," he stressed, the "information in a given natural system is irretrievable" (vi). This gave a sense of urgency to the work: "We had better begin to understand the history and potential of the landscape while it is at least partially intact," he argued (vi). Although the term—or a similar one, salvage biology—appeared in several other places (see, e.g., Janzen, "Science Is Forever"; Benson, "Endangered Birds and Epistemic Concerns"; Raby, "Ark and Archive"; De Bont, "A World Laboratory"), it was never widely adopted. The term *salvage* has also been used to refer to the harvesting of dead birds for museum collections, rather than active hunting (see Sweet, "Collection Building through Salvage").

14. Bolton, "Animals Worth Their Weight in Gold," 400.

15. See Lucas, "Official Extinction," 104. Elizabeth Hennessy, in "Saving Species," makes a similar point about scientists' concerns over the loss of data.

16. Karen Wonders notes that the National Museum in Washington and the American Museum in New York had adopted the group display strategy by the late 1880s and early 1890s, respectively. See Wonders, *Habitat Dioramas*, 123.

17. Akeley, *In Brightest Africa*, 152–53. It should be noted here that in the late nineteenth century sportsmen were doing more for wildlife conservation than naturalists were. See Reiger, *American Sportsmen and the Origins of Conservation*. The literature on Roosevelt is very large. A recent book about his life as a naturalist and explorer is Canfield, *Theodore Roosevelt in the Field*. Roosevelt's personal account of his own African safari, inspired by a White House conversation with Akeley, is Roosevelt, *African Game Trails*.

18. C. E. Akeley, *In Brightest Africa*, 56–57.

19. D. G. Elliot to F. J. V. Skiff, July 30, 1896, and April 15, 1901, DGC, FMA.

20. The literature on this question is large. Among the earliest defenses of scientific collecting came from the pen of American Museum ornithologist Joel A. Allen in 1886 (see his "The Present Wholesale Destruction of Bird-Life in the United States"). The crux of this defense was the question of scale. According to Allen, birds taken as specimens by legitimate scientists were but a pittance compared to the wholesale slaughter of birds for the millinery or market trade, for example. This argument remains the standard defense of scientific collecting even today (see, e.g., Goodman and Lanyon, "Scientific Collecting"; B. D. Patterson, "On the Continuing Need for Scientific Collecting of Mammals"; Remsen, "The Importance of Continued Collecting of Bird Specimens"; Rocha et al., "Specimen Collection"). On the other hand, the potentially damaging role of scientific

collecting on small, remnant populations has been acknowledged by Minteer and colleagues; see "Avoiding (Re)Extinction."

21. A study done in the late 1990s at the Milwaukee Public Museum suggests that dioramas attract more museum visitors than any other kind of exhibit. See Korenic, "Studying the Visitor and the Diorama."

22. The literature on dioramas is also very large. See, e.g., Wonders, *Habitat Dioramas*; Haraway, "Teddy Bear Patriarchy"; Poliquin, *The Breathless Zoo*; Shell, "Skin Deep"; Mendenhall et al., "Diversifying Displays of Biological Sex and Sexual Behavior"; J. E. Jones, *In Search of Brightest Africa*. See also Alberti, introduction, 2; Rothfels, introduction, xi.

Chapter 1

1. The first two quotations come from Burg, *Chicago's White City of 1893*, 113; the last is from Shepp and Shepp, *Shepp's World's Fair Photographed*, 9.

2. "An Historical and Descriptive Account of the Field Columbian Museum," 15. See also Brinkman, "Frederic Ward Putnam."

3. See Brinkman, "Frederic Ward Putnam"; "An Historical and Descriptive Account of the Field Columbian Museum."

4. "An Historical and Descriptive Account of the Field Columbian Museum."

5. Wille, *Forever Open, Clear and Free*, 66. On the building, see Kohlstedt and Brinkman, "Framing Nature." On Director Skiff, see Brinkman, "Frederic Ward Putnam"; Brinkman, "The 'Chicago Idea.'"

6. Brinkman, "Frederic Ward Putnam"; Brinkman, "The 'Chicago Idea.'"

7. Farrington, "Dr. Frederick J. V. Skiff," 198; Skiff, *Classification*, 9.

8. C. B. Cory to E. E. Ayer, June 27, 1894, DGC, FMA; F. J. V. Skiff to C. B. Cory, November 19, 1895, Director's letterbooks (DL), FMA.

9. [Cope], Editor's Table (29, no. 9), 827. For more on Cope and his combativeness, see Brinkman, "Remarking on a Blackened Eye."

10. Skiff, "The Uses of the Museum." For more on women in American museums, see Rossiter, *Women Scientists in America*, 57–60; Kohlstedt, "Innovative Niche Scientists."

11. See contract between Ward's Natural Science Establishment and the Columbian Museum, November 18, 1893, Recorder's Office accession records (ROAR), geology, no. 9, FMA; F. J. V. Skiff to R. Metcalf, March 8, 1894, DL, FMA. See also Ward and McKelvey, *Henry A. Ward*, 253. Other Ward's workers included Edward C. Mirquet, James M. De Laney, Charles Hutchinson, Elbert Potter, and James H. Taylor. See "Back from Chicago," *Rochester (NY) Chronicle*.

12. C. B. Cory to E. E. Ayer, January 20, 1894, DGC, FMA.

13. On Cory, see Osgood, "In Memoriam." Allen is well known as an early proponent of wildlife conservation. See, e.g., his "The American Bisons, Living and Extinct" (175–91), where he argues for legal protections for bison. He was also an advocate of salvage zoology. See, e.g., American Museum of Natural History, *Annual Report [. . .] for the Year 1887–8*, 13.

14. Osgood, "In Memoriam," 154. See also Marconi, "Palm Beach Museum Founder and Golfer." For Cory's collection of short stories, see Cory, *Montezuma's Castle*.

15. Osgood, "In Memoriam."

16. E. E. Ayer to C. B. Cory, December 21, 1893, ROAR, ornithology, no. 15, FMA; C. B. Cory to E. E. Ayer, January 5, 1894, DGC, FMA. In a later letter to E. E. Ayer (January 26, 1894, DGC, FMA), Cory wrote that "Samson without his hair would be better off than C. B. C. without his working collection." See also Osgood, "In Memoriam."

17. Quoted in Osgood, "In Memoriam," 158.

18. C. B. Cory to E. E. Ayer, January 16, 1894, DGC, FMA.

19. Cory to Ayer. Only a few months later, Cory was offered a skeleton and an egg of the extinct great auk for a staggering £750, or about $3,750. See R. Ward to C. B. Cory, May 24, 1894, Zoology offers, no. 15, FMA.

20. Cory to Ayer, January 16, 1894.

21. J. A. Allen to C. B. Cory, June 9, 1894 (two letters with the same date, one official, one personal), Charles B. Cory Papers (Cory Papers, hereafter), Chicago Academy of Sciences / Peggy Notebaert Nature Museum (CAS). Cory's original letter is lost. Its contents can be gleaned from Allen's replies.

22. Allen to Cory, June 9, 1894 (two letters); J. A. Allen to C. B. Cory, July 17, 1894, Cory Papers, CAS.

23. C. B. Cory to E. E. Ayer, June 27, 1894, DGC, FMA. This action was approved by the executive committee at a meeting on July 11, 1894; see "Record of Minutes of the Executive Committee of the Field Columbian Museum," July 11, 1894, FMA.

24. C. B. Cory to E. E. Ayer, July 18, 1894, telegram; E. E. Ayer to C. B. Cory, July 18, 1894, telegram; C. B. Cory to F. J. V. Skiff, August 22, 1894; C. B. Cory to E. E. Ayer, November 8, 1895; all in DGC, FMA. Elliot's misgivings about scientific autonomy were so similar to Allen's that it's interesting to speculate that the issue originated with Cory.

25. "Prof. D. G. Elliot Accepts," *New York Times*; "Curator for the Field Museum," *Chicago Daily Tribune*, 5.

Chapter 2

1. "Daniel Giraud Elliot: A Brief Biographical Sketch," 134. For more on natural science in Philadelphia, see Porter, *The Eagle's Nest*; Peck and Stroud, *A Glorious Enterprise*.

2. Chapman, "Daniel Giraud Elliot."

3. John James Audubon's *The Birds of America* was published in Great Britain.

4. "Daniel Giraud Elliot: A Brief Biographical Sketch."

5. Elliot, "In Memoriam," 5. See also Elliot, *A Monograph of the Paradiseidae*; Elliot, *A Monograph of the Bucerotidae*; Elliot, *A Monograph of the Felidae*.

6. "Daniel Giraud Elliot: A Brief Biographical Sketch." On the American Museum, see Osborn, *The American Museum of Natural History*.

7. American Museum of Natural History, *Annual Report [. . .] for the Year 1887–8*, 13. See also Chapman, "Daniel Giraud Elliot."

8. Chapman, "Daniel Giraud Elliot," 5; see also Allen, "Daniel Giraud Elliot."

9. C. B. Cory to F. J. V. Skiff, June 27, 1894, DGC, FMA; F. J. V. Skiff to D. C. Davies, June 26, 1894, DL, FMA. See also Frank Collins Baker, biographical

notes, Institutional Records, Personnel, Staff, CAS. The *Auk* was an ornithological journal edited by Allen. For more on Baker at the Field Columbian Museum, see Brinkman, "Valuable So Far as It Goes." Two letters from Robert Ridgway, curator of birds at the US National Museum, to Henry K. Coale suggest that the latter was not doing state-of-the-art systematic ornithology. See R. Ridgway to H. K. Coale, June 11 and June 13, 1894, Division of Birds, National Museum of Natural History, record unit 105, box 4, Smithsonian Institution Archives. By July, Baker held a new position as curator of the Chicago Academy of Sciences.

10. C. B. Cory to E. E. Ayer, January 16, 1894; C. B. Cory to F. J. V. Skiff, August 22, 1894; D. G. Elliot to F. J. V. Skiff, October 6, 1894; all in DGC, FMA. For more on Hay and his experiences at the Field Columbian Museum, see Brinkman, "Establishing Vertebrate Paleontology at Chicago's Field Columbian Museum"; Brinkman, *The Second Jurassic Dinosaur Rush*; Brinkman "Edward Drinker Cope's Final Feud."

11. C. B. Cory to F. J. V. Skiff, June 19, 1895, DGC, FMA; F. J. V. Skiff to C. B. Cory, June 26, 1895, DL, FMA.

12. C. B. Cory to F. J. V. Skiff, June 27, 1894, DGC, FMA; C. B. Cory to E. E. Ayer, June 27, 1894, DGC, FMA.

13. C. B. Cory to F. J. V. Skiff, November 10, 1894; C. B. Cory to F. J. V. Skiff, November 8, 1895; C. B. Cory to F. J. V. Skiff, August 22, 1894; all in DGC, FMA. See also C. B. Cory to E. E. Ayer, November 8, 1895, DGC, FMA; F. J. V. Skiff to C. B. Cory, November 19, 1895, DL, FMA.

14. *Guide to the Field Columbian Museum*, 1st ed., 103; Field Columbian Museum, "Annual Report [. . .] for the Year 1895–1896," 99. See also Brinkman, "Frederic Ward Putnam," 89–90.

15. F. J. V. Skiff to H. N. Higinbotham, January 29, 1895, DL, FMA. For more on Elliot's first full year at the museum, see Brinkman, "The Strongest Kind of Competition."

16. D. G. Elliot to F. J. V. Skiff, December 18, 1894, DGC, FMA. Elliot's remark is so similar to one made by Cory to Ayer (January 16, 1894, DGC, FMA) that they probably had the same source. For more on Rowland Ward, see Jackson, "The Ward Family of Taxidermists"; K. Jones, "The Rhinoceros and the Chatham Railway."

17. Elliot to Skiff, December 18, 1894.

18. Elliot to Skiff.

19. F. J. V. Skiff to H. N. Higinbotham, January 29, 1895, DL, FMA.

20. Skiff to Higinbotham.

21. Skiff to Higinbotham. For the sake of comparison: Skiff recommended $6,000 for anthropology, $3,500 for geology, $2,500 for botany, $1,500 for economic geology, and nothing for industrial arts. On the history of dioramas in museums, see Rader and Cain, *Life on Display*, 51–53; Wonders, *Habitat Dioramas*.

22. Skiff to Higinbotham, January 29, 1895.

23. M. Field to J. G. Pangborn, June 16, 1894, DGC, FMA.

24. D. G. Elliot to F. J. V. Skiff, December 24, 1894, DGC, FMA. Rowland Ward had asked $2,000 for the rhinoceros skin; see D. G. Elliot to J. A. Allen, January 14, 1895, Daniel G. Elliot Correspondence (Elliot Correspondence, hereafter),

Mammalogy Departmental Library and Archives (MDLA), American Museum of Natural History (AMNH), New York.

25. D. G. Elliot to F. J. V. Skiff, December 26, 1894, DGC, FMA; D. G. Elliot to F. M. Chapman, April 9, 1895, Daniel G. Elliot Correspondence (Elliot Correspondence, hereafter), Ornithological Department Library and Archives (ODLA), AMNH. See also Field Columbian Museum, "Annual Report [. . .] for the Year 1895–1896"; Brinkman, "Establishing Vertebrate Paleontology at Chicago's Field Columbian Museum."

26. D. G. Elliot to F. J. V. Skiff, n.d., DGC, FMA; D. G. Elliot to J. A. Allen, May 9, 1895, Elliot Correspondence, MDLA, AMNH; F. J. V. Skiff to E. E. Ayer, July 5, 1895, DL, FMA; D. G. Elliot to F. M. Chapman, October 28, 1895, Elliot Correspondence, ODLA, AMNH.

27. D. G. Elliot to F. J. V. Skiff, July 30, 1896; D. G. Elliot to F. J. V. Skiff, May 25, 1896; D. G. Elliot to F. J. V. Skiff, July 19, 1895; all in DGC, FMA.

28. D. G. Elliot to J. A. Allen, October 9, 1895, Elliot Correspondence, ODLA, AMNH. See also Elliot, *North American Shore Birds*; Cory, *How to Know the Ducks, Geese and Swans of North America*.

29. D. G. Elliot to F. M. Chapman, April 9, 1895, Elliot Correspondence, ODLA, AMNH. Cory's *Hunting and Fishing in Florida* (44–49), which includes a lengthy account of hunting panthers in Florida, makes no mention of his dangerous encounter.

30. D. G. Elliot to F. J. V. Skiff, December 18, 1894; D. G. Elliot to D. C. Davies, December 4, 1895; O. F. Aldis to M. Field, March 26, 1894; O. F. Aldis to E. E. Ayer, November 16, 1894; R. Metcalf to F. J. V. Skiff, February 1, 1896; all in DGC, FMA. For more on the museum's troubled occupation of the Palace of Fine Arts, see Kohlstedt and Brinkman, "Framing Nature."

31. C. B. Cory to E. E. Ayer, November 8, 1895, DGC, FMA.

32. F. J. V. Skiff to H. N. Higinbotham, November 19, 1895, DL, FMA; F. J. V. Skiff to C. B. Cory, November 19, 1895, DL, FMA.

33. E. E. Ayer to C. M. Higginson, c. December 11, 1895, DGC, FMA; D. G. Elliot to E. E. Ayer and H. N. Higinbotham, December 18, 1895, DGC, FMA.

Chapter 3

1. D. G. Elliot to F. J. V. Skiff, December 27, 1895, DGC, FMA.

2. On African colonization, see Pakenham, *The Scramble for Africa*. The literature on museums and imperialism is very large. Two excellent volumes are MacKenzie, *Museums and Empire*; Barringer and Flynn, *Colonialism and the Object*.

3. D. J. Elliot to F. J. V. Skiff, December 27, 1895, DGC, FMA. For more on the complex motives of museum patrons in funding collecting expeditions, see Kohler, *All Creatures*, 107–17.

4. Elliot journal, 156.

5. There's an extensive literature on the history of sport hunting and its crossover with conservation. Karl Jacoby's *Crimes against Nature*, for example, examines the effects on local inhabitants of criminalizing traditional practices like hunting and fishing in newly protected park spaces. Edward I. Steinhart's *Black Poachers, White Hunters* argues that the ban on hunting in Kenya in 1977 created a new

class of outlaws, vilified by the press for stealing wildlife from posterity. Louis S. Warren's *The Hunter's Game* argues that early wildlife conservation was driven by the interests of elite, recreational hunters and their ideas about sportsmanship. And John F. Reiger's *American Sportsmen and the Origins of Conservation* suggests that recreational hunters and sportsmen—rather than naturalists—were the instigators of the American conservation movement in the late nineteenth century.

6. D. J. Elliot to F. J. V. Skiff, December 27, 1895, DGC, FMA.

7. "Record of Minutes," December 31, 1895, FMA.

8. F. J. V. Skiff to D. G. Elliot, January 1, 1896, DL, FMA.

9. Bodry-Sanders, *African Obsession*, 2–14.

10. Bodry-Sanders, 15–16; C. E. Akeley, *In Brightest Africa*, 1–4.

11. C. E. Akeley, *In Brightest Africa*, 9. On Ward's, see Kohlstedt, "Henry A. Ward"; Barrow, "The Specimen Dealer."

12. C. E. Akeley to Board of Trustees, c. July 1889; C. E. Akeley to Henry Nehrling, March 31, 1892; both in Carl E. Akeley Papers, Milwaukee Public Museum, Wisconsin. See also Bodry-Sanders, *African Obsession*; Wheeler, "Carl Akeley's Early Work and Environment." Several sources claim that Akeley opened a business in DeKalb, Illinois, but there seems to be no evidence for this. On the muskrat diorama, which is still on display, see Rosenberg, "Carl Akeley's Revolution in Exhibit Design." For an excellent account of Akeley's decorative taxidermy work, see Alvey, Resleure, and Gnoske, "Carl Akeley's 'Lost' Decorative Taxidermy and Anthropomorophic Groups."

13. Wheeler, "Carl Akeley's Early Work and Environment," 139; C. E. Akeley, *In Brightest Africa*, 10. Akeley dissolved his company on January 6, 1896. See Certificate of Dissolution of the Carl E. Akeley Company, January 10, 1896, Office of the Secretary of State, Mss-0742, Milwaukee County Incorporations, vol. K, p. 152, Milwaukee County Historical Society, Wisconsin. Exactly how and when Akeley became associated with the Field Columbian Museum isn't known. He began working on mounted animals on a contract basis for the museum by May 1895, probably earlier. The anthropologist William Henry Holmes, who had commissioned Akeley's mustangs for the World's Columbian Exposition and who was now a curator at the museum, possibly played a part in bringing the taxidermist to Chicago. See C. E. Akeley, *In Brightest Africa*, 10–11; Hough, "Chicago and the West," 369. Elliot first received permission to hire a taxidermist on a contract basis in January, after acquiring African mammals from Rowland Ward; see D. G. Elliot to F. J. V. Skiff, January 7, 1895, DGC, FMA. Akeley's own account of his association with the Field Museum is apocryphal.

14. D. G. Elliot to F. J. V. Skiff, January 17, 1896; D. G. Elliot to F. J. V. Skiff, July 31, 1895, DGC, FMA; D. G. Elliot to F. M. Chapman, December 23, 1895; Elliot Correspondence, ODLA, AMNH.

15. Field Columbian Museum, "Annual Report [. . .] for the Year 1894–1895," 20.

16. D. G. Elliot to F. J. V. Skiff, October 9, 1895, DGC, FMA.

17. Elliot to Skiff, October 9, 1895.

18. F. J. V. Skiff to H. N. Higinbotham, November 19, 1895, DL, FMA; Bodry-Sanders, *African Obsession*, 45.

19. F. J. V. Skiff to D. G. Elliot, January 8, 1896, DL, FMA.

20. D. G. Elliot to F. J. V. Skiff, January 9, 1896, DGC, FMA. In fact, Akeley had gone to Milwaukee for the board meeting at which the Carl E. Akeley Company was formally dissolved. See note 13 in this chapter.

21. See Field Columbian Museum, "Annual Report [. . .] for the Year 1895–1896," 109. For more on Akeley's taxidermy, see Haraway, "Teddy Bear Patriarchy"; Kirk, *Kingdom under Glass*.

22. [D. G. Elliot], c. January 18, 1896, memorandum, DGC, FMA; F. J. V. Skiff to E. E. Ayer, April 26, 1896, DL, FMA.

23. F. J. V. Skiff to D. G. Elliot, January 23, 1896, DL, FMA.

24. Quoted in "To Hunt Wild Beasts in Africa," *New York Times*, 25.

25. D. G. Elliot to F. J. V. Skiff, May 20, 1896, DGC, FMA.

26. D. G. Elliot to F. M. Chapman, February 18, 1896, Elliot Correspondence, ODLA, AMNH. See also D. G. Elliot to J. A. Allen, February 8, 1896, Elliot Correspondence, MDLA, AMNH. Elliot brought Cyclone traps of various sizes. For more on these innovative traps, see Genoways, McLaren, and Timm, "Innovations That Changed Mammalogy."

27. D. G. Elliot to F. J. V. Skiff, February 21, 1896, DGC, FMA.

28. Elliot to Skiff.

29. D. G. Elliot to F. J. V. Skiff, March 1, 1896, DGC, FMA; Voucher file, African Expedition, 1896, FMA. See also D. G. Elliot to J. A. Allen, February 22, 1896, Elliot Correspondence, MDLA, AMNH.

30. Elliot to Skiff, March 1, 1896.

31. Elliot to Skiff. For more on Smith, see Imperato, *Arthur Donaldson Smith and the Exploration of Lake Rudolf.*

32. Elliot to Skiff, March 1, 1896.

33. Elliot to Skiff.

34. Elliot to Skiff.

Chapter 4

1. A. H. Elliot to F. J. V. Skiff, March 10, 1896, DGC, FMA. For more details on the ship, see Flayhart, *The American Line.*

2. F. J. V. Skiff to C. B. Cory, March 6, 1896, DL, FMA; F. J. V. Skiff to D. G. Elliot, March 3, 1896, telegram, DL, FMA.

3. D. G. Elliot to F. J. V. Skiff, March 15, 1896 (copy), DGC, FMA; Voucher file, FMA.

4. Elliot to Skiff, March 15, 1896. See also J. E. Jones, *In Search of Brightest Africa*, 138–39.

5. See Fargher, "Hunter in Somaliland."

6. Fargher, 123.

7. Fargher, 124.

8. Fargher, 125.

9. D. G. Elliot to F. J. V. Skiff, March 15, 1896 (copy), DGC, FMA. An entry in Dodson's journal dated March 16, 1896, reads: "Started work with Proff Elliot." See Dodson, "Ted's Trunk," Natural History Museum, Library and Archives, London (NHM). On the Victorian lure of Somaliland, see Herne, *White Hunters*, 3.

10. Elliot to Skiff, March 15, 1896.

11. Elliot to Skiff.

12. D. G. Elliot to F. J. V. Skiff, March 26, 1896, DGC, FMA.

13. D. G. Elliot to the president and chairman and members of the Executive Committee of the Field Columbian Museum, c. December 1896, ROAR, zoology, no. 151 (hereafter, Elliot's final report), FMA. This twenty-five-page letter served as a final report on the expedition. For more on Akeley as a camera hunter, see Alvey, "The Cinema as Taxidermy"; Haraway, "Teddy Bear Patriarchy," 38–42.

14. D. G. Elliot to F. J. V. Skiff, March 26, 1896, DGC, FMA.

15. [D. G. Elliot to F. J. V. Skiff], c. March 26, 1896, telegram, DGC, FMA; Elliot to Skiff, March 26, 1896.

16. Elliot to Skiff, March 26, 1896.

17. D. G. Elliot to F. J. V. Skiff, March 26, 1896, personal letter, DGC, FMA. (There are two letters with the same date, one official and one personal.)

18. Elliot to Skiff, March 26, 1896, personal letter.

19. D. G. Elliot to F. J. V. Skiff, April 4, 1896, DGC, FMA.

20. Dodson, "Ted's Trunk," NHM; Elliot to Skiff, April 4, 1896.

21. Elliot to Skiff, April 4, 1896; Chapman, "Daniel Giraud Elliot," 9; Dodson, "Ted's Trunk," NHM.

22. On Akeley's personality, see Bodry-Sanders, *African Obsession*.

23. See Beolens, Watkins, and Grayson, *The Eponym Dictionary of Birds*, 155; Dodson, "Ted's Trunk," NHM.

24. D. G. Elliot to F. J. V. Skiff, April 4, 1896, DGC, FMA; D. G. Elliot to F. J. V. Skiff, April 5, 1896, DGC, FMA.

25. Burton, *First Footsteps in East Africa*, 78–79. See also Hunter, *An Account of the British Settlement of Aden in Arabia*; *Red Sea and Gulf of Aden Pilot*, 348.

26. D. G. Elliot to F. J. V. Skiff, April 5, 1896, DGC, FMA.

27. D. G. Elliot to F. J. V. Skiff, April 15, 1896, DGC, FMA.

28. Elliot to Skiff.

29. Elliot to Skiff.

30. D. G. Elliot to F. J. V. Skiff, April 19, 1896, DGC, FMA.

31. Elliot to Skiff.

32. Elliot to Skiff. Demand for ivory from British East Africa peaked in the 1890s. See Brown, *Hunter Away*.

33. Elliot to Skiff, April 19, 1896.

34. Elliot to Skiff; D. G. Elliot to F. J. V. Skiff, June 22, 1896, DGC, FMA; Dodson, "Ted's Trunk," NHM.

35. Elliot to Skiff, April 19, 1896.

36. Elliot journal, 7. Fishes collected in Aden included: *Carangoides bajad, Caranx affinis, C. carangus, C. ferdau, C. hippos, C. ignobilis, C. sem, Chirocentrus dorab, Cypselurus heterurus, Diagramma griseum, Epinephelus stoliczkae, E. suillus, Genyoroge bengalensis, G. gibba, Gerres oyena, Lethrinus harak, L. nebulosus, Lutjanus fulviflamma, L. gibbus, L. rivulatus, Mugil scheli, Plectorhinchus gaterinus, P. schotaf, Pomadasys argenteus, P. hasta, P. maculatum, P. punctulatum, P. stridens, Pristipoma punctulatum, Rastrelliger kanagurta, Rhabdosargus sarba, Scomber microlepidotus, Scomberoides sanctipetri, Selar crumenophthalmus, Siganus canaliculatus, Therapon*

jarbua, Trachinotus africanus, Tylosurus crocodilus, and *Upeneus macronemus*; see Zoological collections database, Field Museum. See also Meek, "List of Fishes and Reptiles Obtained by Field Columbian Museum East African Expedition."

37. D. G. Elliot to F. J. V. Skiff, April 19, 1896, DGC, FMA.

38. Dodson, "Ted's Trunk," NHM. See also A. D. Smith, *Through Unknown African Countries,* 170, 359.

39. A. D. Smith, *Through Unknown African Countries,* 193–94.

40. F. L. James, *The Unknown Horn of Africa,* 23–24. See also D. G. Elliot to F. J. V. Skiff, April 19, 1896, DGC, FMA.

Chapter 5

1. Elliot journal, 1. Except where noted, the main source for chapters 5–7, 9–11, and 13–19 is Elliot's journal. Only quotations from the journal are cited by page number in these chapters. On the *Tuna,* see *Board of Trade Wreck Report for "Tuna."* See also Voucher file, FMA.

2. Elliot journal, 1.

3. On Dualla, see Höhnel, *Discovery of Lakes Rudolf and Stefanie,* 11–12; Bodry-Sanders, *African Obsession,* 50; Prestholdt, *Domesticating the World,* 98. See also A. D. Smith, *Through Unknown African Countries* (99–100), for a curiously unflattering portrayal of the headman. On Elliot's interview, see Elliot's final report, FMA. Langton Prendergast Walsh, a former colonial official for Britain who once lived in Berbera, called Dualla "the best known and most capable" headman. Yet, he and his supervisor, Frederick M. Hunter, also suspected Dualla of unspecified political "intrigues" and "machinations" that resulted in many of the difficulties that Britain had with some Somali tribes. Walsh admitted: "I dare say there was a substantial foundation of truth in these accusations, but, personally, I had no direct or positive evidence of them." See Walsh, *Under the Flag and Somali Coast Stories,* 234–35.

4. Elliot journal, 2; A. D. Smith, *Through Unknown African Countries,* 99; Elliot journal, 2.

5. Elliot journal, 2–3.

6. Elliot journal, 2–3. See also D. G. Elliot to F. J. V. Skiff, April 27, 1896, DGC, FMA.

7. Elliot journal, 3.

8. Elliot journal, 4.

9. See Metz, *Somalia;* Pankhurst, "Ethiopia and Somalia," 406; Walsh, *Under the Flag and Somali Coast Stories,* 214–15. See also Reclus, *The Earth and Its Inhabitants,* 413–14; Njoku, *The History of Somalia,* 54–56.

10. Fargher, "Hunter in Somaliland," 124–26.

11. Fargher, 127–28. See also Walsh, *Under the Flag and Somali Coast Stories.*

12. Burton, *First Footsteps in East Africa,* xxxiii; Kirk, *Kingdom under Glass,* 81. See also Reclus, *The Earth and Its Inhabitants,* 413–14; Njoku, *The History of Somalia,* 54–56; Pankhurst, "Ethiopia and Somalia," 406.

13. Elliot journal, 4. See also Swayne, *Seventeen Trips through Somáliland,* 361–62.

14. Elliot journal, 5.

15. Elliot journal, 11–12. Somaliland was administered as a part of British India

and thus used the Indian rupee as its official unit of currency. During Elliot's expedition, the rupee was equal to a little less than fifty cents.

16. D. G. Elliot to F. J. V. Skiff, April 27, 1896, DGC, FMA. On Merewether, see Welch, *The Harrow School Register*, 457; Potocki, *Sport in Somaliland*.

17. A small fee for "House at Berbera" and a larger donation to the Berbera hospital might have been Elliot's obligations for the house. See Voucher file, FMA.

18. Elliot journal, 6. On Barclay, see "Meetings of the Royal Geographical Society," 437; Sclater and Thomas, *The Book of Antelopes*, 131. American novelist Henry James once described Algie Sartoris as a "drunken idiot" and "disreputable" figure, yet his cousin still boasted of their kinship. H. James, *The Letters of Henry James*, 212. For more on the disastrous Sartoris-Grant marriage, see Chernow, *Grant*, 772–75.

19. See Pankhurst, "Ethiopia and Somalia," 383; Jonas, *The Battle of Adwa*.

20. Elliot journal, 8–9.

21. Elliot journal, 8.

22. Elliot journal, 16. No spoonbills or herons are listed in Elliot, "Catalogue of a Collection of Birds Obtained by the Expedition into Somali-Land."

23. Dodson, "Ted's Trunk," NHM. See also G. James, "Britain's Black-Powder .303"; Elliot, "Catalogue of a Collection of Birds Obtained by the Expedition into Somali-Land," 59–60; Meek, "List of Fishes and Reptiles Obtained by Field Columbian Museum East African Expedition," 181–82; Dall, "List of a Collection of Shells from the Gulf of Aden."

24. Elliot journal, 9–10, 37; Dodson, "Ted's Trunk," NHM.

25. On the history of the mission, see Geshekter, "The Missionary Factor in Somali Dervish History."

26. Elliot journal, 16–18.

27. Elliot journal, 12–13.

28. Elliot journal, 13–14. See also Swayne, *Seventeen Trips through Somáliland*, 341.

29. Phoofolo, "Epidemics and Revolutions."

30. Elliot journal, 15.

31. Elliot journal, 16.

32. Elliot journal, 18.

33. Elliot journal, 18–19.

34. Elliot journal, 17–18. See also Reclus, *The Earth and Its Inhabitants*, 414.

35. Elliot journal, 20–22.

36. Elliot journal, 14–15. There were about sixteen annas to a rupee.

37. See Elliot, "List of Mammals from Somali-Land Obtained by the Museum's East African Expedition," 153.

38. Elliot journal, 21.

39. Elliot journal, 20. See also Elliot, "List of Mammals from Somali-Land Obtained by the Museum's East African Expedition," 150–51.

40. M. L. J. Akeley, *The Wilderness Lives Again*, 17. See also Hornaday, *Taxidermy and Zoological Collecting*. Akeley almost certainly taught this unorthodox method of skinning to Dodson, insisting he use it on Field Museum specimens.

41. Hornaday, *Taxidermy and Zoological Collecting*.

42. Hornaday.

43. Hornaday.

44. C. E. Akeley, *In Brightest Africa*, 17.

45. Elliot journal, 22. See also Meek, "List of Fishes and Reptiles Obtained by Field Columbian Museum East African Expedition," 181.

46. Elliot, "Catalogue of a Collection of Birds Obtained by the Expedition into Somali-Land," 65, uses *Pteroclurus exustus*. The bird is now considered a subspecies, *Pterocles exustus ellioti*. See Bogdanov, "Bemerkungen über die Gruppe der Pterocliden," 167–68.

47. Elliot, "Catalogue of a Collection of Birds Obtained by the Expedition into Somali-Land." See also Swayne, *Seventeen Trips through Somáliland*, 339.

48. Elliot journal, 22.

49. C. E. Akeley, *In Brightest Africa*.

50. For more on the safari's dependence on local knowledge, see Ritvo, *The Animal Estate*, 261–62.

51. On Akeley's racial attitudes, see Bodry-Sanders, *African Obsession*, 54, 218–26.

52. Elliot journal, 24. This is a reference to Falstaff's threadbare regiment in Shakespeare's *Henry IV*.

53. Elliot journal, 24.

54. Elliot's journal never mentions khat, but all other Somali hunting narratives suggest it was ubiquitous. For more on khat, see Carrier and Gezon, "Khat in the Western Indian Ocean."

55. Elliot journal, 25.

56. Elliot journal, 25.

57. Elliot journal, 26.

58. Elliot journal, 26.

Chapter 6

1. Elliot journal, 27.

2. Elliot journal, 27.

3. Elliot journal, 27–28; Elliot, "List of Mammals from Somali-Land Obtained by the Museum's East African Expedition," 119.

4. Elliot, "List of Mammals from Somali-Land Obtained by the Museum's East African Expedition," 139–40.

5. Elliot journal, 29.

6. Elliot journal, 30. See also Elliot, "List of Mammals from Somali-Land Obtained by the Museum's East African Expedition," 140.

7. Elliot journal, 30.

8. Elliot journal, 31.

9. Elliot journal, 278–79.

10. Elliot journal, 31.

11. Elliot journal, 32.

12. Elliot journal, 32. No habitat group of Pelzeln's gazelles was ever made.

13. Elliot journal, 33. See also Elliot, "List of Mammals from Somali-Land Obtained by the Museum's East African Expedition." Possibly Elliot's interest in

the mechanics of bird flight was connected to an interest in the problem of human flight. German glider enthusiast Otto Lilienthal, who died later that same summer in a crash, was convinced that the secret of human flight would be found by observing birds. Wilbur Wright, too, began watching birds fly in 1896. See Mc-Cullough, *The Wright Brothers*, 27–30.

14. Elliot journal, 33. H. G. C. Swayne, in *Seventeen Trips through Somáliland* (339), recommended the British-made .577 Snider-Enfield breech-loading rifle as a useful weapon for a Somali escort. See also Elliot, "List of Mammals from Somali-Land Obtained by the Museum's East African Expedition."

15. Elliot journal, 33–34.

16. Elliot journal, 34–35. See also Meek, "List of Fishes and Reptiles Obtained by Field Columbian Museum East African Expedition," 181.

17. Elliot journal, 35. On Midgans, see Reclus, *The Earth and Its Inhabitants*, 399.

18. Elliot journal, 37.

19. Elliot journal, 34.

20. Elliot journal, 36.

21. C. E. Akeley, *In Brightest Africa*, 115.

22. Akeley, 115.

23. Akeley, 115–16.

24. Akeley.

25. Akeley, 117.

26. Akeley.

27. Elliot journal, 38.

28. C. E. Akeley, *In Brightest Africa*, 118.

29. Akeley, 118; Elliot journal, 37–38. See also Elliot, "List of Mammals from Somali-Land Obtained by the Museum's East African Expedition," 122.

Chapter 7

1. Elliot journal, 38–39. For more on Victorian notions of disciplined, self-restrained manliness, see Bederman, *Manliness and Civilization*.

2. Elliot journal, 39–40.

3. Elliot journal, 40–41.

4. Elliot journal, 42–43. To solve this problem, Elliot later changed his policy.

5. Elliot journal, 42–43.

6. Elliot journal, 43.

7. Elliot journal, 43–44.

8. Elliot journal, 41.

9. Elliot journal, 45–46. See Outram, "New Spaces in Natural History," for an interesting discussion of the kind of comprehensive vision one can get of nature from museum study, as opposed to field study.

10. Kingsley, *Glaucus*, 31; and see Conniff, *The Species Seekers*, 12.

11. Elliot journal, 44, 46–47.

12. Elliot journal, 47–48.

13. Elliot journal, 55.

14. Dodson, "Ted's Trunk," NHM.

15. Elliot journal, 48; Dodson, "Ted's Trunk," NHM.

16. Elliot journal, 49, 51; and see Meek, "List of Fishes and Reptiles Obtained by Field Columbian Museum East African Expedition," 181.

17. Elliot journal, 49.

18. Elliot journal, 50. Formaldehyde solutions, which delay but do not prevent decay, were commonly used as a biological preservative in the nineteenth century. See Elliot, "List of Mammals from Somali-Land Obtained by the Museum's East African Expedition," 115–17; Meek, "List of Fishes and Reptiles Obtained by Field Columbian Museum East African Expedition," 181.

19. Elliot journal, 50–52. See also Elliot, "Catalogue of a Collection of Birds Obtained by the Expedition into Somali-Land," 41.

20. Elliot journal, 53.

21. Elliot journal, 56.

22. Elliot journal, 58.

23. Elliot, "Catalogue of a Collection of Birds Obtained by the Expedition into Somali-Land," 60.

24. Elliot journal, 59. See also Elliot, "Catalogue of a Collection of Birds Obtained by the Expedition into Somali-Land," 51.

25. Elliot journal, 64; see also ROAR, anthropology, no. 176, FMA. I. M. Lewis, in "The So-Called 'Galla Graves' of Northern Somaliland" (104), notes that an apron of stones is a common feature of both ancient Galla cairns and modern Somali graves.

26. Elliot journal, 61–62.

27. Elliot journal, 62.

28. Elliot journal, 63. See also Elliot, *A Monograph of the Bucerotidae*.

29. See Elliot, "Catalogue of a Collection of Birds Obtained by the Expedition into Somali-Land," 58; Elliot, "List of Mammals from Somali-Land Obtained by the Museum's East African Expedition," 152–53; Meek, "List of Fishes and Reptiles Obtained by Field Columbian Museum East African Expedition," 179.

30. Elliot journal, 65.

31. Elliot journal, 65. Regarding the frogs, see Meek, "List of Fishes and Reptiles Obtained by Field Columbian Museum East African Expedition," 175–78.

32. Elliot journal, 66.

33. Elliot journal, 66.

34. Elliot journal, 66–67.

Chapter 8

1. D. G. Elliot to F. J. V. Skiff, February 7, 1896 (there are three letters with the same date), DGC, FMA; F. J. V. Skiff to D. C. Davies, February 28, 1896 (there are two letters with the same date), DL, FMA.

2. Lull, "Memorial of Oliver Perry Hay," 31–32; see also Nieuwland, *American Dinosaur Abroad*, 162–63.

3. For more on Hay, see Lull, "Memorial of Oliver Perry Hay"; Brinkman, "Establishing Vertebrate Paleontology at Chicago's Field Columbian Museum," 91–92. On Baur, see Hay, "George Baur"; Wheeler, "George Baur's Life and Writings." On Baur's struggles at Chicago, see Rainger, "Biology, Geology, or Neither, or Both." For Hay's interest in working with Whitman, see C. O. Whitman to O. P.

Hay, May 5, 1895, William Perry Hay Papers, Special and Area Studies Collection, University of Florida, Smathers Libraries, Gainesville.

4. S. C. Eastman to O. P. Hay, October 10, 1893; F. J. V. Skiff to O. P. Hay, March 3, 1894; Acting Director [W. H. Holmes] to O. P. Hay, July 26, 1894; all in DGC, FMA.

5. D. G. Elliot to F. J. V. Skiff, October 25, 1894, DGC, FMA; D. G. Elliot to F. J. V. Skiff, December 18, 1894, DGC, FMA.

6. Elliot to Skiff, December 18, 1894. The segregation of exhibit and study collections had been championed previously by John Edward Gray, keeper of zoology at the British Museum. See Gray, "On Museums, Their Use and Improvement."

7. D. G. Elliot to F. J. V. Skiff, December 18, 1894. On problems with the Ward's cases, see R. Metcalf to H. A. Ward, February 7, 1894, DL, FMA.

8. *Guide to the Field Columbian Museum*, 2nd ed., 170–71. Upton Sinclair's novel *The Jungle* vividly describes the horrible conditions found in a Chicago slaughterhouse in the early twentieth century. One wonders whether Hall 22 would have remained intact had one or more of Chicago's meatpacking princes, Philip Armour or Gustavus Swift, for example, been more active in the museum's organization.

9. *Guide to the Field Columbian Museum*, 2nd ed., 173–74.

10. Field Columbian Museum, "Annual Report [. . .] for the Year 1894–1895," 22.

11. D. G. Elliot to F. J. V. Skiff, June 20, 1895, DGC, FMA. See also ROAR, obsolete-miscellaneous, no. 115, FMA. On the terra-cotta pavilion, see F. J. V. Skiff to E. Walker, March 11, 1896, DL, FMA.

12. D. G. Elliot to F. J. V. Skiff, March 1, 1896, DGC, FMA; F. J. V. Skiff to E. E. Ayer, April 26, 1896, DL, FMA; D. G. Elliot to J. A. Allen, February 1, 1896, MDLA, AMNH; D. G. Elliot to J. A. Allen, February 8, 1896, MDLA, AMNH. Some biographical details about Brandler and O'Brien can be found in Storrs, "Visitin' 'Round in Minneapolis"; Stoddard, *Memoirs of a Naturalist*. Little is known about Fischer.

13. F. J. V. Skiff to H. C. Ives, January 30, 1896, DL, FMA. The Anthropology Department also received a large share of museum resources.

14. See Field Columbian Museum, "Annual Report [. . .] for the Year 1895–1896," 103, 128.

15. F. J. V. Skiff to C. B. Cory, March 6, 1896, DL, FMA.

16. Skiff to Cory.

17. See Field Columbian Museum, "Annual Report [. . .] for the Year 1895–1896," 107; *Guide to the Field Columbian Museum*, 4th ed., 115–25. On the construction of cases, see F. J. V. Skiff to C. L. Wheeler, April 25, 1896, DL, FMA.

18. F. J. V. Skiff to E. E. Ayer, April 1, 1896, DL, FMA. See also Field Columbian Museum, "Annual Report [. . .] for the Year 1895–1896," 107, 109.

19. F. J. V. Skiff to D. C. Davies, February 28, 1896; F. J. V. Skiff to C. B. Cory, March 6, 1896, DL, FMA; O. P. Hay to F. J. V. Skiff, May 22, 1896, DGC, FMA. For more on the Carpenter collection, see Palmer, *Type Specimens of Marine Mollusca Described by P. P. Carpenter*, 31–32.

20. O. P. Hay to F. J. V. Skiff, June 9, 1896, DGC, FMA. See also F. J. V. Skiff to O. P. Hay, June 8, 1896, DL, FMA; F. J. V. Skiff to E. R. Cooper, April 6,

1896, DL, FMA; O. P. Hay to F. C. Baker, June 18, 1896, Frank Collins Baker Papers, CAS.

21. These papers were: Hay, "On the Structure and Development of the Vertebral Column of *Amia*"; Hay, "On Certain Portions of the Skeleton of *Protostega gigas*" (which makes no reference to any particular museum specimens, although Geology Department records show that specimen number UR 79 was the object of Hay's description); Hay, "Description of a New Species of *Petalodus*" (the specimen on which Hay established this species should have become the museum's first fossil type specimen, but Hay's paper gives no information on the provenance of the specimen, and it cannot now be identified in the museum's collection); and Hay, "On the Skeleton of *Toxochelys latiremis*" (again, no reference is made to any particular museum material). For more details, see Brinkman, "Establishing Vertebrate Paleontology at Chicago's Field Columbian Museum."

22. [Cope], Editor's Table (30, no. 5), 385–86.

23. McVicker, "Buying a Curator," 41, 49.

24. F. J. V. Skiff to G. F. Kunz, c. February 18, 1896, George Kunz Papers, box 27, folder 30, Central Archives, AMNH.

25. G. Kunz to F. J. V. Skiff, March 2, 1896, DGC, FMA.

26. Quoted in D. J. Meltzer, "When Destiny Takes a Turn for the Worse," 205. See also Brinkman, "The 'Chicago Idea'"; Brinkman, "Edward Drinker Cope's Final Feud." For more on Holmes at the Field Museum, see Fernlund, *William Henry Holmes and the Rediscovery of the American West*.

Chapter 9

1. Elliot journal, 67–68.

2. D. G. Elliot to F. J. V. Skiff, May 20, 1896, DGC, FMA.

3. Elliot to Skiff.

4. Elliot journal, 67, 73.

5. When Berbera was ruled by the Egyptians, it was nominally a part of the Ottoman Empire.

6. Elliot journal, 70–71.

7. Elliot journal, 60, 68.

8. Elliot journal, 68–69.

9. Elliot journal, 70.

10. Elliot journal, 68–70.

11. Elliot journal, 72. Jamal Gabobe, in "European Travel Writing in Somaliland" (86–87), argues that the trope of the master and servant was a common narrative device used by European and American explorers to devalue their non-European collaborators.

12. Elliot journal, 73.

13. Elliot journal, 68, 74. See also Pearce, *Rambles in Lion Land*, 25.

14. Elliot journal, 74–75.

15. Elliot journal, 75; Dodson, "Ted's Trunk," NHM.

16. Elliot journal, 75–76.

17. See Elliot, "Catalogue of a Collection of Birds Obtained by the Expedition

into Somali-Land," 57; Elliot, "List of Mammals from Somali-Land Obtained by the Museum's East African Expedition," 113–15.

18. Elliot journal, 76.

19. Elliot journal, 76.

20. Elliot journal, 77.

21. Elliot journal, 77–78. According to Elliot, "Catalogue of a Collection of Birds Obtained by the Expedition into Somali-Land," the birds collected at Laferug included *Bradyornis pumillus*, *Campothera nubica*, *Coracias lorti*, *Corvus edithae*, *Hedydipna metallica*, *Lophotis gindiana*, *Melierax poliopterus*, *Nilaus capensis minor*, and *Spreo superbus*. See also Elliot, "List of Mammals from Somali-Land Obtained by the Museum's East African Expedition," 141.

22. See Elliot, "List of Mammals from Somali-Land Obtained by the Museum's East African Expedition," 120–22, 126–29.

23. Elliot journal, 78. See also Elliot, "List of Mammals from Somali-Land Obtained by the Museum's East African Expedition," 134–35.

24. Elliot journal, 81–82.

25. Elliot journal, 86. See also Meek, "List of Fishes and Reptiles Obtained by Field Columbian Museum East African Expedition," 183.

26. Elliot journal, 82; Dodson, "Ted's Trunk," NHM.

27. Elliot journal, 86–87. See also Elliot, "List of Mammals from Somali-Land Obtained by the Museum's East African Expedition," 134.

28. Elliot journal, 84.

29. Elliot journal, 84–85. See also Elliot, "List of Mammals from Somali-Land Obtained by the Museum's East African Expedition," 109–11.

30. Elliot journal, 85; Elliot, "List of Mammals from Somali-Land Obtained by the Museum's East African Expedition," 145.

31. Elliot journal, 87.

32. Elliot journal, 88.

33. Elliot journal, 90.

34. Dodson, "Ted's Trunk," NHM.

35. Dodson.

36. Elliot journal, 90–91. See also Elliot, "Catalogue of a Collection of Birds Obtained by the Expedition into Somali-Land," 59–60.

37. Elliot journal, 92. See also Dodson, "Ted's Trunk," NHM; Elliot, "List of Mammals from Somali-Land Obtained by the Museum's East African Expedition," 132–34.

Chapter 10

1. See "Jackson Park's Future," *New York Times*; Hines, *Burnham of Chicago*, 313–14. More on Burnham and the exposition can be found in E. Larson, *The Devil in the White City*.

2. F. J. V. Skiff to E. R. Graham, March 18, 1896, DL, FMA; F. J. V. Skiff to E. E. Ayer, March 23, 1896, DL, FMA.

3. See F. J. V. Skiff to E. E. Ayer, April 1, 1896, DL, FMA. Charles B. Atwood, who was the original architect of the building and who also worked at Burnham's firm, died in December 1895. See "Work of the Late Charles B. Atwood," *New York Times*.

4. Field Columbian Museum, "Annual Report [. . .] for the Year 1895–1896," 103.

5. Biographical details on Aldis can be found in "Obituary Record of Yale Graduates."

6. On staff and exposition construction techniques, see Kidder, *Building Construction and Superintendence*, 800–803; Graff, "Dream City, Plaster City."

7. O. F. Aldis to F. J. V. Skiff, June 2, 1896, DGC, FMA. After working with Goodman closely for several exasperating days, Aldis confessed that he had no confidence in the building superintendent's energy or skill.

8. Field Columbian Museum, "Annual Report [. . .] for the Year 1895–1896," 103–4.

9. Field Columbian Museum, 104. See also F. J. V. Skiff to J. F. Foster, June 4, 1896, DL, FMA.

10. Field Columbian Museum, "Annual Report [. . .] for the Year 1896–1897," 194.

11. Field Columbian Museum, "Annual Report [. . .] for the Year 1895–1896," 104; Field Columbian Museum, "Annual Report [. . .] for the Year 1896–1897," 193. Elliot complained bitterly and often of the cold temperatures that prevailed inside the museum's building.

12. "Rats," *Chicago Evening Post*. See also, Field Columbian Museum, "Annual Report [. . .] for the Year 1895–1896," 105.

13. O. F. Aldis to F. J. V. Skiff, June 2, 1896, DGC, FMA. There is a large literature on the Great Chicago Fire. See, e.g., Sawislak, *Smoldering City*. On the fire and civic leadership, see McCarthy, *Noblesse Oblige*, 64–65. For a general account of the fires at the fairgrounds, see Burg, *Chicago's White City of 1893*, 287–88; Miller, *City of the Century*, 549–51. For more details, see "Origin of Fair Fires," *Chicago Daily Tribune*, on the February fire; and "Fair Buildings in Ashes," *New York Times*, on the July fire.

14. See "Origin of Fair Fires," *Chicago Daily Tribune*. See also Bragdon, "Letter from Chicago," 140.

15. See "Origin of Fair Fires," *Chicago Daily Tribune*.

16. Field Columbian Museum, "Annual Report [. . .] for the Year 1895–1896," 110–11.

17. Field Columbian Museum, 104. See also Field Columbian Museum, "Annual Report [. . .] for the Year 1896–1897," 194; F. J. V. Skiff to George Manierre, April 30, 1897, DL, FMA. Bailey didn't last long. He became obsessively involved as an expert witness in the sensational Luetgert murder case of 1897 and left the museum early in 1899; see Loerzel, *Alchemy of Bones*, 126–29, 132; D. G. Elliot to F. J. V. Skiff, September 22, 1899, DGC, FMA.

18. See Field Columbian Museum, "Annual Report [. . .] for the Year 1895–1896," 107; Field Columbian Museum, "Annual Report [. . .] for the Year 1897–1898," 285; Field Columbian Museum, "Annual Report [. . .] for the Year 1898–1899," 329.

19. Field Columbian Museum, "Annual Report [. . .] for the Year 1895–1896," 104–5.

20. Field Columbian Museum, 104.

Chapter 11

1. Elliot journal, 96.

2. Elliot journal, 93, 95. The expedition acquired four oxpecker specimens at two other localities. See Elliot, "Catalogue of a Collection of Birds Obtained by the Expedition into Somali-Land," 33, 53.

3. Elliot journal, 97. See also Elliot, "Catalogue of a Collection of Birds Obtained by the Expedition into Somali-Land," 31–32.

4. Elliot journal, 94.

5. Elliot journal, 94.

6. Elliot journal, 95.

7. Elliot journal, 106–7. See also Elliot, "List of Mammals from Somali-Land Obtained by the Museum's East African Expedition," 142–43. Elliot was wrong. A second species, *Heterocephalus phillipsi*, was described by Oldfield Thomas in 1885. See Thomas, "Notes on the Rodent Genus *Heterocephalus*."

8. D. G. Elliot to F. J. V. Skiff, February 11, 1897, ROAR, zoology, no. 151, FMA.

9. Elliot journal, 92.

10. Elliot journal, 101–2.

11. Elliot journal, 99.

12. Elliot journal, 100.

13. Elliot journal, 103.

14. Elliot journal, 105.

15. Elliot journal, 101, 105.

16. Elliot journal, 107.

17. Elliot, "List of Mammals from Somali-Land Obtained by the Museum's East African Expedition," 140–42.

18. Elliot journal, 108–9. Elliot's view about Somali avarice comes directly from Swayne, *Seventeen Trips through Somáliland*.

19. Elliot journal, 109.

20. Elliot journal, 111–12.

21. Elliot journal, 112.

22. Elliot journal, 111. See also Elliot, "List of Mammals from Somali-Land Obtained by the Museum's East African Expedition," 130–31.

23. Elliot journal, 113.

24. Elliot journal, 113.

25. Elliot journal, 113–14.

26. Elliot journal, 114.

27. Elliot journal, 115.

28. Elliot journal, 115–16.

29. Elliot journal, 116–17.

30. Elliot journal, 117. See also Elliot, "List of Mammals from Somali-Land Obtained by the Museum's East African Expedition," 112–13.

31. Elliot journal, 118.

32. Elliot journal, 119.

33. Elliot journal, 120.

Chapter 12

1. Elliot journal, 120–21.
2. Elliot journal, 128.
3. Elliot journal, 124–25, 130.
4. Elliot journal, 128–29.
5. Elliot journal, 132.
6. Elliot journal, 126. See also Elliot, "List of Mammals from Somali-Land Obtained by the Museum's East African Expedition," 143.
7. Elliot journal, 130; Elliot, "Catalogue of a Collection of Birds Obtained by the Expedition into Somali-Land," 30–31.
8. Elliot journal, 130–31.
9. Elliot journal, 133.
10. Elliot journal, 134–35.
11. Elliot journal, 135.
12. Elliot journal, 135–36, 138.
13. C. E. Akeley, *In Brightest Africa* (152–53), tells this story without naming names. Interestingly, Norman Smith himself wrote in "Big Game Shooting" (39) that "it is quality, and not quantity, that makes a fine bag, and that every real sportsman will avoid the slightest chance of being called a butcher."
14. Elliot journal, 136.
15. Dodson, "Ted's Trunk," NHM.
16. Elliot journal, 143.
17. Elliot journal, 142.
18. Elliot journal, 145.
19. Elliot journal, 146, 149. Aoul was the local name for Soemmerring's gazelle.
20. Dodson, "Ted's Trunk," NHM. See also Elliot, "List of Mammals from Somali-Land Obtained by the Museum's East African Expedition," 143–45.
21. D. G. Elliot to F. J. V. Skiff, June 22, 1896, DGC, FMA; Dodson, "Ted's Trunk," NHM.
22. Elliot journal, 147–48.

Chapter 13

1. F. J. V. Skiff to O. P. Hay, June 8, 1896, DL, FMA.
2. O. P. Hay to F. J. V. Skiff, June 9, 1896, Director's general correspondence, FMA. Hay published a series of papers in 1896 and 1897 in which he differed with several other naturalists over the mode of development of the vertebral column in fishes. This was the controversy that Skiff declined to publish in the museum's journal.
3. F. J. V. Skiff to H. N. Higinbotham, July 28, 1896, DL, FMA; F. J. V. Skiff to M. Fischer, September 15, 1896, DL, FMA. For more on Hay's peripatetic career, see Rainger, "Collectors and Entrepreneurs."
4. O. P. Hay to G. Baur, July 1, 1896, Georg Baur Records, Ernst Mary Library, Archives, Harvard University, Museum of Comparative Zoology (MCZ), Cambridge, Massachusetts.
5. Baur, "On the Origin of the Galapagos Islands (Concluded)," 318. See also

M. J. James, *Collecting Evolution*, 52. For more on Adams, see "Mr. Charles F. Adams"; Gill, "Charles Francis Adams."

6. See G. Baur to F. J. V. Skiff, March 27, 1894, ROAR, zoology, no. 10, FMA; F. J. V. Skiff to G. Baur, March 28, 1894, ROAR, zoology, no. 10, FMA. For details of the exhibit, see *Guide to the Field Columbian Museum*, 1st ed., 119–20. On Baur's controversial theory, see Baur, "The Differentiation of Species on the Galapagos Islands and the Origin of the Group." The more orthodox view can be found in many places, including Darwin, *Journal and Remarks*, 453; and, Darwin, *Geological Observations on the Volcanic Islands*, 97–116.

7. G. Baur to F. J. V. Skiff, June 23, 1894, DGC, FMA; F. J. V. Skiff to G. Baur, June 27, 1894, DL, FMA.

8. O. P. Hay to G. Baur, July 1, 1896, Georg Baur Records, MCZ.

Chapter 14

1. Elliot journal, 148.

2. Elliot journal, 149.

3. D. G. Elliot to F. J. V. Skiff, June 22, 1896, DGC, FMA.

4. Elliot to Skiff.

5. Elliot journal, 150.

6. Elliot journal, 151.

7. Elliot journal, 151. See also Elliot, "List of Mammals from Somali-Land Obtained by the Museum's East African Expedition," 124–26.

8. Elliot journal, 152.

9. See Sharpe, "On a Collection of Birds Made by Dr. A. Donaldson Smith in Western Somaliland," 503.

10. Elliot journal, 154.

11. Elliot journal, 153–54.

12. See Potocki, *Sport in Somaliland*, 5–6; Pease, *Travel and Sport in Africa*, 51. Dysentery in the field could be very dangerous. Dodson's younger brother, William—also an adventurer and taxidermist—died of dysentery in Aden while on an expedition, in 1899. See Ogilvie-Grant, "On the Birds of Southern Arabia," 243.

13. Elliot journal, 155.

14. C. E. Akeley, *In Brightest Africa*, 75–76.

15. Akeley, 75–76.

16. D. G. Elliot to F. J. V. Skiff, May 25, 1896, DGC, FMA; Elliot journal, 156.

17. Elliot journal, 156–58.

18. Elliot journal, 158–59.

19. Elliot journal, 159.

20. Elliot journal, 159–60.

21. Elliot journal, 162.

22. Elliot journal, 163–64.

23. Elliot journal, 164.

24. See Elliot, "Catalogue of a Collection of Birds Obtained by the Expedition into Somali-Land," 53–54.

25. Elliot journal, 166.

26. Elliot journal, 167.

27. Elliot journal, 167–68.

28. Elliot journal, 168–69. Much of Elliot's data about the relative abundance of certain animals was anecdotal—it came from local guides or the accounts of big-game hunters.

29. Elliot journal, 171. See also Elliot, "List of Mammals from Somali-Land Obtained by the Museum's East African Expedition," 151–52.

30. Elliot journal, 171.

31. Elliot journal, 171–72.

Chapter 15

1. D. G. Elliot to F. J. V. Skiff, May 20, 1896, DGC, FMA.

2. D. G. Elliot to F. J. V. Skiff, May 25, 1896, DGC, FMA.

3. Elliot to Skiff.

4. See D. G. Elliot to F. J. V. Skiff, July 7, 1896, DGC, FMA; F. J. V. Skiff to D. G. Elliot, June 26, 1896, DGC, FMA; F. J. V. Skiff to US Consul, Aden, June 26, 1896, DL, FMA; see also "Record of Minutes," May 5, 1896, FMA.

5. Skiff to Elliot, June 26, 1896.

6. F. J. V. Skiff to US Consul, June 26, 1896, DL, FMA.

7. F. J. V. Skiff to E. L. Burchard, August 29, 1896, DL, FMA.

Chapter 16

1. Elliot journal, 173.

2. Elliot journal, 174. Ultimately, the museum mounted only one habitat group of gerenuks.

3. Elliot journal, 174–75.

4. Elliot journal, 175.

5. Elliot journal, 176.

6. Elliot journal, 177.

7. Elliot journal, 178.

8. Elliot journal, 178–79.

9. Elliot journal, 179.

10. Elliot, "Catalogue of a Collection of Birds Obtained by the Expedition into Somali-Land," 31.

11. Elliot journal, 181.

12. Elliot journal, 181. See also Elliot's final report, FMA.

13. Elliot journal, 182–83. The curator wrote at length about the doglike qualities of cheetahs in Elliot, *A Monograph of the Felidae.*

14. Elliot journal, 182.

15. Elliot journal, 183. See also D. G. Elliot to F. J. V. Skiff, July 7, 1896, DGC, FMA.

16. Elliot journal, 184–85.

17. Elliot journal, 183–84. See also Elliot, "List of Mammals from Somali-Land Obtained by the Museum's East African Expedition," 148–49.

18. Elliot journal, 183–84.

19. Elliot journal, 185.

20. Elliot journal, 187.

21. Elliot journal, 189.

22. Elliot journal, 189.

23. Elliot journal, 189–90.

24. Elliot journal, 190. Birds collected at Legud included *Anthothreptes orientalis, Buchanga assimilis, Buphaga erythrorhynchus, Chalcopelia afra, Cinnyris osiris, Coracias lorti, Dryoscopus funebris, Francolinus granti, Granatina ianthinogastra, Irrisor erythrorhynchus, Nilaus capensis minor, Parus thruppi, Poeocepahlus rufiventris, Pternistes infuscatus, Tricholaema stigmatothorax,* and *Turtur damarensis.* Elliot later described two new species from Legud, including a warbler, *Sylviella isabellina,* and a flycatcher, *Pachyprora bella.* See Elliot, "Catalogue of a Collection of Birds Obtained by the Expedition into Somali-Land."

25. Elliot journal, 191–92.

26. Elliot journal, 191–92.

27. Elliot journal, 193.

28. Elliot journal, 193–94.

29. Elliot journal, 195, 202.

30. Elliot journal, 196.

31. Elliot journal, 196–97.

32. Elliot journal, 197–98.

33. Elliot journal, 198. See also Elliot, "Catalogue of a Collection of Birds Obtained by the Expedition into Somali-Land," 34, where he noted that these birds were "as nice a morsel as is our own reed or rice bird."

34. Elliot journal, 198–99.

35. Elliot journal, 199.

36. Elliot journal, 200. These skins don't appear in Elliot, "List of Mammals from Somali-Land Obtained by the Museum's East African Expedition."

37. Elliot journal, 201. See also Elliot, "Catalogue of a Collection of Birds Obtained by the Expedition into Somali-Land," 52.

38. Elliot journal, 202.

39. Elliot journal, 203–4.

40. Elliot journal, 203.

Chapter 17

1. Elliot journal, 207. The term *white hunter,* however, originated in British East Africa sometime after this expedition, in the early twentieth century. See Herne, *White Hunters,* 3.

2. Elliot journal, 205; D. G. Elliot to F. J. V. Skiff, July 30, 1896, DGC, FMA.

3. Elliot journal, 205–6.

4. Elliot journal, 206–7.

5. Elliot journal, 208.

6. Elliot journal, 209.

7. Elliot journal, 209.

8. Elliot journal, 210–11.

9. Elliot journal, 211.

10. Elliot journal, 211.

11. Elliot, "Catalogue of a Collection of Birds Obtained by the Expedition into Somali-Land," 47, 54–55.

12. Elliot journal, 214.

13. Elliot journal, 214. See also Elliot, "Catalogue of a Collection of Birds Obtained by the Expedition into Somali-Land," 66.

14. See Meek, "List of Fishes and Reptiles Obtained by Field Columbian Museum East African Expedition," 175.

15. Elliot, "List of Mammals from Somali-Land Obtained by the Museum's East African Expedition," 136.

16. Elliot, 136–37.

17. Elliot journal, 216–17.

18. Elliot, "List of Mammals from Somali-Land Obtained by the Museum's East African Expedition," 137; Elliot journal, 219.

19. Elliot journal, 218.

20. Elliot journal, 219.

21. Elliot journal, 222.

22. Elliot journal, 221.

23. See Elliot, "List of Mammals from Somali-Land Obtained by the Museum's East African Expedition," 147–48.

24. Elliot journal, 222.

25. Elliot journal, 225.

26. Elliot, "List of Mammals from Somali-Land Obtained by the Museum's East African Expedition," 137.

27. C. E. Akeley, *In Brightest Africa*, 120–22. See also Elliot, "List of Mammals from Somali-Land Obtained by the Museum's East African Expedition," 141.

28. Elliot journal, 224.

Chapter 18

1. Swayne, *Seventeen Trips through Somáliland*, 7; Burton, *First Footsteps in East Africa*. On Burton's injury, see Millard, *River of the Gods*, 55.

2. Elliot journal, 227.

3. Elliot journal, 213. By this point, Elliot had begun to suspect that his kittens were leopards and not cheetahs.

4. Elliot journal, 215, 226.

5. Elliot journal, 220.

6. Elliot journal, 227–28.

7. Elliot journal, 228.

8. Elliot journal, 228, 240.

9. Elliot journal, 220–21.

10. Elliot journal, 229.

11. Elliot journal, 229–30; D. G. Elliot to F. J. V. Skiff, July 30, 1896, DGC, FMA.

12. Elliot journal, 231–32.

13. Elliot journal, 30.

14. Elliot journal, 231.

15. Elliot journal, 233.

16. Elliot journal, 233; Elliot, "Catalogue of a Collection of Birds Obtained by the Expedition into Somali-Land," 47–48, 58.

17. Elliot, "List of Mammals from Somali-Land Obtained by the Museum's East African Expedition," 137–38; Elliot journal, 233.

18. Elliot journal, 233–34.

19. Elliot journal, 235.

20. Elliot journal, 235–36.

21. Elliot journal, 235–36. See also Dodson, "Ted's Trunk," NHM.

22. *Sheffield Independent*, "The Latest Egyptian Developments"; "The Threatened Extinction of the African Elephant," *Huddersfield Chronicle*. On Crispi, see, e.g., "Francesco Crispi," *Pall Mall Gazette*. On the Uganda Railway, see, e.g., Whitehouse, "The Building of the Kenya and Uganda Railway." A pair of maneless, man-eating lions shot near the Tsavo River in 1898 during construction of this railroad would later be added to the Field Museum's collections. See J. H. Patterson, *The Man-Eaters of Tsavo and Other East African Adventures*. Today, these notorious lions are displayed together with many of the mammals hunted in Somaliland in the museum's Carl E. Akeley Memorial Hall.

Chapter 19

1. Elliot journal, 236; Potocki, *Sport in Somaliland*, 29. In his journal, Elliot recorded the man's name incorrectly as Sheik Mulla.

2. Elliot journal, 236–37.

3. Elliot journal, 237.

4. Elliot journal, 239.

5. Elliot journal, 239–40.

6. Elliot journal, 241.

7. Elliot journal, 241.

8. Elliot journal, 244.

9. Elliot journal, 244.

10. Elliot journal, 242.

11. Dodson, "Ted's Trunk," NHM.

12. Elliot journal, 244–45.

13. Elliot journal, 246–47.

14. Elliot journal, 247–48.

15. Elliot journal, 248.

16. Elliot journal, 249.

17. Elliot journal, 250.

18. See Fisher and Bolton, *Ants of Africa and Madagascar*, 308–10.

19. Elliot journal, 253–54.

20. Elliot journal, 255.

21. Elliot journal, 255–56.

22. Elliot journal, 257.

23. Elliot journal, 256–57.

24. Elliot journal, 257–58.

25. Elliot journal, 258.

Chapter 20

1. Elliot, "Catalogue of a Collection of Birds Obtained by the Expedition into Somali-Land," 42–43, 49.

2. Elliot journal, 260–61.

3. Elliot journal, 261.

4. Elliot journal, 261–62. The wells were actually made by the Borana, a group of seminomadic pastoralists. See Eshelby, "The Way of the Borana."

5. Elliot journal, 262.

6. Elliot journal, 263.

7. Elliot journal, 263.

8. Elliot journal, 263–64.

9. Elliot journal, 264. See also Elliot, "Catalogue of a Collection of Birds Obtained by the Expedition into Somali-Land," 37.

10. Elliot journal, 265.

11. Elliot journal, 266. See also Elliot, "Catalogue of a Collection of Birds Obtained by the Expedition into Somali-Land," 30–31.

12. Elliot journal, 266.

13. Elliot journal, 267.

14. Elliot journal, 267–68. See also Elliot, "List of Mammals from Somali-Land Obtained by the Museum's East African Expedition," 146.

15. Elliot journal, 268.

16. See Elliot, "Catalogue of a Collection of Birds Obtained by the Expedition into Somali-Land," 41, 48, 55; Meek, "List of Fishes and Reptiles Obtained by Field Columbian Museum East African Expedition," 179.

17. Elliot journal, 269.

18. Elliot journal, 270.

19. Elliot journal, 271.

20. Elliot journal, 271.

21. Elliot journal, 271–72. See also Elliot, "List of Mammals from Somali-Land Obtained by the Museum's East African Expedition," 152, which notes that only one fox was added to the museum's collections, so one of these specimens was probably discarded or exchanged.

22. C. E. Akeley, *In Brightest Africa*, 94–95.

23. Akeley, 95–96.

24. Elliot journal, 272–73; C. E. Akeley, *In Brightest Africa*, 99.

25. C. E. Akeley, *In Brightest Africa*, 100; Elliot journal, 274.

26. C. E. Akeley, *In Brightest Africa*, 100; Elliot journal, 274.

27. Elliot journal, 275.

28. Elliot journal, 275. See also C E. Akeley, *In Brightest Africa*; D. G. Elliot to F. J. V. Skiff, September 11, 1896, Recorder, personnel matters (RPM), FMA.

29. Elliot journal, 275. See also Elliot, "List of Mammals from Somali-Land Obtained by the Museum's East African Expedition," 145–46.

Chapter 21

1. Elliot journal, 276.

2. Elliot journal, 276.

3. Elliot journal, 276–77; D. G. Elliot to F. J. V. Skiff, September 11, 1896, RPM, FMA; C. E. Akeley, *In Brightest Africa*, 97.

4. Elliot journal, 277.

5. Elliot journal, 278.

6. Elliot journal, 277. On the Boran fight, see A. D. Smith, *Through Unknown African Countries*, 194–204.

7. Elliot journal, 280.

8. Elliot journal, 281.

9. Elliot, "Catalogue of a Collection of Birds Obtained by the Expedition into Somali-Land," 40.

10. Elliot journal, 281.

11. Elliot journal, 281–82.

12. Bull, *Safari*, 188.

13. Elliot, "Catalogue of a Collection of Birds Obtained by the Expedition into Somali-Land," 56–57.

14. Elliot journal, 283.

15. Elliot journal, 283–84.

16. Elliot journal, 284, 287–89. Two months later, an English big-game hunter, Alfred E. Pease, gave a tobe to an Ogaden man who visited his camp bringing a chit from Elliot that explained that the man had "shown great kindness" to Elliot when he was sick. Perhaps it was this same man. See Pease, *Travel and Sport in Africa*, 56.

17. Elliot journal, 283, 286, 290[b] (the journal has two pages numbered 290). Arsenious compounds have a long history of use as both medicines and poisons. See Jolliffe, "A History of the Use of Arsenicals in Man."

18. Elliot journal, 285. See also Elliot, "Catalogue of a Collection of Birds Obtained by the Expedition into Somali-Land," 62; Elliot, "List of Mammals from Somali-Land Obtained by the Museum's East African Expedition," 117–18, 152.

19. Elliot journal, 286.

20. Elliot, "List of Mammals from Somali-Land Obtained by the Museum's East African Expedition," 147.

21. Elliot journal, 287–88.

22. Elliot journal, 290[a].

23. Elliot journal, 290[b].

24. Elliot journal, 293. See also Elliot, "Catalogue of a Collection of Birds Obtained by the Expedition into Somali-Land," 45.

25. Elliot journal, 296.

26. Elliot journal, 296.

27. Elliot journal, 296.

28. Elliot journal, 299.

29. Elliot journal, 301–2.

30. Elliot journal, 303. See also Mosley, *Burke's Peerage, Baronetage and Knightage*; notice of Lieutenant Andrew's resignation, *London Gazette*.

31. Elliot journal, 303–4.

32. Elliot journal, 304; Cavendish, "Through Somaliland and around and South of Lake Rudolf," 373.

33. Elliot journal, 305.

34. Elliot journal, 305.

35. Elliot journal, 306.

36. Elliot journal, 306–7.

37. Elliot journal, 307.

38. Elliot journal, 309. Meek, "List of Fishes and Reptiles Obtained by Field Columbian Museum East African Expedition," does not list a specimen that seems to fit this description.

39. Elliot journal, 310; Elliot, "Catalogue of a Collection of Birds Obtained by the Expedition into Somali-Land," 50, and see 33, 58. Elliot recorded the locality for these birds as Robeleh, but this is probably a mistake.

40. Elliot journal, 310.

41. Elliot journal, 310–12.

42. Elliot journal, 312.

43. See Meek, "List of Fishes and Reptiles Obtained by Field Columbian Museum East African Expedition," 178; Elliot, "Catalogue of a Collection of Birds Obtained by the Expedition into Somali-Land," 65.

Chapter 22

1. Elliot journal, 314. See also Cavendish, "Through Somaliland and around and South of Lake Rudolf."

2. Elliot journal, 314.

3. Elliot journal, 313–14.

4. Elliot journal, 314.

5. Elliot journal, 289, 315. One hundred and sixty-one objects are listed in ROAR, anthropology, no. 300, FMA. Akeley turned these objects over to Holmes on December 22, 1896. A selection of them—including many of the weapons—was placed on display in Hall 6 the following day; see W. H. Holmes to F. J. V. Skiff, December 23, 1896, ROAR, anthropology, no. 300, FMA.

6. D. G. Elliot to F. J. V. Skiff, June 14, 1896, DGC, FMA. On salvage anthropology, see Gruber, "Ethnographic Salvage and the Shaping of Anthropology"; Brantlinger, *Dark Vanishings*; Redman, *Prophets and Ghosts*.

7. Elliot journal, 315. See also Voucher file, FMA. There is no further record of Elliot's pet cats beyond Berbera.

8. D. G. Elliot to F. J. V. Skiff, July 30, 1896, DGC, FMA; D. G. Elliot to F. J. V. Skiff, October [actually November] 4, 1896, RPM, FMA; F. J. V. Skiff to D. G. Elliot, September 14, 1896, DL, FMA. Elliot estimated this cost at about $400. See Voucher file, FMA.

9. D. G. Elliot to F. J. V. Skiff, October 6, 1896[b] (one of two letters with the same date), DGC, FMA.

10. *Board of Trade Wreck Report for "Tuna."* Few details about Dualla's death are available, but see Pease, *Travel and Sport in Africa*, 17, 20; Johnston, *The Story of My Life*, 130.

11. D. G. Elliot to F. J. V. Skiff, October 6, 1896[a], DGC, FMA. See also Henry J. Stadelman to Marshall Field & Co., Chicago, October 23, 1896 (copy), DGC, FMA.

12. D. G. Elliot to F. J. V. Skiff, October 6, 1896[b], DGC, FMA; Field Columbian Museum, "Annual Report [. . .] for the Year 1895–1896," 101.

13. Elliot to Skiff, October 6, 1896[b].

14. D. G. Elliot to F. J. V. Skiff, October 26, 1896, DGC, FMA. See also D. G. Elliot to F. J. V. Skiff, October [actually November] 4, 1896, RPM, FMA.

15. Elliot mentioned the baira antelope in his journal (231), noting: "I doubt if it is that animal." On the dik-diks, see Elliot, "Catalogue of a Collection of Birds Obtained by the Expedition into Somali-Land"; Elliot, "List of Mammals from Somali-Land Obtained by the Museum's East African Expedition," 113–18, 135–39; Menges, "Eine neue Antilope des Somalilandes." Interestingly, in 1898, zoologist Oldfield Thomas named a new species of dik-dik, *Madoqua cavendishi*, from a specimen collected by Henry Cavendish, who probably used one of Elliot's rifles to collect it. See Thomas, "Description of a New Dik-Dik Antelope." English zoologist William Edward de Winton described a new species of giraffe from Somaliland in 1899. Specimens of this animal had been collected by Arthur Donaldson Smith, Henry Cavendish, and Arthur Neumann. See de Winton, "On the Giraffe of Somaliland."

16. D. G. Elliot to F. J. V. Skiff, October 26, 1896, DGC, FMA. See also D. G. Elliot to F. J. V. Skiff, October [actually November] 4, 1896, RPM, FMA; "Cable dated Nov 2," note taken from a telegram, DGC, FMA.

17. D. G. Elliot to F. J. V. Skiff, April 24, 1897, Memos, all departments, box 1, distribution no. 327, FMA.

18. D. G. Elliot to F. J. V. Skiff, October [actually November] 4, 1896, Recorder, historical documents, FMA.

19. D. G. Elliot to F. J. V. Skiff, September 11, 1896, Recorder, historical documents, FMA.

20. D. G. Elliot to F. J. V. Skiff, October 6, 1896, DGC, FMA. Unfortunately, Skiff's letter has not been located.

21. F. J. V. Skiff to D. G. Elliot, October 17, 1896, RPM, FMA.

22. D. G. Elliot to F. J. V. Skiff, October [actually November] 4, 1896, RPM, FMA.

23. [Cope], Editor's Table (30, no. 10), 806.

24. R. McMurdy to F. J. V. Skiff, November 9, 1896, DGC, FMA; F. J. V. Skiff to R. McMurdy, November 10, 1896, DL, FMA.

25. D. G. Elliot to F. J. V. Skiff, October [actually November] 4, 1896, RPM, FMA; D. G. Elliot to F. J. V. Skiff, October 6, 1896, DGC, FMA.

26. "Denies the Elliott-Ackley Story," *Chicago Daily Tribune*, 3.

27. D. G. Elliot to F. J. V. Skiff, October [actually November] 4, 1896, RPM, FMA. See also A. H. Elliot to F. J. V. Skiff, November 5, 1896, DGC, FMA. See also "Returned from Africa," *New York Times*.

28. "Denies the Elliott-Ackley Story," *Chicago Daily Tribune*, 3.

Chapter 23

1. Field Columbian Museum, "Annual Report [. . .] for the Year 1895–1896," 99.

2. Field Columbian Museum, "Annual Report [. . .] for the Year 1896–1897," 183–85.

3. Field Columbian Museum, 185. On dioramas as the motivation for survey collecting, see Kohler, *All Creatures*.

4. Field Columbian Museum, "Annual Report [. . .] for the Year 1896–1897," 186, and see 183. Fieldwork wasn't the only way Elliot worked to expand the zoological collections. The museum purchased rare mammals whenever possible, including a collection of fifty-six skins from Akeley in January 1897, which included an African elephant. Other specimens arrived as donations. The Ringling brothers, for example, donated several specimens of exotic African circus animals, including a giraffe. See D. G. Elliot to F. J. V. Skiff, January 6, 1897, DGC, FMA; F. J. V. Skiff to D. G. Elliot, January 13, 1897, DL, FMA.

5. Field Columbian Museum, "Annual Report [. . .] for the Year 1896–1897," 186. See also Voucher file, FMA.

6. Elliot's final report, 23–24, FMA; D. G. Elliot to F. J. V. Skiff, July 29, 1897, DGC, FMA.

7. D. G. Elliot to F. J. V. Skiff, November 11, 1897, DGC, FMA. The first lecture, "The East African Expedition—London to South of Toyo Plain," debuted April 17, 1897. The second, "The East African Expedition—Toyo to Ogaden and Berbera," followed a week later. Both were repeated by popular demand the following October. Elliot consolidated his lectures into a single performance that he reprised in November 1905. The new lecture was titled: "A Naturalist in Africa—Field Columbian Museum Expedition." See Field Columbian Museum, "Annual Report [. . .] for the Year 1896–1897," 176; Field Columbian Museum, "Annual Report [. . .] for the Year 1897–1898," 264–65; Field Columbian Museum, "Annual Report [. . .] for the Year 1904–1905," 338. On Elliot's book proposal, see F. J. V. Skiff to D. G. Elliot, July 7, 1897, DGC, FMA.

8. D. G. Elliot to R. B. Sharpe, April 14, 1897, R. B. Sharpe Papers, NHM. See also Elliot journal, 293; Field Columbian Museum, "Annual Report [. . .] for the Year 1896–1897," 177; Field Columbian Museum, "Annual Report [. . .] for the Year 1897–1898," 266–67; Elliot, "Catalogue of a Collection of Birds Obtained by the Expedition into Somali-Land," 37–38, 45, 58–61; Elliot, "List of Mammals from Somali-Land Obtained by the Museum's East African Expedition." Two additional articles on the fishes, reptiles, amphibians, and shells collected in Africa also appeared: Meek, "List of Fishes and Reptiles Obtained by Field Columbian Museum East African Expedition"; Dall, "List of a Collection of Shells from the Gulf of Aden."

9. D. G. Elliot to F. J. V. Skiff, February 26, 1898, DGC, FMA; D. G. Elliot to F. J. V. Skiff, September 13, 1898, DGC, FMA.

10. Elliot to Skiff, February 26, 1898; D. G. Elliot to F. J. V. Skiff, August 5, 1898, DGC, FMA.

11. Field Columbian Museum, "Annual Report [. . .] for the Year 1897–1898," 279–80; Field Columbian Museum, "Annual Report [. . .] for the Year 1898–1899," 365; Elliot, "Preliminary Descriptions of New Rodents from the Olympic Mountains," 226–27. For more on Elliot's participation in the Harriman expedition, see Goetzmann and Sloan, *Looking Far North*. Conservationist John Muir, who was also on the Harriman expedition, later referred to Elliot fondly—but disapprovingly—as "our grand big-game Doctor." See Badè, *The Life and Letters of John Muir*, 332.

12. Field Columbian Museum, "Annual Report [. . .] for the Year 1899–1900," 444; D. G. Elliot to F. J. V. Skiff, c. late 1900, DGC, FMA. Surber didn't last long, either. He resigned early in 1901 and was succeeded by Edmund Heller.

13. D. G. Elliot to F. J. V. Skiff, April 15, 1901, DGC, FMA.

14. *Guide to the Field Columbian Museum*, 5th ed., 125. See also Field Columbian Museum, "Annual Report [. . .] for the Year 1896–1897," 193.

15. Field Columbian Museum, "Annual Report [. . .] for the Year 1896–1897," 193; D. G. Elliot to F. J. V. Skiff, September 25, 1897, DGC, FMA.

16. Field Columbian Museum, "Annual Report [. . .] for the Year 1897–1898," 285, 287.

17. Field Columbian Museum, 287; Field Columbian Museum, "Annual Report [. . .] for the Year 1898–1899," 365.

18. Field Columbian Museum, "Annual Report [. . .] for the Year 1898–1899," 370–71.

19. Field Columbian Museum, 370–71. On Mason, see D. G. Elliot to F. J. V. Skiff, April 23, 1901, DGC, FMA.

20. Field Columbian Museum, "Annual Report [. . .] for the Year 1900–1901," 21, 24; Field Columbian Museum, "Annual Report [. . .] for the Year 1899–1900," 451–52. The spotted hyena group is not mentioned in the annual report, but it does appear in 1900 for the first time in the *Guide to the Field Columbian Museum*, 6th ed., 89.

21. Field Columbian Museum, "Annual Report [. . .] for the Year 1901–1902," 105. See also Brinkman, "Frederic Ward Putnam," 96–97; Field Columbian Museum, "Annual Report [. . .] for the Year 1899–1900," 451; Field Columbian Museum, "Annual Report [. . .] for the Year 1900–1901," 21; *Guide to the Field Columbian Museum*, 6th ed.

22. Elliot, "List of Mammals from Somali-Land Obtained by the Museum's East African Expedition," 112–13, 124–26.

23. D. G. Elliot to F. J. V. Skiff, July 30, 1896, DGC, FMA; and see *Guide to the Field Columbian Museum*, 7th ed., 92–93.

24. "Scientific Notes and News," 991; Hornaday, "America Leads the World in the Taxidermist's Art"; Bryan is quoted in Andrei, *Nature's Mirror*, 120.

25. Ernest Hemingway, "A Trip to the Field Museum," unpublished manuscript, c. 1910, "The Early Years—Ernest and Marcelline Hemingway in Oak Park," Collection, Oak Park Public Library, Illinois. Akeley's elephant group went on display in 1909.

Chapter 24

1. Field Columbian Museum, "Annual Report [. . .] for the Year 1901–1902," 85; Field Columbian Museum, "Annual Report [. . .] for the Year 1902–1903," 167. On Akeley's concrete spraying device, see Teichert, "Carl Akeley"; Dewey, "My Friend 'Ake.'"

2. O. C. Farrington to F. J. V. Skiff, June 30, 1905, DGC, FMA; F. J. V. Skiff to E. E. Ayer, November 22, 1905, DL, FMA.

3. F. J. V. Skiff to H. N. Higinbotham, March 29, 1905, DGC (Akeley), FMA.

4. "Record of Minutes," July 24, 1905, FMA; Field Columbian Museum, "Annual Report [. . .] for the Year 1904–1905," 357.

5. "Record of Minutes," July 24, 1905, FMA; H. N. Higinbotham to D. G. Elliot, July 27, 1905; D. G. Elliot to H. N. Higinbotham, August 2, 1905; H. N. Higinbotham to D. G. Elliot, August 10, 1905; all letters in DGC, FMA.

6. "Trouble in Field Museum," *Pittsburgh (PA) Dispatch*; "Burchard Leaves the Field Museum," *Chicago Daily Tribune*. For more details of the revolt, see Brinkman, "The 'Chicago Idea'"; Brinkman, "Edward Drinker Cope's Final Feud."

7. McVicker, "Buying a Curator," 48; Fernlund, *William Henry Holmes and the Rediscovery of the American West*, 161.

8. "Burchard Leaves the Field Museum," *Chicago Daily Tribune*. See also Fernlund, *William Henry Holmes and the Rediscovery of the American West*; Brinkman, "The 'Chicago Idea'"; H. N. Higinbotham to E. E. Ayer, January 9, 1897, Recorder, historical documents, box 1, ed. L. Burchard, FMA. For more on Holmes's experience in Chicago, see D. J. Meltzer, "When Destiny Takes a Turn for the Worse."

9. D. C. Davies to F. J. V. Skiff, March 24, 1903, DL, FMA; H. N. Higinbotham to D. G. Elliot, December 28, 1903, DGC, FMA; D. G. Elliot to H. N. Higinbotham, December 29, 1903, DGC, FMA. On Elliot's health, see A. H. Elliot to D. C. Davies, letters dated March 25, 1902, April 4, 1902, April 15, 1902, and n.d., DGC, FMA; H. N. Higinbotham, January 27, 1904, memorandum, RPM, FMA.

10. H. N. Higinbotham, January 25, 1904, memorandum; H. N. Higinbotham, January 27, 1904, memorandum; both in RPM, FMA.

11. Higinbotham, January 25, 1904, memorandum.

12. F. J. V. Skiff to H. N. Higinbotham, February 8, 1904, RPM, FMA. Skiff may have had a point with regard to Elliot's temperament. During the Harriman expedition, Elliot arranged to give a formal lecture on Somaliland to his shipmates and colleagues. Much to his surprise, he was greeted with cheers, hoots, and cries of "What's the matter with Elliot? He's all right! Who's all right?" See Wolfe, *John of the Mountains*, 418.

13. D. G. Elliot to H. N. Higinbotham, August 2, 1905, DGC, FMA.

14. D. G. Elliot to F. J. V. Skiff, October 2, 1905, DGC, FMA.

15. F. J. V. Skiff to D. G. Elliot, January 6, 1906, RPM, FMA; D. G. Elliot to F. J. V. Skiff, January 6, 1906, RPM, FMA.

16. Field Museum of Natural History, "Annual Report [. . .] for the Year 1906," 9–10.

17. Osgood, "In Memoriam," 159–60. See also employment card for Charles B. Cory, FMA.

18. See Laubacher, "Ready to Migrate." Akeley gave his six months' notice on August 1, 1907. See C. E. Akeley to F. J. V. Skiff, August 1, 1907, DGC, FMA.

19. S. Field to F. J. V. Skiff, January 10, 1907, RPM, FMA; F. J. V. Skiff to D. G. Elliot, January 22, 1907, DGC, FMA.

Conclusion

1. Lyell, *The Principles of Geology*; Darwin, *On the Origin of Species by Means of Natural Selection*, 475; Sepkoski, *Catastrophic Thinking*, 81. A comprehensive history of extinction and biodiversity is beyond the scope of this book. For more

on these ideas, see Barrow, *Nature's Ghosts*; Farnham, *Saving Nature's Legacy*. On the early history of the conservation movement, see also Bowler, *The Earth Encompassed*, 318–23; Grove, *Green Imperialism*.

2. "Opening of the Museum," *Chicago Daily Tribune*; "An Historical and Descriptive Account of the Field Columbian Museum," 13–14. See also Brinkman, "Frederic Ward Putnam"; Conn, *Museums and American Intellectual Life*, 78.

3. *Guide to the Field Columbian Museum*, 2nd ed., 170–71, 173. See also Brinkman, "Valuable So Far as It Goes"; Brinkman, "The Strongest Kind of Competition."

4. Wallace, "On the Physical Geography of the Malay Archipelago," 234.

5. Turvey and Cheke, "Dead as a Dodo," 159. See also Strickland and Melville, *The Dodo and Its Kindred*.

6. Salvin, "Notes on Captain Markham's 'Visit to the Galapagos Islands,'" 757; Baur, "On the Origin of the Galapagos Islands (Concluded)," 318; Rothschild quoted in Hennessy, "Mythologizing Darwin's Islands," 70; "The First of the Series of Expeditions in Search of Specimens for the Academy of Sciences," *San Francisco Chronicle*, 16. See also E. J. Larson, *Evolution's Workshop*, 117–18. For much more on salvage collecting in the Galapagos Islands, see M. J. James, *Collecting Evolution*. Rothschild was sometimes accused of overcollecting. "I can't agree with you in thinking that Zoology is best advanced by collectors of the kind you employ," wrote a Cambridge professor, "no doubt they answer admirably the purpose of stocking a Museum; but they unstock the world—and that is a terrible consideration." Quoted in Conniff, *The Species Seekers*, 332. See also Rothschild, *Dear Lord Rothschild*.

7. Quammen, *The Song of the Dodo*, 308. See also Lorimer, "On Auks and Awkwardness," 200; Newton, "Abstract of Mr. J. Wooley's Researches in Iceland Respecting the Gare-Fowl or Great Auk"; Greenberg, *A Feathered River across the Sky*. For a cautionary tale on extinction, see Thornhill, *The Tragic Tale of the Great Auk*.

8. Hornaday, *Taxidermy and Zoological Collecting*, i, 2, and see 157–58. See also Hornaday, "The Extermination of the American Bison, with a Sketch of Its Discovery and Life History"; Shell, "Skin Deep."

9. D. G. Elliot to F. J. V. Skiff, July 30, 1896, DGC, FMA.

10. Ritvo, *The Animal Estate*, 284. For Chapman's advocacy, see, e.g., Chapman, "The Whitney South Sea Expedition." On Hornaday's ethical awakening, see Bechtel, *Mr. Hornaday's War*.

11. See Merriam, *Birds through an Opera Glass*; Mitman, *Reel Nature*; Barrow, *Nature's Ghosts*, 286–87; Benson, "Endangered Birds and Epistemic Concerns," 180–82; Grinnell, "Conserve the Collector."

12. Reiger, *American Sportsmen and the Origins of Conservation*, 97; Bryden, *The Great and Small Game of Africa*, v; Elliot, *The Gallinaceous Game Birds of North America*. Elliot wasn't the only museum zoologist of his generation to fail to make the transition to conservationist. His Field Columbian Museum colleague Charles B. Cory, for example, wrote in 1902 in reply to an invitation to attend a meeting of the Audubon Society, "I do not protect birds. I kill them." Quoted in Barrow, *A Passion for Birds*, 141.

13. Elliot, "The World's Hunting Fields and Disappearance of Big Game."

14. See Barrow, *Nature's Ghosts*, 154; *Three Addresses Delivered at the Meeting Commemorating the Fiftieth Anniversary of Field Museum of Natural History*, 9; Bowler, *The Earth Encompassed*, 436; Kohler, *All Creatures*, 274–78.

15. Quoted in Kohler, *All Creatures*, 277.

16. On shooting pandas, see Roosevelt and Roosevelt, *Trailing the Giant Panda*; on harpooning manatees, see Barber and Barber, *Recollections of a Museum Collector*.

17. See Barrow, *Nature's Ghosts*, 139, 148–52; *Three Addresses Delivered at the Meeting Commemorating the Fiftieth Anniversary of Field Museum of Natural History*, 11; Asma, *Stuffed Animals and Pickled Heads*, 43–45.

18. S. Meltzer, *Theatres of Nature*, 12–13.

19. C. E. Akeley, *In Brightest Africa*, 153. See also Bodry-Sanders, *African Obsession*.

20. Elliot, "Address Given before the Linnaean Society of New York," n.p.

21. Elliot, n.p.

BIBLIOGRAPHY

ARCHIVAL SOURCES

American Museum of Natural History, New York (AMNH)

Elliot, Daniel G. Correspondence. Mammalogy Departmental Library and Archives (MDLA).
———. Correspondence. Ornithological Department Library and Archives (ODLA).
Kunz, George. Papers. Central Archives.

Chicago Academy of Sciences / Peggy Notebaert Nature Museum (CAS)

Baker, Frank Collins. Biographical notes. Institutional Records, Personnel, Staff.
———. Papers. Manuscripts and Institutional Archives.
Cory, Charles B. Papers. Manuscripts and Institutional Archives.

Field Museum, Chicago

Akeley, Carl E. Field notes. 1896. Zoology Department Archives.
Director's general correspondence (DGC). Field Museum Archives (FMA).
Director's letterbooks (DL). Field Museum Archives (FMA).
Elliot, Daniel Giraud. Manuscript journal on the Field Columbian Museum's zoological expedition to Africa. 1896. Mary W. Runnells Rare Book Room.
Employment cards. Field Museum Archives (FMA).
Memos, all departments. Field Museum Archives (FMA).
"Minutes of the Meetings of the Board of Trustees of the Field Columbian Museum." September 1893–December 1912. Field Museum Archives (FMA).
Recorder, historical documents. Field Museum Archives (FMA).
Recorder, personnel matters (RPM). Field Museum Archives (FMA).
Recorder's Office accession records (ROAR). Anthropology. Field Museum Archives (FMA).
———. Geology. Field Museum Archives (FMA).
———. Obsolete-Miscellaneous. Field Museum Archives (FMA).
———. Ornithology. Field Museum Archives (FMA).
———. Zoology. Field Museum Archives (FMA).
"Record of Minutes of the Executive Committee of the Field Columbian Museum." May 1894–December 1913. Field Museum Archives (FMA).

Voucher file, African Expedition. 1896. Field Museum Archives (FMA).
Zoological collections database. Online.
Zoology offers. Field Museum Archives (FMA).

Harvard University, Museum of Comparative Zoology, Cambridge, Massachusetts (MCZ)

Baur, Georg. Records. Ernst Mayr Library, Archives.

Milwaukee County Historical Society, Wisconsin

Office of the Secretary of State. Mss-0742. Milwaukee County Incorporations, Vol. K.

Milwaukee Public Museum, Wisconsin

Akeley, Carl E. Papers.

Natural History Museum, Library and Archives, London (NHM)

Dodson, Edward. "'Ted's Trunk': Transcriptions of the Diaries, Letters and Other Documentation of Edward Dodson, 1873–1948, Naturalist and Taxidermist . . . from His 'Expeditions' Trunk to Lake Rudolph, Somaliland, Morocco, Tripolitana and Patagonia between 1894 and 1901." Transcribed by John P. Williamson. 2001.
Sharpe, R. B. Papers.

Oak Park Public Library, Illinois

"The Early Years—Ernest and Marcelline Hemingway in Oak Park." Collection.

Smithsonian Institution Archives, Washington, DC

Division of Birds, National Museum of Natural History. 1854–1959. Record Unit 105, Box 4.

University of Florida, Smathers Libraries, Gainesville

Hay, William Perry. Papers. Special and Area Collections.

PUBLISHED SOURCES

Akeley, Carl E. *In Brightest Africa*. Garden City, NY: Garden City, 1920.
Akeley, Mary L. Jobe. *The Wilderness Lives Again: Carl Akeley and the Great Adventure*. New York: Dodd, Mead, 1940.
Alberti, Samuel J. M. M. "Introduction: The Dead Ark." In *The Afterlives of Animals: A Museum Menagerie*, edited by Samuel J. M. M. Alberti, 1–16. Charlottesville: University of Virginia Press, 2011.
Allen, J. A. "The American Bisons, Living and Extinct." *Memoirs of the Museum of Comparative Zoology* 4, no. 10 (1976): 1–249.
———. "Daniel Giraud Elliot." *Science* 43, no. 1101 (1916): 159–62.
———. "The Present Wholesale Destruction of Bird-Life in the United States." *Science* 7, no. 16 (1886): 191–95.
Alvey, Mark. "The Cinema as Taxidermy: Carl Akeley and the Preservation

Obsession." *Framework: The Journal of Cinema and Media* 48, no. 1 (2007): 23–45.

Alvey, Mark, Tia Resleure, and Thomas P. Gnoske. "Carl Akeley's 'Lost' Decorative Taxidermy and Anthropomorphic Groups: The European Connection." *Journal of the History of Collections* 35, no. 2 (2023): 333–46.

American Museum of Natural History. *Annual Report of the American Museum of Natural History for the Year 1887–8*. New York: American Museum of Natural History, 1888.

Andrei, Mary Anne. *Nature's Mirror: How Taxidermists Shaped America's Natural History Museums and Saved Endangered Species*. Chicago: University of Chicago Press, 2020.

Arengo, Felicity, Ana L. Porzecanski, Mary E. Blair, George Amato, Christopher Filardi, and Eleanor J. Sterling. "The Essential Role of Museums in Biodiversity Conservation." In *The Future of Natural History Museums*, edited by Eric Dorfman, 82–100. New York: Routledge, 2018.

Armstrong, David M. *Distribution of Mammals in Colorado*. Lawrence: University of Kansas, 1972.

Asma, Stephen T. *Stuffed Animals and Pickled Heads: The Culture and Evolution of Natural History Museums*. Oxford: Oxford University Press, 2001.

Badè, William Frederic. *The Life and Letters of John Muir*. Vol. 2. Boston: Houghton Mifflin, 1924.

Barber, Charles M., with James P. Barber. *Recollections of a Museum Collector*. Warsaw, IN: Other Road, 2017.

Barringer, Tim, and Tom Flynn, eds. *Colonialism and the Object: Empire, Material Culture and the Museum*. London: Routledge, 1998.

Barrow, Mark V., Jr. *Nature's Ghosts: Confronting Extinction from the Age of Jefferson to the Age of Ecology*. Chicago: University of Chicago Press, 2009.

———. *A Passion for Birds: American Ornithology after Audubon*. Princeton, NJ: Princeton University Press, 1998.

———. "The Specimen Dealer: Entrepreneurial Natural History in America's Gilded Age." *Journal of the History of Biology* 33, no. 3 (2000): 493–534.

Baur, G. "The Differentiation of Species on the Galapagos Islands and the Origin of the Group." In *Biological Lectures Delivered at the Marine Biological Laboratory of Wood's Holl in the Summer Session of 1894*, 67–78. Boston: Ginn, 1895.

———. "On the Origin of the Galapagos Islands (Concluded)." *American Naturalist* 25, no. 292 (1891): 307–26.

Bechtel, Stefan. *Mr. Hornaday's War: How a Peculiar Victorian Zookeeper Waged a Lonely Crusade for Wildlife That Changed the World*. Boston: Beacon Press, 2012.

Bederman, Gail. *Manliness and Civilization: A Cultural History of Gender and Race in the United States, 1880–1917*. Chicago: University of Chicago Press, 1996.

Benson, Etienne. "Endangered Birds and Epistemic Concerns: The California Condor." In *Endangerment, Biodiversity and Culture*, edited by Fernando Vidal and Nélia Dias, 175–94. London: Routledge, 2015.

Beolens, Bo, Michael Watkins, and Michael Grayson. *The Eponym Dictionary of Birds*. London: Bloomsbury, 2014.

Board of Trade. *Board of Trade Wreck Report for "Tuna."* No. 5471 (1897).

Bodry-Sanders, Penelope. *African Obsession: The Life and Legacy of Carl Akeley*. 2nd ed. Jacksonville, FL: Batax Museum, 1998.

Bogdanov, Modest. "Bemerkungen über die Gruppe der Pterocliden." *Bulletin de l'Académie Impériale des Sciences de Saint Pétersbourg* 27 (1881): 164–68.

Bolton, Gambier. "Animals Worth Their Weight in Gold." *Royal Magazine* 2 (1899): 400–405.

Bowler, Peter J. *The Earth Encompassed: A History of the Environmental Sciences*. New York: W. W. Norton, 1992.

Bragdon, Claude F. "Letter from Chicago." *American Architecture and Building News* 43, no. 952 (1894): 139–42.

Brantlinger, Patrick. *Dark Vanishings: Discourse on the Extinction of Primitive Races, 1800–1930*. Ithaca, NY: Cornell University Press, 2003.

Brinkman, Paul D. "The 'Chicago Idea': Patronage, Authority, and Scientific Autonomy at the Field Columbian Museum, 1893–1897." *Museum History Journal* 8, no. 2 (2015): 168–87.

———. "Edward Drinker Cope's Final Feud." *Archives of Natural History* 43, no. 2 (2016): 305–20.

———. "Establishing Vertebrate Paleontology at Chicago's Field Columbian Museum, 1893–1898." *Archives of Natural History* 27, no. 1 (2000): 81–114.

———. "Frederic Ward Putnam, Chicago's Cultural Philanthropists, and the Founding of the Field Museum." *Museum History Journal* 2, no. 1 (2009): 73–100.

———. "Remarking on a Blackened Eye: Persifor Frazer's Blow-by-Blow Account of a Fistfight with His Dear Friend Edward Drinker Cope." *Endeavour* 39, nos. 3–4 (2015): 188–92.

———. *The Second Jurassic Dinosaur Rush: Museums and Paleontology in America at the Turn of the Twentieth Century*. Chicago: University of Chicago Press, 2010.

———. "The Strongest Kind of Competition: Expanding Zoology at Chicago's Field Columbian Museum, 1894–1895." *Colligo* 1, no. 1 (2018): 53–68.

———. "Valuable So Far as It Goes: Establishing Zoology at Chicago's Field Columbian Museum, 1893–1894." *Journal of the History of Collections* 31, no. 1 (2019): 93–109.

Brown, Monty. *Hunter Away: The Life and Times of Arthur Henry Neumann*. London: Privately printed, 1993.

Bryden, H. A., ed. *The Great and Small Game of Africa: An Account of the Distribution, Habits, and Natural History of the Sporting Mammals, with Personal Hunting Experiences*. London: Rowland Ward, 1899.

Bull, Bartle. *Safari: A Chronicle of Adventure*. New York: Viking, 1992.

Burg, David F. *Chicago's White City of 1893*. Lexington: University Press of Kentucky, 1976.

Burton, Richard Francis. *First Footsteps in East Africa, or An Exploration of Harar*. Vol. 2. 1856. London: Longman, Brown, Green, and Longmans, 1894.

Canfield, Michael R. *Theodore Roosevelt in the Field*. Chicago: University of Chicago Press, 2015.

Carrier, Neil, and Lisa Gezon. "Khat in the Western Indian Ocean: Regional Linkages and Disjunctures." *Études Océan Indien* 42–43 (2009): 271–97.

Cavendish, H. S. H. "Through Somaliland and around and South of Lake Rudolf." *Geographical Journal* 11, no. 4 (1898): 372–93.

Chapman, Frank M. "Daniel Giraud Elliot." *Auk* 34, no. 10 (1917): 1–10.

———. "The Whitney South Sea Expedition." *Science* 81, no. 2091 (1935): 95–97.

Chernow, Ron. *Grant*. New York: Penguin Books, 2017.

Chicago Daily Tribune. "Burchard Leaves the Field Museum." February 3, 1897.

———. "Curator for the Field Museum." November 23, 1894.

———. "Denies the Elliott-Ackley Story." November 29, 1896.

———. "Opening of the Museum." June 3, 1894.

———. "Origin of Fair Fires." February 20, 1894.

Chicago Evening Post. "Rats." February 20, 1894.

Conn, Steven. *Museums and American Intellectual Life, 1876–1926*. Chicago: University of Chicago Press, 1998.

Conniff, Richard. *The Species Seekers: Heroes, Fools, and the Mad Pursuit of Life on Earth*. New York: W. W. Norton, 2011.

[Cope, E. D.]. Editor's Table. *American Naturalist* 29, no. 9 (1895): 827.

———. Editor's Table. *American Naturalist* 30, no. 5 (1896): 385–86.

———. Editor's Table. *American Naturalist* 30, no. 10 (1896): 806.

Cory, Charles B. *How to Know the Ducks, Geese and Swans of North America: All the Species Being Grouped According to Size and Color*. Boston: Little, Brown, 1897.

———. *Hunting and Fishing in Florida: Including a Key to the Water Birds Known to Occur in the State*. Boston: Estes and Lauriat, 1896.

———. *Montezuma's Castle and Other Weird Tales*. Boston: Rockwell and Churchill, 1899.

Dall, W. H. "List of a Collection of Shells from the Gulf of Aden Obtained by the Museum's East African Expedition." *Field Columbian Museum, Zoological Series* 1, no. 9 (1899): 187–89.

"Daniel Giraud Elliot: A Brief Biographical Sketch on the Occasion of His Eightieth Birthday to Emphasize His Long Devotion to Scientific Work and His Services to the Museum." *American Museum Journal* 14 (1915): 133–41.

Darwin, Charles. *Geological Observations on the Volcanic Islands Visited During the Voyage of H. M. S. Beagle, Together with Some Brief Notices of the Geology of Australia and the Cape of Good Hope* [. . .]. London: Smith Elder, 1844.

———. *Journal and Remarks, 1832–1836*. London: Henry Colburn, 1839.

———. *On the Origin of Species by Means of Natural Selection, or the Preservation of Favoured Races in the Struggle for Life*. London: John Murray, 1859.

De Bont, Raf. "A World Laboratory: Framing the Albert National Park." *Environmental History* 22 (2017): 404–32.

Dewey, C. L. "My Friend 'Ake.'" *Nature Magazine* 10 (1927): 387–91.

De Winton, William Edward. "On the Giraffe of Somaliland." *Annals and Magazine of Natural History* 4 (1899): 211–12.

Elliot, Daniel Giraud. "Address Given before the Linnaean Society of New York on March 24th, 1914, in Acknowledgement of the Bestowal of the Linnaean Medal of Honor for Distinguished Work in the Fields of Mammalogy and Ornithology." Privately printed, 1914.

———. "Catalogue of a Collection of Birds Obtained by the Expedition into

Somali-Land." *Field Columbian Museum, Ornithological Series* 1, no. 2 (1897): 29–67.

———. *The Gallinaceous Game Birds of North America: Including the Partridges, Grouse, Ptarmigan, and Wild Turkeys* [. . .]. London: Suckling, 1897.

———. "In Memoriam: Philip Lutley Sclater." *Auk* 31, no. 1 (1914): 1–12.

———. "List of Mammals from Somali-Land Obtained by the Museum's East African Expedition." *Field Columbian Museum, Zoological Series* 1, no. 6 (1897): 109–55.

———. *A Monograph of the Bucerotidae, or Family of the Hornbills.* London: privately published, 1876–82.

———. *A Monograph of the Felidae, or Family of the Cats.* London: privately published, 1883.

———. *A Monograph of the Paradiseidae, or Birds of Paradise.* London: privately published, 1873.

———. *North American Shore Birds: A History of the Snipes, Sandpipers, Plovers and Their Allies, Inhabiting the Beaches and Marshes of the Atlantic and Pacific Coasts* [. . .]. New York: Francis P. Harper, 1895.

———. "Preliminary Descriptions of New Rodents from the Olympic Mountains." *Field Columbian Museum, Zoological Series* 1, no. 11 (1899): 225–28.

———. "The World's Hunting Fields and Disappearance of Big Game." *New York Times*, April 15, 1900.

Eshelby, K. "The Way of the Borana." *New African* 441 (2005): 32–34.

Fargher, James. "Hunter in Somaliland: Consul Frederick M. Hunter and the Creation of the British Somaliland Protectorate." *Northeast African Studies* 21, no. 1 (2021): 119–36.

Farnham, Timothy J. *Saving Nature's Legacy: Origins of the Idea of Biological Diversity.* New Haven, CT: Yale University Press, 2007.

Farrington, Oliver C. "Dr. Frederick J. V. Skiff." *Museum Work, Including the Proceedings of the American Association of Museums* 3, nos. 7–8 (1921): 197–98.

———. "The Rise of Natural History Museums." *Proceedings of the American Association of Museums* 9 (1915): 36–53.

Fernlund, Kevin J. *William Henry Holmes and the Rediscovery of the American West.* Albuquerque: University of New Mexico Press, 2000.

Field Columbian Museum. "Annual Report of the Director to the Board of Trustees for the Year 1894–1895." *Publications of the Field Columbian Museum, Report Series* 1, no. 1 (1895).

———. "Annual Report of the Director to the Board of Trustees for the Year 1895–1896." *Publications of the Field Columbian Museum, Report Series* 1, no. 2 (1896).

———. "Annual Report of the Director to the Board of Trustees for the Year 1896–1897." *Publications of the Field Columbian Museum, Report Series* 1, no. 3 (1897).

———. "Annual Report of the Director to the Board of Trustees for the Year 1897–1898." *Publications of the Field Columbian Museum, Report Series* 1, no. 4 (1898).

———. "Annual Report of the Director to the Board of Trustees for the Year 1898–1899." *Publications of the Field Columbian Museum, Report Series* 1, no. 5 (1899).

———. "Annual Report of the Director to the Board of Trustees for the Year 1899–1900." *Publications of the Field Columbian Museum, Report Series* 1, no. 6 (1900).

———. "Annual Report of the Director to the Board of Trustees for the Year 1900–1901." *Publications of the Field Columbian Museum, Report Series* 2, no. 1 (1901).

———. "Annual Report of the Director to the Board of Trustees for the Year 1901–1902." *Publications of the Field Columbian Museum, Report Series* 2, no. 2 (1902).

———. "Annual Report of the Director to the Board of Trustees for the Year 1902–1903." *Publications of the Field Columbian Museum, Report Series* 2, no. 3 (1903).

———. "Annual Report of the Director to the Board of Trustees for the Year 1904–1905." *Publications of the Field Columbian Museum, Report Series* 2, no. 5 (1905).

Field Museum of Natural History. "Annual Report of the Director to the Board of Trustees for the Year 1906." *Publications of the Field Columbian Museum, Report Series* 3, no. 1 (1907).

Fisher, Brian L., and Barry Bolton. *Ants of Africa and Madagascar: A Guide to the Genera*. Oakland: University of California Press, 2016.

Flayhart, William Henry, III. *The American Line, 1871–1902*. New York: W. W. Norton, 2001.

Gabobe, Jamal. "European Travel Writing on Somaliland: The Rhetoric of Empire and the Emergence of the Somali Subject." *Bildhaan: An International Journal of Somali Studies* 19 (2019): article 9.

Genoways, Hugh H., Suzanne B. McLaren, and Robert M. Timm. "Innovations That Changed Mammalogy: The Cyclone Trap." *Journal of Mammalogy* 101, no. 2 (2020): 325–27.

Geshekter, Charles. "The Missionary Factor in Somali Dervish History, 1890–1910." *Bildhaan: An International Journal of Somali Studies* 20 (2020): article 5.

Gill, B. J. "Charles Francis Adams: Diary of a Young American Taxidermist Visiting New Zealand, 1884–1887." *Archives of Natural History* 41, no. 1 (2014): 1–16.

Goetzmann, William H., and Kay Sloan. *Looking Far North: The Harriman Expedition to Alaska, 1899*. Princeton, NJ: Princeton University Press, 1983.

Goodman, Steven Michael, and Scot Lanyon. "Scientific Collecting." *Conservation Biology* 8, no. 1 (1994): 314–15.

Graff, Rebecca S. "Dream City, Plaster City: Worlds' Fairs and the Gilding of American Material Culture." *International Journal of Historical Archaeology* 16, no. 4 (2012): 696–716.

Gray, J. E. "On Museums, Their Use and Improvement, and on the Acclimatization of Animals." *Annals and Magazine of Natural History* 14 (1864): 283–97.

Greenberg, Joel. *A Feathered River across the Sky: The Passenger Pigeon's Flight to Extinction*. New York: Bloomsbury, 2014.

Grinnell, Joseph. "Conserve the Collector." *Science* 41, no. 1050 (1915): 229–32.

Grove, Richard H. *Green Imperialism: Colonial Expansion, Tropical Island Edens and the Origins of Environmentalism, 1600–1860*. Cambridge: Cambridge University Press, 1995.

Gruber, Jacob W. "Ethnographic Salvage and the Shaping of Anthropology." *American Anthropologist* 72, no. 6 (1970): 1289–99.

Guide to the Field Columbian Museum. 6th ed. Chicago: Field Columbian Museum, 1900.

Guide to the Field Columbian Museum. 7th ed. Chicago: Field Columbian Museum, 1903.

Guide to the Field Columbian Museum with Diagrams and Descriptions. 1st ed. Chicago: Field Columbian Museum, 1894.

Guide to the Field Columbian Museum with Diagrams and Descriptions. 2nd ed. Chicago: Field Columbian Museum, 1894.

Guide to the Field Columbian Museum with Diagrams and Descriptions. 4th ed. Chicago: Field Columbian Museum, 1896.

Guide to the Field Columbian Museum with Diagrams and Descriptions. 5th ed. Chicago: Field Columbian Museum, 1897–98.

Haraway, Donna. "Teddy Bear Patriarchy: Taxidermy in the Garden of Eden, New York City, 1908–1936." *Social Text* 11 (1984–85): 20–64.

Hay, O. P. "Description of a New Species of *Petalodus* (*P. securiger*) from the Carboniferous of Illinois." *Journal of Geology* 3, no. 5 (1895): 561–64.

———. "George Baur." *Science* 8, no. 185 (1898): 68–71.

———. "On Certain Portions of the Skeleton of *Protostega gigas.*" *Field Columbian Museum, Zoological Series* 1, no. 3 (1895): 57–62.

———. "On the Skeleton of *Toxochelys latiremis.*" *Field Columbian Museum, Zoological Series* 1, no. 5 (1896): 101–6.

———. "On the Structure and Development of the Vertebral Column of *Amia.*" *Field Columbian Museum, Zoological Series* 1, no. 1 (1895): 1–54.

Hennessy, Elizabeth. "Mythologizing Darwin's Islands." In *Darwin, Darwinism and Conservation in the Galapagos Islands: The Legacy of Darwin and Its New Applications*, edited by Diego Quiroga and Ana Sevilla, 65–90. Basel: Springer International, 2017.

———. "Saving Species: The Co-Evolution of Tortoise Taxonomy and Conservation in the Galapagos Islands." *Environmental History* 25 (2020): 263–86.

Herne, Brian. *White Hunters: The Golden Age of African Safaris.* New York: Henry Holt, 1999.

Hines, Thomas S. *Burnham of Chicago: Architect and Planner.* 2nd ed. Chicago: University of Chicago Press, 1979.

"An Historical and Descriptive Account of the Field Columbian Museum." *Field Columbian Museum, Publication 1* 1, no. 1 (1894): 1–91.

Höhnel, Ludwig von. *Discovery of Lakes Rudolf and Stefanie: A Narrative of Count Samuel Teleki's Exploring and Hunting Expedition in Eastern Equatorial Africa in 1887 and 1888.* 2 vols. London: Longmans, Green, 1894.

Hornaday, William T. "America Leads the World in the Taxidermist's Art." *New York Herald*, March 21, 1901.

———. "The Extermination of the American Bison, with a Sketch of Its Discovery and Life History." In *Report of the United States National Museum, under the Direction of the Smithsonian Institution, 1887*, 369–548. Washington, DC: Government Printing Office, 1889.

———. *Taxidermy and Zoological Collecting: A Complete Handbook for the Amateur Taxidermist, Collector, Osteologist, Museum-Builder, Sportsman, and Traveller.* New York: Charles Scribner's Sons, 1891.

Hough, Emerson. "Chicago and the West." *Field and Stream* 44 (1895): 369.

Huddersfield Chronicle (West Yorkshire). "The Threatened Extinction of the African Elephant." April 22, 1896.

Hunter, F. M. *An Account of the British Settlement of Aden in Arabia*. London: Trubner, 1877.

Imperato, Pascal James. *Arthur Donaldson Smith and the Exploration of Lake Rudolf*. Lake Success: Medical Society of the State of New York, 1987.

Jackson, Christine E. "The Ward Family of Taxidermists." *Archives of Natural History* 45, no. 1 (2018): 1–13.

Jacoby, Karl. *Crimes against Nature: Squatters, Poachers, Thieves, and the Hidden History of American Conservation*. Berkeley: University of California Press, 2001.

James, F. L. *The Unknown Horn of Africa: An Exploration from Berbera to the Leopard River*. 2nd ed. London: George Philip, 1890.

James, Garry. "Britain's Black-Powder .303, the Lee-Metford." *Gun Digest* 32, no. 3 (2015).

James, Henry. *The Letters of Henry James*. Vol. 3, *1883–1895*. Edited by Leon Edel. Cambridge, MA: Belknap Press of Harvard University Press, 1980.

James, Matthew J. *Collecting Evolution: The Galapagos Expedition That Vindicated Darwin*. Oxford: Oxford University Press, 2017.

Janzen, Daniel H. "Science Is Forever." *Oikos* 46, no. 3 (1986): 281–83.

Johnston, Harry H. *The Story of My Life*. Garden City, NY: Garden City, 1923.

Jolliffe, D. M. "A History of the Use of Arsenicals in Man." *Journal of the Royal Society of Medicine* 86 (1993): 287–89.

Jones, Jeannette Eileen. *In Search of Brightest Africa: Reimagining the Dark Continent in American Culture, 1884–1936*. Athens: University of Georgia Press, 2010.

Jones, Karen. "The Rhinoceros and the Chatham Railway: Taxidermy and the Production of Animal Presence in the 'Great Indoors.'" *History* 101, no. 348 (2016): 710–35.

Jonas, Raymond. *The Battle of Adwa: African Victory in the Age of Empire*. Cambridge, MA: Harvard University Press, 2011.

Kidder, Frank Eugene. *Building Construction and Superintendence*. 9th ed. New York: William T. Comstock, 1909.

Kingsley, Charles. *Glaucus, or The Wonders of the Shore*. 5th ed. London: MacMillan, 1873.

Kirk, Jay. *Kingdom under Glass: A Tale of Obsession, Adventure, and One Man's Quest to Preserve the World's Great Animals*. New York: Henry Holt, 2010.

Kohler, Robert E. *All Creatures: Naturalists, Collectors, and Biodiversity, 1850–1950*. Princeton, NJ: Princeton University Press, 2006.

Kohlstedt, Sally G. "Henry A. Ward: The Merchant Naturalist and American Museum Development." *Journal of the Society for the Bibliography of Natural History* 9, no. 4 (1980): 647–61.

———. "Innovative Niche Scientists: Women's Role in Reframing North American Museums, 1880–1930." *Centaurus* 55, no. 2 (2013): 153–74.

Kohlstedt, Sally G., and Paul D. Brinkman. "Framing Nature: Reflections on the Formative Years of Natural History Museum Development in the United States." *Proceedings of the California Academy of Sciences* 55, no. S1 (2004): 7–33.

Korenic, Mary. "Studying the Visitor and the Diorama, Part I." *Informal Science Review* 1 (1997): 12–15.

Kuklick, H. and Kohler, R. E., eds. "Science in the Field." Special issue, *Osiris* 11 (1997).

Larson, Edward J. *Evolution's Workshop: God and Science on the Galapagos Islands.* New York: Basic Books, 2001.

Larson, Erik. *The Devil in the White City: Murder, Magic, and Madness at the Fair That Changed America.* New York: Crown, 2003.

Laubacher, Matthew. "'Ready to Migrate': The Field Museum Exodus of 1907– 1908." *Museum History Journal* 11, no. 1 (2018): 42–56.

Lewis, I. M. "The So-Called 'Galla Graves' of Northern Somaliland." *Man* 61 (1961): 103–6.

Loerzel, Robert. *Alchemy of Bones: Chicago's Luetgert Murder Case of 1897.* Urbana: University of Illinois Press, 2003.

London Gazette. Notice of Lieutenant Andrew's resignation. September 29, 1896, 5379.

Lorimer, Jamie. "On Auks and Awkwardness." *Environmental Humanities* 4 (2014): 195–205.

Lucas, F. A. "Official Extinction." *Forest and Stream* 28 (1887): 104.

Lull, Richard Swann. "Memorial of Oliver Perry Hay." *Bulletin of the Geological Society of America* 42 (1931): 30–48.

Lyell, Charles. *The Principles of Geology.* Vol. 2. London: John Murray, 1832.

MacKenzie, John M. *Museums and Empire: Natural History, Human Cultures and Colonial Identities.* Manchester: Manchester University Press, 2009.

Marconi, Richard A. "Palm Beach Museum Founder and Golfer: The Forgotten Charles Barney Cory." *Tustenegee* 3, no. 1 (2012): 20–28.

McCarthy, Kathleen D. *Noblesse Oblige: Charity and cultural philanthropy in Chicago, 1849–1929.* Chicago: University of Chicago Press, 1982.

McCullough, David. *The Wright Brothers.* New York: Simon and Schuster, 2015.

McVicker, Donald. "Buying a Curator: Establishing Anthropology at Field Columbian Museum." In *Assembling the Past: Studies in the Professionalization of Archaeology,* edited by Alice B. Kehoe and Mary Beth Emmerichs, 37–52. Albuquerque: University of New Mexico Press, 1999.

Meek, Seth Eugene. "List of Fishes and Reptiles Obtained by Field Columbian Museum East African Expedition to Somali-Land in 1896." *Field Columbian Museum, Zoological Series* 1, no. 8 (1897): 163–84.

"Meetings of the Royal Geographical Society, Session 1895–96." *Geographical Journal* 7, no. 4 (1896): 436–37.

Meltzer, David J. "When Destiny Takes a Turn for the Worse: William Henry Holmes and, Incidentally, Franz Boas in Chicago, 1892–97." *Histories of Anthropology Annual* 6 (2010): 171–224.

Meltzer, Sally. *Theatres of Nature: Dioramas at the Field Museum.* Chicago: Field Museum, 2007.

Mendenhall, Chase D., Vivienne G. Hayes, Jay R. Margolis, and Eric Dorfman. "Diversifying Displays of Biological Sex and Sexual Behavior in a Natural History Museum." *Museum International* 72, nos. 1–2 (2020): 152–61.

Menges, J. "Eine neue Antilope des Somalilandes." *Zoologischer Anzeiger* 17 (1894): 130–31.

Merriam [Bailey], Florence. *Birds through an Opera Glass*. New York: Chatauqua, 1889.

Metz, Helen Chapin, ed. *Somalia: A Country Study*. Washington, DC: Government Printing Office, 1992.

Millard, Candice. *River of the Gods: Genius, Courage, and Betrayal in the Search for the Source of the Nile*. New York: Doubleday, 2022.

Miller, Donald L. *City of the Century: The Epic of Chicago and the Making of America*. New York: Simon and Schuster, 1996.

Minteer, Ben A., James P. Collins, Karen E. Love, and Robert Puschendorf. "Avoiding (Re)Extinction." *Science* 344 (2014): 260–61.

Mitman, Gregg. *Reel Nature: America's Romance with Wildlife on Film*. Cambridge, MA: Harvard University Press, 1999.

Moehlman, P. D., F. Kebede, and H. Yohannes. "African Wild Ass: *Equus africanus*." IUCN Red List of Threatened Species. Published 2015.

Morgan, Lewis H. *Ancient Society, or Researches in the Lines of Human Progress from Savagery through Barbarism to Civilization*. New York: Henry Holt, 1877.

Moskovits, Debby. "1994: A Grand Experiment for Conservation Begins." In *125 Moments in the Natural History of the Field Museum*, edited by Franck Mercurio, Erin Hogen, and Chandler Garland, 282–85. Chicago: Field Museum, 2018.

Mosley, Charles, ed. *Burke's Peerage, Baronetage and Knightage*. 107th ed. 3 vols. Wilmington, DE: Burke's Peerage, 2003.

"Mr. Charles F. Adams." *Auk* 10, no. 4 (1893): 385–86.

Newton, Alfred. "Abstract of Mr. J. Wooley's Researches in Iceland Respecting the Gare-Fowl or Great Auk (*Alca impennis*, Linn.)." *Ibis* 3 (1861): 374–99.

New York Times. "Fair Buildings in Ashes." July 6, 1894.

———. "Jackson Park's Future." May 24, 1896.

———. "Prof. D. G. Elliot Accepts." November 23, 1894.

———. "Returned from Africa." November 22, 1896.

———. "To Hunt Wild Beasts in Africa." February 2, 1896.

———. "Work of the Late Charles B. Atwood." January 1, 1896.

Nieuwland, Ilja. *American Dinosaur Abroad: A Cultural History of Carnegie's Plaster Diplodocus*. Pittsburgh, PA: University of Pittsburgh Press, 2019.

Njoku, Raphael Chijioke. *The History of Somalia*. Santa Barbara, CA: Greenwood, 2013.

"Obituary Record of Yale Graduates: Owen Franklin Aldis." *Bulletin of Yale University*, 22nd ser., no. 22 (1926): 68–69.

Ogilvie-Grant, W. R. "On the Birds of Southern Arabia." *Novitates Zoologicae* 7 (1900): 243–73.

Osborn, Henry Fairfield. *The American Museum of Natural History, Its Origin, Its History, the Growth of Its Departments to December 31, 1909*. New York: Irving Press, 1911.

Osgood, Wilfred H. "In Memoriam: Charles Barney Cory." *Auk* 39, no. 2 (1922): 151–66.

Outram, Dorinda. "New Spaces in Natural History." In *Cultures of Natural History*, edited by N. Jardine, J. A. Secord, and E. C. Spary, 249–65. Cambridge: Cambridge University Press, 1996.

Pakenham, Thomas. *The Scramble for Africa: White Man's Conquest of the Dark Continent from 1876 to 1912*. New York: Random House, 1990.

Pall Mall Gazette (London). "Francesco Crispi." March 28, 1896.

Palmer, Katherine Van Winkle. *Type Specimens of Marine Mollusca Described by P. P. Carpenter from the West Coast (San Diego to British Columbia)*. Denver, CO: Geological Society of America, 1958.

Pankhurst, Richard. "Ethiopia and Somalia." In *General History of Africa*, vol. 6, *Africa in the Nineteenth Century until the 1880s*, edited by J. F. Ade Ajayi, 376–411. Paris: UNESCO, 1989.

Patterson, Bruce D. "Dr. Bill Stanley: A Tribute." *In the Field* 87, no. 1 (2016): 8.

———. "On the Continuing Need for Scientific Collecting of Mammals." *Mastozoología Neotropical* 9, no. 2 (2002): 253–62.

Patterson, John Henry. *The Man-Eaters of Tsavo and Other East African Adventures*. London: MacMillan, 1908.

Pearce, Francis B. *Rambles in Lion Land: Three Months' Leave Passed in Somaliland*. London: Chapman and Hall, 1898.

Pease, A. E. *Travel and Sport in Africa*. London: Arthur L. Humphreys, 1902.

Peck, Robert McCracken, and Patricia Tyson Stroud. *A Glorious Enterprise: The Academy of Natural Sciences of Philadelphia and the Making of American Science*. Philadelphia: University of Pennsylvania Press, 2012.

Phoofolo, Pule. "Epidemics and Revolutions: The Rinderpest Epidemic in Late Nineteenth-Century Southern Africa." *Past and Present* 138, no. 1 (1993): 112–43.

Pittsburgh (PA) Dispatch. "Trouble in Field Museum." February 3, 1897.

Poliquin, Rachel. *The Breathless Zoo: Taxidermy and the Cultures of Longing*. University Park: Pennsylvania State University Press, 2012.

Porter, Charlotte M. *The Eagle's Nest: Natural History and American Ideas, 1812–1842*. Tuscaloosa: University of Alabama Press, 1986.

Potocki, Joseph. *Sport in Somaliland: Being an Account of a Hunting Trip to That Region*. 1900. Camden, SC: Briar Patch Press, 1988.

Powell, Miles A. *Vanishing America: Species Extinction, Racial Peril, and the Origins of Conservation*. Cambridge, MA: Harvard University Press, 2016.

Prestholdt, Jeremy. *Domesticating the World: African Consumerism and the Genealogies of Globalization*. Berkeley: University of California Press, 2008.

Quammen, David. *The Song of the Dodo: Island Biogeography in an Age of Extinctions*. New York: Simon and Schuster, 1996.

Raby, Megan. "Ark and Archive: Making a Place for Long-Term Research on Barro Colorado Island, Panama." *Isis* 106, no. 4 (2015): 798–824.

Rader, Karen A., and Victoria E. M. Cain. *Life on Display: Revolutionizing U.S. Museums of Science and Natural History in the Twentieth Century*. Chicago: University of Chicago Press, 2014.

Rainger, Ronald. "Biology, Geology, or Neither, or Both: Vertebrate Paleontology at the University of Chicago, 1892–1950." *Perspectives on Science* 1, no. 1 (1993): 478–519.

———. "Collectors and Entrepreneurs: Hatcher, Wortman, and the Structure of American Vertebrate Paleontology circa 1900." *Earth Sciences History* 9, no. 1 (1990): 14–21.

Reclus, Élisée. *The Earth and Its Inhabitants: Africa*. Vol. 4, *South and East Africa*. New York: D. Appleton, 1890.

Records of Big Game, with Their Distribution, Characteristics, Dimensions, Weights, and Measurements of Horns, Antlers, Tusks, and Skins. 3rd ed. London: Rowland Ward, 1899.

Redman, Samuel J. *Prophets and Ghosts: The Story of Salvage Anthropology*. Cambridge, MA: Harvard University Press, 2021.

Red Sea and Gulf of Aden Pilot. 5th ed. London: Hydrographic Office, 1900.

Reiger, John F. *American Sportsmen and the Origins of Conservation*. 3rd ed. Corvallis: Oregon State University Press, 2001.

Remsen, J. V., Jr. "The Importance of Continued Collecting of Bird Specimens to Ornithology and Bird Conservation." *Bird Conservation International* 5 (1995): 145–80.

Rideout, Henry Milner. *William Jones: Indian, Cowboy, American Scholar, and Anthropologist in the Field*. New York: Frederick A. Stokes, 1912.

Ritvo, Harriet. *The Animal Estate: The English and Other Creatures in the Victorian Age*. Cambridge, MA: Harvard University Press, 1987.

Rocha, L. A., A. Aleixo, G. Allen, F. Almeda, C. C. Baldwin, M. V. L. Barclay, J. M. Bates et. al. "Specimen Collection: An Essential Tool." *Science* 344 (2014): 814–15.

Rochester (NY) Chronicle. "Back from Chicago." February 1, 1894.

Roosevelt, Theodore. *African Game Trails: An Account of the African Wanderings of an American Hunter-Naturalist*. New York: Charles Scribner's Sons, 1910.

Roosevelt, Theodore, and Kermit Roosevelt. *Trailing the Giant Panda*. New York: Charles Scribner's Sons, 1929.

Rosenberg, Gary D. "Carl Akeley's Revolution in Exhibit Design at the Milwaukee Public Museum." In *Museums at the Forefront of the History and Philosophy of Geology: History Made, History in the Making*, edited by Gary D. Rosenberg and Renee M. Clary, 191–202. Boulder, CO: Geological Society of America, 2018.

Rossiter, Margaret W. *Women Scientists in America*. Vol. 1, *Struggles and Strategies to 1940*. Baltimore, MD: Johns Hopkins University Press, 1982.

Rothfels, Nigel. Introduction to *Representing Animals*, edited by Nigel Rothfels, vii–xv. Bloomington: Indiana University Press, 2002.

Rothschild, Miriam. *Dear Lord Rothschild: Birds, Butterflies and History*. Glenside, PA: Balaban, 1983.

Salvin, Osbert. "Notes on Captain Markham's 'Visit to the Galapagos Islands.'" *Proceedings of the Royal Geographical Society* 2, no. 12 (1880): 755–58.

San Francisco Chronicle. "The First of the Series of Expeditions in Search of Specimens for the Academy of Sciences Is Back from Socorro." August 14, 1903.

Sawislak, Karen. *Smoldering City: Chicagoans and the Great Fire, 1871–1874*. Chicago: University of Chicago Press, 1995.

"Scientific Notes and News." *Science* 6, no. 157 (1897): 991.

Sclater, Philip Lutley, and Oldfield Thomas. *The Book of Antelopes*. Vol. 3. London: R. H. Porter, 1897–98.

Sepkoski, David. *Catastrophic Thinking: Extinction and the Value of Diversity from Darwin to the Anthropocene*. Chicago: University of Chicago Press, 2020.

————. "Extinction and Biodiversity: A Historical Perspective." In *The Routledge Handbook of Philosophy of Biodiversity*, edited by Justin Garson, Anya Plutynski, and Sahotra Sarkar, 26–39. London: Routledge, 2017.

Sharpe, R. Bowdler. "On a Collection of Birds Made by Dr. A. Donaldson Smith in Western Somaliland." *Proceedings of the Zoological Society of London*, 1895, 457–520.

Sheffield Independent. "The Latest Egyptian Developments." June 9, 1896.

Shell, Hanna Rose. "Skin Deep: Taxidermy, Embodiment, and Extinction in W. T. Hornaday's Buffalo Group." *Proceedings of the California Academy of Sciences* 55, no. S1 (2004): 88–112.

Shepp, James W., and Daniel B. Shepp. *Shepp's World's Fair Photographed.* Chicago: Globe Bible, 1893.

Sinclair, Upton. *The Jungle.* New York: Doubleday, Jabber, 1906.

Skiff, Frederick J. V. *Classification: Address by Frederick J. V. Skiff before the Graduating Class of Colorado College, Commencement 1905.* Colorado Springs: Colorado College, 1905.

————. "The Uses of the Museum." *Chicago Times Herald*, April 29, 1895.

Smith, Arthur Donaldson. *Through Unknown African Countries: The First Expedition from Somaliland to Lake Lamu.* London: E. Arnold, 1897.

Smith, Norman B. "Big Game Shooting." In *The "House" on Sport*, vol. 2, edited by W. A. Morgan, 1–40. London: Gale and Polden, 1899.

Steinhart, Edward I. *Black Poachers, White Hunters: A Social History of Hunting in Colonial Kenya.* Athens: Ohio University Press, 2006.

Stoddard, Herbert L. *Memoirs of a Naturalist.* Norman: University of Oklahoma Press, 1969.

Storrs, Caryl B. "Visitin' 'Round in Minneapolis." *Minneapolis Star Tribune*, March 6, 1917.

Strickland, H. E., and A. G. Melville. *The Dodo and Its Kindred.* London: Reeve, Benham and Reeve, 1848.

Swayne, H. G. C. *Seventeen Trips through Somáliland: A Record of Exploration and Big Game Shooting, 1885 to 1893.* London: Rowland Ward, 1895.

Sweet, Paul R. "Collection Building through Salvage: A Review." In *Collections in Context: Proceedings of the 5th International Meeting of European Bird Curators*, edited by Ernst Bauernfeind, Anita Gamauf, Hans-Martin Berg, and Yoko Muraoka, 157–68. Vienna: Vienna Natural History Museum, 2010.

Teichert, Pietro. "Carl Akeley: A Tribute to the Founder of Shotcrete." *Shotcrete Magazine* 4, no. 3 (2002): 10–12.

Thomas, Oldfield. "Description of a New Dik-Dik Antelope (*Madoqua*) discovered in N.E. Africa by Mr. H. S. H. Cavendish." *Proceedings of the Zoological Society of London*, 1898, 278–79.

————. "Notes on the Rodent Genus *Heterocephalus*." *Proceedings of the Zoological Society of London*, 1885, 845–48.

Thornhill, Jan. *The Tragic Tale of the Great Auk.* Toronto: Groundwood Books, 2016.

Three Addresses Delivered at the Meeting Commemorating the Fiftieth Anniversary of Field Museum of Natural History, September 15, 1943. Chicago: Field Museum Press, 1943.

Turvey, Samuel T., and Anthony S. Cheke. "Dead as a Dodo: The Fortuitous Rise to Fame of an Extinction Icon." *Historical Biology* 20, no. 2 (2008): 149–63.

Vetter, Jeremy. *Field Life: Science in the American West during the Railroad Era*. Pittsburgh, PA: University of Pittsburgh Press, 2016.

Wallace, Alfred Russell. "On the Physical Geography of the Malay Archipelago." *Journal of the Royal Geographical Society* 33 (1863): 217–34.

Walsh, Langton Prendergast. *Under the Flag and Somali Coast Stories*. London: Andrew Melrose, 1932.

Ward, Roswell, and Blake McKelvey. *Henry A. Ward: Museum Builder to America*. Rochester, NY: Rochester Historical Society, 1948.

Warren, Louis S. *The Hunter's Game: Poachers and Conservationists in Twentieth-Century America*. New Haven, CT: Yale University Press, 1997.

Welch, R. Courtenay, ed. *The Harrow School Register, 1801–1893*. London: Longmans, Green, 1894.

Wheeler, William Morton. "Carl Akeley's Early Work and Environment." *Natural History* 27, no. 2 (1927): 133–41.

———. "George Baur's Life and Writings." *American Naturalist* 33, no. 385 (1899): 15–29.

Whitehouse, G. C. "The Building of the Kenya and Uganda Railway." *Uganda Journal* 12, no. 1 (1948): 1–15.

Wille, Lois. *Forever Open, Clear and Free: The Struggle for Chicago's Lakefront*. 2nd ed. Chicago: University of Chicago Press, 1991.

Wolfe, Linnie Marsh, ed. *John of the Mountains: The Unpublished Journals of John Muir*. Madison: University of Wisconsin Press, 1938.

Wonders, Karen. *Habitat Dioramas: Illusions of Wilderness in Museums of Natural History*. Uppsala: Acta Universitatis Upsaliensis, 1993.

INDEX

Abdi Kareem (tent boy), 65, 100, 118

Abyssinia/Abyssinians, 59, 219, 230

Abyssinian scimitarbill, *Rhinopomastus minor*, 202

Academy of Natural Sciences (Philadelphia), 20

accessory materials, for habitat group displays, 33, 38, 269, 272

Adadleh, 145–46, 149, 152–53, 155, 157, 159–62, 167, 173–74, 176, 182, 184–88, 187 (fig. 16.2), 189, 193–94, 207, 212, 256–57

Adams, Charles F., 164, 317n5

Aden, 44, 48, 50–51, 53–56, 58–61, 66, 153, 166–67, 178–80, 218–19, 224, 254, 259–63, 305n36, 317n12

Adwa, Battle of, 64, 219

Africa, 7–8, 32. *See also place names*

African expedition (1896): banner/flags, 62, 183, 184 (fig. 16.1); Berbera headquarters, 61 (fig. 5.3), 61–62, 63 (fig. 5.4); departure, 40–42; first specimens obtained, 64, 69–70; funding for, 39–40, 42, 46–48, 50–53, 60, 120, 166–67, 179; homeward bound, 261–67; planning/preparation, 38–40; productivity, 115–16, 129, 143, 207, 212, 220, 224, 238, 249, 261; proposal for, 32–33; report of results, 268–70; size of company, 222 (fig. 19.1); stakes of, 89; supplies and equipment, 46–48, 53–54, 102, 119; surplus items, disposal of, 254–55, 259;

voyage, 48–58. *See also* Akeley, Carl Ethan; Elliot, Daniel Giraud; Skiff, Frederick J. V.

African mammals, 27–28, 32–33, 37 (fig. 3.1), 47, 67, 69–70, 167, 303n13. *See also names of mammals*

African pygmy falcon, *Poliohierax semitorquatus*, 217

Ahmet Hirsi (gun bearer and tent boy), 84, 93, 96, 118, 212–13, 215–16, 218–19, 224, 239, 249

Akeley, Carl Ethan (taxidermist), 2, 5–10, 41, 43, 46–47, 49, 62, 64–65, 90 (fig. 7.1), 94–96, 100–103, 107, 109–10, 117, 121, 123–25, 125 (fig. 9.5), 127–31, 137, 140–43, 145–46, 149–51, 153–58, 160–61, 166, 168–75, 177, 181–83, 186–88, 187 (fig. 16.2), 190–92, 194–205, 200 (fig. 17.1), 207–14, 216–18, 221, 223–26, 229, 233–45, 247–49, 251–57, 259, 261, 263–81, 284–85, 294, 303n12, 303n13, 304n20, 307n40, 316n13, 326n4; and armed robbery for camel's milk, 86–87; bagging first mammals on African expedition, 69–72; and Belgian Congo expedition (1921), 294; birthday (32nd) on African expedition, 101–3; and British East Africa expedition (1905), 281; early experience, 34–36; and Elliot, 266–67, 269–71; expedition attendants, 73 (fig. 5.6); and habitat group

14–17, 24–27, 112–14, 163–65; founding of, 11–13; funding for African expedition, 39–40, 42, 46–48, 50–53, 60, 120, 166–67, 179; Hall 19, 110, 164–65; Hall 20, 110, 111 (fig. 8.2); Hall 22, 106–7; Hall 23, 107; Hall 24, 276; hall of mammals, 110; history of zoology at, 8–10; lecture series, 28; management, 13–14 (*see also* Skiff, Frederick J. V.); museum publications, 15, 17, 22, 163–64; naming of, 12; need for full-time collector, 271; new taxidermy building, 138–39, 273; Ornithology Department, 280–81, 285, 288; permanent home for, 27; pest control, 137; reinstallation of exhibits, 33; repair/restoration, of exhibits, 110; repairs and renovations, 133–39; restructuring, 276; Section of Animal Industries, 106–7; Section of Fishery Industries, 107; Section of Taxidermy, 272; sourcing of collections/displays, 11–13, 15; South Court, 276; and taxidermy fire hazards, 36–38; telephone system, 139; West Court/Pavilion, 37 (fig. 3.1), 38, 107, 108 (fig. 8.1), 272, 276, 277 (fig. 23.5), 288; Zoology Department, 104–14, 108 (fig. 8.1), 108–9, 139, 163–65, 268, 281, 288. *See also* Skiff, Frederick J. V.

field observation, museum zoology and, 292–93

fieldwork, 32–33, 269, 284. *See also* African expedition (1896)

Finley, William, 291

firefighters, 138

Fischer, Moritz (preparator), 108, 112, 164

fish collection, 53, 305n36

fishing expedition to Florida and Gulf of Mexico, 109

flash flood, on African expedition, 152

flora/fauna, African, colonization of, 32

Flower, William H., 35

food and cooking, on African expedition, 91, 128–30, 184–86, 203, 227. *See also* meat, in expedition diet

food supply, on African expedition, 91–92, 128–30, 173–74, 222

fossils, 112–13; type specimens, 312n21

fossil turtle *(Protostega gigas; Toxochelys latiremis)*, 113

fox, *Otocyon megalotis*, 142, 238, 248, 322n21

France, and Somaliland, 63

frog, *Rana delalandii*, 203

Galápagos Islands, 164–65, 289

Galla, 99, 100 (fig. 7.3), 233

gazelle. *See* dibatag; gerenuk; Pelzeln's gazelle; Soemmerring's gazelle; Speke's gazelle

Gelalo, 151, 187

gerenuk, *Lithocranius walleri*, 123, 131, 166, 181, 196, 198, 203, 207–9, 214–16, 220, 229, 234–35, 254; habitat group display, 181, 229, 272, 318n2

giraffe, 238, 264, 325n15

glossy starling, *Dilophus carunculatus*, 235

golf, game of, 260

Golis Range, 97, 121, 257

Goodman, John B., 109, 135, 314n7

gorillas, 294

Graham, Ernest R., 133–34

graves, Somali, 99, 100 (fig. 7.3)

Gray, John Edward, 311n6

grazing, 74, 115, 186, 226, 232

great auk, 290

Great Chicago Fire (October 1871), 137

greater kudu, *Strepsiceros kudu*, 130–32, 140, 142–43, 199, 200 (fig. 17.1), 209–10, 216; habitat group display, 199–200, 216, 275, 275 (fig. 23.3)

great spotted hyena, *Hyena crocuta*, 188

"great white hunter" figure, Elliot and, 198–99, 206

Grinnell, Joseph, 291, 293

Milmil, 224

Milne-Edwards, Alphonse, 265

Milwaukee Public Museum, 34–35, 107, 299n21

mirage, over African plain, 150

morale, issues of, 117–19

Muir, John, 326n11

Musa Mohammed (head cook), 65, 78 (fig. 6.1), 118, 187–88

Musée national d'histoire naturelle, Paris, 265

museum anthropologists, 260

museum zoology, broadening scope of, 292–94

music, on African expedition, 116–17, 117 (fig. 9.1)

Nabokov, Vladimir, 93

Nagal, field camp at, 193

naked mole rat, *Heterocephalus glaber*, 142

naming, of newly discovered species, 49–50, 270. *See also* new species

Nasr Hablod, 199, 203, 204 (fig. 17.2), 205, 208 (fig. 17.3), 209

National Museum, Washington, DC, 21, 298n16

national parks, 292

Neumann, Arthur, 325n15

newspapers, English, 219

new species, 49–50, 92–93, 167, 182, 195, 203–6, 204 (fig. 17.2), 206, 248, 263–64, 270; *Bufo viridis somalacus*, 102; chameleon, *Rhampholeon manders*, 126; dik-dik *(Madoqua cavendishi)*, 325n15; flycatcher *(Pachyprora bella)*, 319n24; *Lygosoma akeleyi*, 65; *Phrynobatrachus hailiensis*, 102; *Pterois ellioti*, 65; warbler *(Sylviella isabellina)*, 319n24

nightjar, *Caprimulgus donaldsoni*, 168–69

Nile Valley sunbird, *Hedydipna metallica*, 95

Nur, Sultan (ruler of Habr Yunis), 2–6, 3 (fig. 1.1), 153

O'Brien, Capt. Michael, 138

O'Brien, William, 109

obsolescence of nature, 8–10

offensive terms, treatment of, 297n5

Ogaden, 173, 183, 207, 230, 232, 247, 252, 260, 323n16

Old World babbler, *Argya aylmeri*, 217

Olmsted, Frederick Law, 133

Olympic Mountains expedition (1898), 271

oral tradition, Somali, 218

oriole, *Oriolus larvatus*, 256

ornithology, 15–21, 25–26. *See also* bird collection; *names of birds*

oryx, *Oryx beisa*, 149, 172–73, 182, 191, 195, 201–2, 221, 228–29, 253

ostrich, 228, 237, 239, 240 (fig. 20.2)

overcollecting, 329n6

oversize specimens, display of, 108 (fig. 8.1)

owl, *Scops capensis*, 247

paleontology, 112–13

paradise flycatcher, *Terpsiphone cristata*, 237

passenger pigeon, 290

Pease, Alfred E., 323n16

Pectinator spekei, 146

Pelzeln's gazelle, *Gazella pelzelni*, 78, 82–83, 92, 96, 103, 121

Peter Redpath Museum, Montreal, Philip P. Carpenter Collection, 112

Philippines, xi

photographing specimens *vs.* collecting, 291

photography, on African expedition, 46–47, 73 (fig. 5.6), 95, 146, 151, 173, 269. *See also* Akeley, Carl Ethan

pipit, *Anthus sordidus*, 246

poisoning/contamination, fear of, 187–89

poor planning, instance of, 85–87

Potocki, Count, expedition of 1895–96, 169